PROBABILITY THEORY AND PROBABILITY LOGIC

P. ROEPER AND H. LEBLANC

Probability Theory and Probability Logic

UNIVERSITY OF TORONTO PRESS

Toronto Buffalo London

Toronto Buffalo London
Printed in Canada

ISBN 0-8020-0807-0

Printed on acid-free paper

Toronto Studies in Philosophy
Editors: James R. Brown and Calvin Normore

Canadian Cataloguing in Publication Data

Roeper, Peter
Probability theory and probability logic

(Toronto studies in philosophy)
Includes bibliographical references and index.
ISBN 0-8020-0807-0

1. Probabilities. 2. Semantics (Philosophy). 3. Logic. I. Leblanc, Hugues, 1924– . II. Title. III. Series.

BC141.R66 1999 160 C99-931586-2

University of Toronto Press acknowledges the financial assistance to its publishing program of the Canada Council for the Arts and the Ontario Arts Council.

Contents

Acknowledgments

Thanks are due to the FOUNDATION UQAM and to the SSHRCC for their generous support during the writing of the book.

H. Leblanc and P. Roeper

Introduction

The Theory of Probability, as we understand it, connects with logic in two ways. For one, it involves a study of the logic of probability judgments and the logical relations among probability judgments; secondly, it investigates the place of probability vis-a-vis logic itself. The logic of probability judgments, encapsulated in systems of constraints, is pursued in Part I, called *Probability Theory*, while Part II, *Probability Logic*, is devoted to locating probability itself within logic as a semantic notion in the same family as logical consequence and consistency. The two enterprises are of course closely connected; results from Part I are made use of in Part II, and consideration in Part II serve to justify constraints adopted in Part I.

An issue to be settled at the outset is what the items are that we should attribute probability to. The main options are: sets of various kinds on the one hand, and statements or propositions on the other. The axiomatic treatment of the former is mainly due to Kolmogorov, while Keynes originated the attribution of probabilities to statements, Carnap and Popper being major contributors to this approach. From a formal point of view, the two approaches are not all that different. Formal constraints and results can largely be translated from one idiom into the other. We opt for statements, because of our aim to trace the connection between probability and logic, to treat probability as a semantic notion. So we take $P(A)$, the probability of statement A, to be a number which is the value of the *probability function* P for statement A as argument. In addition to such unary, or absolute, probability functions we study binary, or relative, ones. If P is a relative probability function then $P(A,B)$, the probability of A relative to, or given, B, is its value for statements A and B as arguments. Rényi, Carnap, and Popper are the main originators of the study of relative probability functions.

Another respect in which treatments of probability theory differ is the interpretation of probability judgments. There is the possibility of a purely mathematical approach which eschews any stance concerning the meaning of probability. In that case though, formal codifications of probability have to be taken as given and there is no room for criticising or justifying proposed systems of constraints. Then there are subjective accounts of probability which have it that the probability of A is the extent to which A is believed. But more common perhaps is an account which focuses on the role of probability in inductive reasoning and interprets the probability of A as the degree of confirmation of A. This is the interpretation that we adopt in this book. Moreover, following Carnap, we argue that degree of confirmation can be equated with degree of logical consequence. This enables us to mount justifying arguments for the adoption of systems of constraints to characterise types of probability functions, e.g., intuitionistic probability functions.

Part I is devoted largely to formal matters, which can be dealt with without appeal to the interpretation of probability. We present systems of constraints for probability functions appropriate for finitary and infinitary propositional logic, for quantificational logic, and for intuitionistic logic. We then turn to central results about the relationship between absolute and relative probability functions, in particular the representation of relative probability functions by classes of absolute probability functions, the recursive definability of absolute probability functions, and the classification of relative probability functions.

Part II is concerned with the link between probability functions and semantic notions. We consider the program of *Probability Semantics*, which seeks to define the semantic notions in terms of probability functions. We argue that the program is misconceived and that the link in question is to be found in understanding degree of confirmation as degree of logical consequence. The consequences of this identification are pursued for probability functions for both classical and intuitionistic logic. If one takes the probability of a statement to be its degree of confirmation or its degree of consequence, an immediate question is what it is that the statement receives its (partial) confirmation from or that it is a (partial) consequence of. In the case of absolute probability functions we identify a set of statements associated with the probability function in question, calling the set the *assumption set* of the function. We then argue that the probability of a statement is the degree to which it is confirmed by, and hence is a consequence of, the assumption set. In the case of relative probability functions we argue that $P(A,B)$ is the degree to which A is confirmed by (and hence a consequence of) B together with a set of assumptions which depends not only on the function P, but on the statement B as well.

On the basis of this understanding of probability, justifications are provided for the systems of constraints for the various classes of probability functions, i.e. those for finitary and infinitary propositional languages L_P and L_ω, for the quantificational language L_Q and for the language L_I of intuitionistic propositional logic. We also identify and justify language-independent constraints for absolute and relative probability functions.

Part I

Probability Theory

Introduction to Part I

Probability values can be attributed to a variety of items such as event types, properties, sets of various kinds, and statements. From a mathematical perspective sets are the preferred arguments of probability functions. From a logical point of view it is primarily statements which can be said to be more or less probable and, since we are concerned here with probability logic, we opt for statements. This choice is even more appropriate in so far as a matter particular interest for us is the relationship between probability and logical notions like logical consequence and consistency. Part II is devoted to the investigation of these relationships in connection with the probability functions introduced in Part I. The gist of our account can be conveyed by saying that probability is degree of consequence, where it is then an aim of our inquiry to determine what the assumptions are in each case which confer a degree of logical implication on the statement in question. Since both of the logical notions mentioned concern not only individual statements but also sets of statements we extend the investigation of probability functions to include functions which take sets of statements rather than statements as arguments.

The families of functions studied here differ from one another in various ways: in the number of arguments, the language to which the statements belong, and the semantic interpretation of the statements. There are absolute (unary) and relative (binary) functions. As to languages, propositional ones, being the most widely studied, are also at the centre of our investigations. Probability functions defined on propositional languages are characterised in Chapter 1 with some indication of the origins of different sets of constraints. In Chapter 2 we extend the discussion to quantificational languages, as the reader would expect, and also, perhaps unexpectedly, to infinitary propositional languages. As to probability functions for quantificational logic, we formulate constraints appropriate for the objectual and for the substitutional interpretation of quantification. We establish in the process that no more than denumerably many additional terms are required in order to generate all distinct probability functions that are possible on the objectual understanding of the quantifiers.

The reason for including infinitary propositional languages is that when probability functions are defined on the sets in a field of sets it is usually assumed that the field contains with every family of sets belonging to the filed also the union and intersection of all the sets in the family (i.e. the field is *complete*) or, at least, that this is so for every countable family of sets (i.e. the field is a σ-*field*). Since we have opted for statements as the arguments of probability functions, results in probability theory which depend on the existence of infinite unions and intersections of sets cannot be reproduced unless infinite conjunctions and disjunctions of statements are at hand. Hence we characterise and study in Chapter 2 also probability functions for infinitary languages with countable conjunctions and disjunctions. These functions play a role in Chapter 4 when we investigate the representation of relative probability functions by ordered sets of absolute ones, and whenever it is advantageous to represent an infinite set of statements by a single statement.

We also include an account of intuitionistic probability functions for reasons which have to do with the interpretation of probability that we have primarily in mind, namely the interpretation according to which $P(A)$ expresses the degree of support or confirmation that A receives from the so-called background knowledge or evidence. Since our own view identifies degree of confirmation with degree of logical consequence, the study of intuitionistic probability functions is of particular interest. While in classical logic degree of support, however understood, supervenes upon the truth-conditional semantics of the statements under consideration, the intuitively most informative semantics for intuitionistic logic employs conclusive confirmation as its central notion and explains the confirmation of complex statements with reference to the confirmation of constituents. Therefore the partial confirmation of a statement, its probability, should be analogously related to the probability of its constituents, and the constraints for intuitionistic probability functions should be generalisations of the constraints for intuitionistic consequence. The characterisation of intuitionistic probability functions is included in Chapter 1, while the justification of the constraints employed and link with logical consequence are studied in Part II.

Following on the characterisation of probability functions in the first two chapters, Chapter 3 is a systematic study of the relationship between absolute and relative probability functions. Chapter 4, as already mentioned, deals with the representation of relative probability functions by ordered sets of absolute ones. Chapter 5 addresses the question of recursive definability of probability functions and establishes that absolute probability functions can be recursively defined from probabilities of atomic statements, not indeed the probabilities of individual atomic statements alone, but of conjunctions of atomic statements. This is a significant result, which reveals a much larger degree of similarity between truth-value functions and absolute probability functions than is commonly recognised. Chapter 6, the last chapter of Part I, provides a systematic survey of families of relative probability functions for propositional languages in terms of certain equivalence relations among statements which can be formulated probabilistically.

Chapter 1

Probability Functions for Propositional Logic

In a language of finitary propositional logic only finite conjunctions and disjunctions are recognised; indeed the connectives '∧' and '∨' are 2-place connectives. Best known of course are languages for classical propositional logic, and probability functions defined on the statements of such a language L_P occupy a central place in probability theory. But we also introduce here probability functions defined on sets of statements rather than statements and probability functions for intuitionistic rather than classical propositional logic. The reason for including such functions in our study is in both cases our interest in tracing the connection between semantics and probability theory.

Unary functions for the statements of L_P constitute the core of probability theory. A variety of formalisations of absolute probability are known, usually invoking, in addition to conjunctions and negations of statements, the notion of logical equivalence. This notion is one of a family which also includes logical consequence and inconsistency, and these notions can be presupposed when the definition and study of probability functions are seen as extending the logical theory of L_P. If, however, probability theory is expected to make a contribution to or to supersede the normal account of logical notions, then it is desirable to characterise probability functions in a way that does not presuppose those notions. Such characterisations, labelled *autonomous* and advocated and pioneered by Popper, are preferred by us.

The values of probability functions are usually restricted to the closed interval [0,1]. We give an account also of a more general kind of 1-argument function, called *measure functions*, on whose values no upper limit is placed, but which otherwise behave like absolute probability functions. These functions will play a theoretical role later in the book.

Less well known than absolute probability functions are relative ones, binary functions which ascribe a probability value to the first argument relative to the second one. Again we opt for autonomous constraints, derived from Popper, to characterise these functions. Often what is meant by a relative probability function is a partial function derived from an absolute probability function by the equation

$$P(A,B) = P(A \wedge B) \,/\, P(B), \textit{ when } P(B) \neq 0.$$

When these functions are made total by stipulating

$$P(A,B) = 1, \textit{ when } P(B) = 0,$$

the resulting functions, called *Kolmogorov's relative probability functions*, constitute a subclass of Popper's relative probability functions. And the 2-valued functions among them can be interpreted as having as value the truth-value of the conditional whose antecedent is the second argument and whose consequent is the first argument.

Probability functions that take sets of statements as arguments can also be either unary or binary. Sets of statements are usually interpreted conjunctively, which means that the functions on sets of statements can be correlated with the functions on statements by equating a set of statements, when finite, with the conjunction of its members.

Probability functions for intuitionistic logic, finally, will be characterised by constraints which are largely similar to those for probability functions for L_P. The formulation of divergent constraints, in particular those concerning negation, will be guided by the intuitionistic understanding of the meaning of connectives.

Section 1. Absolute Probability Functions for L_P

Absolute probability functions predate relative ones. Most notable is Kolmogorov's account in (1933), who defined his functions on fields of sets (the sets called "events"). Since Carnap's work on probability[1] it has become common practice among logicians to take statements as the arguments of probability functions. This facilitates the study of the connections between probability and logical notions, which is a major concern of this book. In this chapter the statements in question are those of a propositional language L_P with countably many atomic statements whose primitive connectives are '$\neg$' and '$\wedge$', with '$\vee$', '$\supset$', and '$\equiv$' defined in the usual way.

In simple and familiar versions of probability theory logical notions, in particular that of logical equivalence,[2] are themselves employed in the characterisation of absolute probability functions, as in the following one.

AK1 $0 \leq P(A)$

AK2 $P(\neg(A \wedge \neg A)) = 1$

AK3' *If* $A \wedge B$ *is logically false, then* $P(A \vee B) = P(A) + P(B)$

AK4 *If* A *and* B *are logically equivalent, then* $P(A) = P(B)$

An alternative to **AK3'** is the following constraint which does not invoke logical falsity:

AK3 $P(A) = P(A \wedge B) + P(A \wedge \neg B)$

Popper's Constraints

The use of semantic notions in characterisations of probability functions was regarded by Popper as a shortcoming. *Autonomous* characterisations, i.e. ones that are independent of semantic notions, have been advocated by him since 1938.[3] It is of course essential to characterise probability functions independently of semantic notions if an account of these notions in terms of probability functions is aimed at, as is the case with *probability semantics*, a program to be discussed in Part II. For our own aim, pursued in Part II, of treating probabilities themselves as generalisations of semantic notions, it is also important to develop autonomous characterisations which can be seen to support our contention. We take as our standard axiomatisation the following one, derived from Popper's work.

AP1	$0 \leq P(A)$	(*Non-negativity*)
AP2	$P(\neg(A \wedge \neg A)) = 1$	(*Normality*)
AP3	$P(A) = P(A \wedge B) + P(A \wedge \neg B)$	(*Addition*)
AP4	$P(A \wedge B) \leq P(B \wedge A)$	(*Commutativity*)
AP5	$P(A \wedge (B \wedge C)) \leq P((A \wedge B) \wedge C)$	(*Associativity*)
AP6	$P(A) \leq P(A \wedge A)$	(*Idempotence*)[4]

AP1–AP3 are identical with **AK1**, **AK2**, and **AK3**, and together with **AP4–AP6** yield **AK4**, as is shown later in this section.

Characterised by Carnap in (1950) is a family of absolute probability functions meeting this extra constraint:

AC *If $P(A) = 0$, then A is logically false*,[5]

one which — given constraints **AP2** and **AK4** — guarantees that for Carnap P

$$P(A) = 0 \textit{ if and only if } A \textit{ is logically false}$$

and

$$P(A) = 1 \textit{ if and only if } A \textit{ is logically true.}$$

We shall refer to these functions as Carnap's absolute probability functions.

State-Descriptions

Constraint **AC** is of course non-autonomous. An equivalent autonomous constraint can be extracted from Carnap's own characterisation of absolute probability functions in terms of state-descriptions. Since it is a useful way of explicitly defining individual absolute probability functions, we give an account here of Carnap's method. Let $E_1, E_2, ..., E_n, ...$ be in some prearranged order the atomic statements of L_P; and by a *state-description* in $E_1, E_2, ..., E_n$ understand any of the conjunctions of the sort

$$(...(\pm E_1 \wedge \pm E_2) \wedge ...) \wedge \pm E_n$$

or, for short, of the sort

$$\pm E_1 \wedge ... \wedge \pm E_n,$$

where, for each i from 1 through n, $\pm E_i$ is either E_i or $\neg E_i$. Carnap assigns, for every $n \geq 1$, an absolute probability to each state-description in $E_1, E_2, ..., E_n$ under the constraint that

$$P(E_1) + P(\neg E_1) = 1$$

and

$$P((\pm E_1 \wedge ... \wedge \pm E_n) \wedge E_{n+1}) + P((\pm E_1 \wedge ... \wedge \pm E_n) \wedge \neg E_{n+1}) = P(\pm E_1 \wedge ... \wedge \pm E_n).$$

For any given statement A, let then n be such that E_n is an atomic component of A while, for any $m > n$, E_m is not such a component, and define

$$P(A) = P(W_1) + P(W_2) + ... + P(W_h),$$

where W_1, W_2, ..., and W_h are the h ($h \geq 1$) state-descriptions in E_1, E_2, ..., and E_n to whose disjunction A is logically equivalent. The resulting functions are indeed absolute probability functions.

Returning now to constraint **AC**, the equivalent autonomous constraint alluded to above requires that the probability of each state-description be *larger than* 0.

Measure Functions

Of some, more technical, interest is a generalisation of the notion of an absolute probability function, obtained by dropping constraint **AP2** from **AP1–AP6**. The wider class of functions satisfying the remaining constraints is to be called the class of *measure functions*. Since measure functions do not meet this consequence of **AP2**:

$$P(A) \leq 1,$$

they take as their values non-negative reals plus ∞. However, they do meet

$$P(A \vee B) = P(A) + P(B) - P(A \wedge B),$$

and hence this constraint, known as *the Finite Additivity Law*:

$$\textit{If } P(A \wedge B) = 0, \textit{ then } P(A \vee B) = P(A) + P(B);$$

consequently they are said to be *finitely additive*. We do not attribute any particular logical significance to measure functions. The reason for introducing them here is that they will turn up in Chapter 4 in connection with the representation of relative probability functions (i.e. binary probability functions) by means of unary ones. Consequences of constraints **AP1** and **AP3–AP6** are collected in Lemma 1 of Appendix I, while results which hold for absolute probability functions only are to be found in Lemma 2.

Replacing in **AP1–AP6** '*P*' by '*TV*', the latter short for 'the truth-value of', yields six obvious truths about truth-value functions. Hence truth-value functions are absolute probability functions. Conversely, any two-valued absolute probability function is a truth-value function. By Lemma 1(f) and **AP2** the two values must indeed be 0 and 1; by Lemma 2(a) $P(\neg A) = 1 - P(A)$; and by Lemmas 1(b) and 2(e) $P(A \wedge B) = 1$ if and only if $P(A) = P(B) = 1$. So, two-valued absolute probability functions are but truth-value functions and vice-versa;[6] and in the light of this one may conclude that absolute probability theory is a *generalisation* in some sense of two-valued propositional logic.

Popper's Functions Identical with Kolmogorov's Functions

It still needs to be shown that Popper's autonomously characterised absolute probability functions are the same functions as Kolmogorov's, i.e. that they meet constraint **AK4**. Indeed, as we will see, it holds for all measure functions that logically equivalent statements can be substituted for one another. For proof of this let P be any measure function, suppose $P(A) > 0$, let again $E_1, E_2, ..., E_n, ...$ be the atomic statements of L_P, and define a sequence of statements $\pm E_1, \pm E_2, ..., \pm E_n, ...$ as follows:

(i) *If* $P(E_1 \wedge A) > 0$, *then* $\pm E_1$ *is* E_1, $\neg E_1$ *otherwise,*

and

(ii) *If* $P(((\pm E_1 \wedge ... \wedge \pm E_n) \wedge E_{n+1}) \wedge A) > 0$, *then* $\pm E_{n+1}$ *is* E_{n+1}, $\neg E_{n+1}$ *otherwise.*

By this definition $P((\pm E_1 \wedge ... \wedge \pm E_n) \wedge A) > 0$ for every n and hence by Lemma 1(a) $P(\pm E_1 \wedge ... \wedge \pm E_n) > 0$ for every n. Define next a sequence of measure functions $P_1, P_2, ..., P_n, ...$ as follows:

$$P_n(B) = P((\pm E_1 \wedge ... \wedge \pm E_n) \wedge B) \,/\, P(\pm E_1 \wedge ... \wedge \pm E_n),$$

where, in the event that $P(\pm E_1 \wedge ... \wedge \pm E_n) = \infty$,

$$P((\pm E_1 \wedge ... \wedge \pm E_n) \wedge B) \,/\, P(\pm E_1 \wedge ... \wedge \pm E_n)$$

equals 0 if $P((\pm E_1 \wedge ... \wedge \pm E_n) \wedge B)$ is finite, and 1 otherwise. Finally, let a unary function P^2 be defined as the limit of this sequence of functions, i.e.

$$P^2(B) = \underset{n \to \infty}{\text{limit}}\, P_n(B).$$

Then P^2 constitutes a truth-value function, proof of which – being somewhat lengthy – is to be found in Appendix II. However, since $P((\pm E_1 \wedge ... \wedge \pm E_n) \wedge A) > 0$ for every n, $P_m(A) = 1$ for every m such that the atomic statements that occur in A are among $E_1, ..., E_m$. So, $P^2(A) = 1$.

We have established that if $P(A) > 0$, there then exists a truth-value function P^2 which accords A the value 1, or, in other words, there exists an assignment of truth-values to the atomic statements of L_P under which A has the value 1. But if so, then A is not logically false. Hence, if A is logically false, then $P(A) = 0$. Consequently, if P is an absolute probability function, then by Lemma 2(a) $P(A) = 1$ for any logically true A.

Suppose now that A logically implies B, and hence that $A \wedge \neg B$ is logically false. Then, for any measure function P, $P(A \wedge \neg B) = 0$ and hence $P(A) = P(A \wedge B)$ by **AP3** and $P(A) \leq P(B)$ by Lemma 1(a). If, moreover, B logically implies A, so that A and B are logically equivalent, then $P(A) = P(A \wedge B) = P(B)$ by Lemma 1(d). Thus obtained as a consequence is constraint **AK4**. Theorem 1 summarises these results.

Theorem 1. (a) *If A is logically false, then $P(A) = 0$ for any measure function P for L_P;*
(b) *If A is logically true, then $P(A) = 1$ for any absolute probability function P for L_P;*
(c) *If A logically implies B, then $P(A) \leq P(B)$ for any measure function P for L_P;*
(d) *If A and B are logically equivalent, then $P(A) = P(B) = P(A \wedge B)$ for any measure function P for L_P.*

So, A and B, if logically equivalent, can be interchanged in the argument places of measure functions, hence of absolute probability functions. Since such a step often greatly simplifies proofs we frequently avail ourselves of it, and when doing so declare it justified by *Interchange of Equivalents (IE)*.

Furthermore, let P be a measure function and assume that for two statements A and B the following holds:

$$P(A) = P(A \wedge B) = P(B),$$

which by Lemma 1(k) is equivalent to

$$P(A \wedge C) = P(B \wedge C) \textit{ for every statement } C.$$

By Lemma 3(b) it then follows that A and B can be substituted for one another in any statement D without affecting the value of P for D as argument. We talk in that case of A and B being *indiscernible under P*. Hence Theorem 1(d) can be rephrased thus:

If A and B are logically equivalent, then A and B are indiscernible under every measure function, hence every absolute probability function P.

Section 2. Relative Probability Functions for L_P

Relative probability functions defined on fields of sets were first described by Rényi in (1955). These functions, extensions to two arguments of Kolmogorov's absolute ones, are partial functions. They take as their first arguments the members of a field S_1 of sets, and as their second arguments anywhere from a single one to all the members of S_1 other than the empty set, a set of sets that we shall refer to as S_2.[7] Of most interest here are the cases where S_2 contains every member of S_1 except the empty set.

Carnap's Relative Probability Functions

Statement analogues of Rényi's relative probability functions of this type can be had and have indeed been used by many, Carnap in (1952), for instance. Barred in this case from serving as a second argument must be any logical falsehood of L_P; otherwise **RC3** below would yield $P(C,C) = 0$ for any logical falsehood C, thus contradicting **RC2**. Hence, the following list of constraints for a family of partial binary functions, to be referred to as *Carnap's 1952 functions*, that take all the statements of L_P as their first arguments but only those that are not logically false as their second arguments:

RC1 $0 \leq P(A,B)$

RC2 $P(A,A) = 1$

RC3 *If* $A \wedge B$ *is logically false, then* $P(A \vee B,C) = P(A,C) + P(B,C)$

RC4 $P(A \wedge B,C) = P(A,B \wedge C) \times P(B,C)$

RC5 *If A and A' are logically equivalent, then* $P(A,B) = P(A',B)$

RC6 *If B and B' are logically equivalent, then* $P(A,B) = P(A,B')$[8]

For fixed B and as a function of A, $P(A,B)$ is one of Kolmogorov's absolute probability functions, as we establish in Theorem 2. These functions are linked together to some extent by constraint **RC4**.

It is advantageous to turn Carnap's partial functions into total ones, this by stipulating that $P(A,B) = 1$ when B is logically false, an option discussed by Carnap in (1950) and widely adopted in studies of the subject. One then obtains:

RC1' $0 \leq P(A,B)$, *when B is not logically false*

RC2' $P(A,A) = 1$, *when A is not logically false*

RC3' *If* $A \wedge B$ *is logically false, then* $P(A \vee B,C) = P(A,C) + P(B,C)$, *when C is not logically false*

RC4' $P(A \wedge B,C) = P(A,B \wedge C) \times P(B,C)$, *when neither of C and* $B \wedge C$ *is logically false*

RC5' *If A and A' are logically equivalent, then* $P(A,B) = P(A',B)$, *when B is not logically false*

RC6' *If B and B' are logically equivalent, then* $P(A,B) = P(A,B')$, *when neither of B and B' is logically false*

RC7' *If B is logically false, then* $P(A,B) = 1$ *for every statement A of* L_P

The restrictions in **RC1'**, **RC2'**, and **RC4'** can clearly be lifted in the presence of **RC7'**. And **RC3'** can be replaced by the more familiar *Special Addition Law*. Hence these constraints:

RC1'' $0 \leq P(A,B)$

RC2'' $P(A,A) = 1$

RC3'' *If B is not logically false, then* $P(A,B) + P(\neg A,B) = 1$

RC4'' $P(A \wedge B,C) = P(A,B \wedge C) \times P(B,C)$

RC5'' *If A and A' are logically equivalent, then* $P(A,B) = P(A',B)$

RC6'' *If B and B' are logically equivalent, then* $P(A,B) = P(A,B')$

RC7'' *If B is logically false, then* $P(A,B) = 1$ *for any statement A of* L_P

Popper's Relative Probability Functions

Constraints **RC3''** and **RC2''** yield

If B is not logically false, then $P(\neg B,B) = 0$,

and hence, together with **RC7''**,

B is logically false if and only if $P(A,B) = 1$ *for every statement A of* L_P.

So, for Carnap's 1952 functions logical falsity coincides with what Popper calls *P-abnormality*, a matter we shall return to shortly. This characterisation of relative probability functions, relying as it does on the semantic notions of logical falsity and logical equivalence, is not autonomous. Autonomous characterisations are given by Popper in (1955) and (1959). New with and crucial to them is the notion of *P*-abnormality just mentioned, which is defined for statements of L_P thus:

B is P-abnormal if and only if $P(A,B) = 1$ *for every statement A of* L_P.

Popper's own constraints for his functions are somewhat unwieldy; so we adopt a familiar set of constraints which is autonomous like Popper's but largely parallels the non-autonomous **RC1''–RC7''**.

RP1 $0 \leq P(A,B)$ (*Non-negativity*)

RP2 $P(A,A) = 1$ (*Normality*)

RP3 $P(A,B) + P(\neg A,B) = 1$ *if B is P-normal* (*Addition*)

RP4 $P(A \wedge B,C) = P(A,B \wedge C) \times P(B,C)$ (*Multiplication*)

RP5 $P(A \wedge B,C) \leq P(B \wedge A,C)$ (*Commutation on the Left*)

RP6 $P(A,B \wedge C) \leq P(A,C \wedge B)$ (*Commutation on the Right*)

RP7 *At least one statement of* L_P *is P-normal*

We call the functions meeting these constraints *Popper's relative probability functions* or, simply, *relative probability functions for* L_P. Consequences of the constraints are listed and proved in Lemma 4 of Appendix I.

Constraints for Carnap's Relative Probability Functions

Of particular significance is the condition placed by Popper on *B* if $P(A,B) + P(\neg A,B)$ is to equal 1, i.e. that *B* be *P*-normal. All the logical falsehoods of L_P are of course *P*-abnormal, but for Popper other statements as well may be *P*-abnormal. Hence Carnap's 1952 functions are but some of the functions characterised by **RP1–RP6**, namely those which meet the additional constraint

RC52 *If* $P(A,B) = 1$ *for every statement A of* L_P, *then B is logically false.*

A further family of relative probability functions, studied by Carnap in (1950) and referred to hereafter as *Carnap's 1950 functions*, are those that in addition to **RP1–RP6** meet this extra constraint:

RC50 *If* $P(A,\mathrm{T}) = 0$, *then A is logically false.*

Here, and throughout the book, 'T' is short for an arbitrary but fixed statement of the kind $C \vee \neg C$.

RC52 and **RC50** are obviously not autonomous. Equivalent autonomous constraints, making use of state-descriptions, can be found, but because of their greater simplicity we prefer the present constraints as long as semantic notions are not explicitly our concern.

Kolmogorov's Relative Probability Functions

Yet another family of relative probability functions deserves mention at this point, to wit: the partial ones usually associated with Kolmogorov, who in (1933) defined $P(A,B)$ as the quotient

$$P'(A \wedge B) \,/\, P'(B),$$

when $P'(B) > 0$, Here the unary function P' is any absolute probability function. Total functions can be had if we set $P(A,B)$ at 1 when $P'(B) = 0$. Given that the absolute probability of T always equals 1, $P(A,\mathrm{T}) = P'(A)$ and hence

$$P(A,B) = P(A \wedge B,\mathrm{T}) \,/\, P(B,\mathrm{T}), \textit{ when } P(B,\mathrm{T}) \neq 0$$

and

$$P(A,B) = 1, \textit{ if } P(B,\mathrm{T}) = 0.$$

But for every relative probability function P, $P(A,\mathrm{T})$ as a function of A is an absolute probability function and the equation

$$P(A,B) = P(A \wedge B,\mathrm{T}) \,/\, P(B,\mathrm{T})$$

follows from the Multiplication Law **RP4**. So Kolmogorov's relative probability functions are those among Popper's relative probability functions which meet the constraint

RK *If* $P(B,\mathrm{T}) = 0$, *then* $P(A,B) = 1$ *for every statement* A *of* L_P.

Note also that by Lemmas 4(t) and 4(w) a relative probability function meets constraint **RC50** if and only if it meets both **RC52** and **RK**.

Kolmogorov's definition of relative probability functions points to a close association between absolute and relative probability functions. In particular, if P is a relative probability function and C is P-normal, then $P(A,C)$, regarded as a function of A, is an absolute probability function. We use 'P_C' to symbolise this function and call it the *C-restriction* of P:

$$P_C(A) = P(A,C).$$

Theorem 2. *Let P be a relative probability function of Popper's for L_P, i.e. a function meeting constraints* **RP1–RP7**; *and let C be a P-normal statement of L_P. Then P_C is an absolute probability function of Popper's, i.e. a function meeting* **AP1–AP6**.

Proof:

AP1: $0 \leq P_C(A)$ by **RP1**.

AP2: $P_C(\neg(A \wedge \neg A)) = 1$ by Lemma 4(m).

AP3: $P_C(A) = P_C(A \wedge B) + P_C(A \wedge \neg B)$ by Lemma 4(i).

AP4: $P_C(A \wedge B) \leq P_C(B \wedge A)$ by **RP5**.

AP5: $P_C(A \wedge (B \wedge C)) \leq P_C((A \wedge B) \wedge C)$ by Lemma 4(q).

AP6: $P_C(A) \leq P_C(A \wedge A)$ by Lemma 4(c). ¤

Indiscernibility

This result allows us to establish that logically equivalent statements are interchangeable in relative probability functions as they are in absolute ones. Let P be a relative probability function, C a P-normal statement of L_P, and P_C the restriction of P to C. Since P_C is an absolute probability function by Theorem 2, Theorem 1 implies that $P_C(A) = 1$ if A is logically true; that $P_C(A) \leq P_C(B)$ if A logically implies B; and that $P_C(A) = P_C(B)$ when A is logically equivalent to B. So, taking into account that $P(A,C) = 1$ for every statement A if C is P-abnormal, $P(A,C) = 1$ for every statement C if A is logically true; $P(A,C) \leq P(B,C)$ for every statement C if A logically implies B; and $P(A,C) = P(B,C)$ for every statement C if A and B are logically equivalent. But if $P(A,C) = P(B,C)$ for every statement C, then by Lemma 4(r) $P(C,A) = P(C,B)$ for every statement C. So:

Theorem 3. *Let P be a relative probability function of Popper's for L_P. Then*
(a) *If A is logically true, $P(A,C) = 1$ for every statement C;*
(b) *If A logically implies B, $P(A,C) \leq P(B,C)$ for every statement C;*
(c) *If A and B are logically equivalent, then, for every statement C, both $P(A,C) = P(B,C)$ and $P(C,A) = P(C,A)$.*

Hence logically equivalent statements can be interchanged in the argument places of relative as well as those of absolute probability functions. Where we make use of this possibility in a proof we say that the step in question is justified by *Interchange of Equivalents* (*IE*). Theorem 3 connects semantic notions with autonomously characterised relative probability functions, and so introduces a theme which will occupy us in Part II.

Let now P be a relative probability function and assume that for two statements A and B

$$P(A,C) = P(B,C) \text{ for every statement } C.$$

We say in this case that A and B are *indiscernible under P*. For, as we have just seen, it follows that

$$P(C,A) = P(C,B) \text{ for every statement } C,$$

which means that A and B can substitute for one another both in the first and in the second argument place without affecting the value of P. Moreover, by Lemma 6 they can be substituted for one another when they are merely constituents of the arguments of P. Hence Theorem 3(c) can be rephrased thus:

If A and B are logically equivalent, then A and B
are indiscernible under every relative probability function P.

Probabilities and Truth-Values

Relative probability functions are also called *conditional* ones. The appellation proves to be a particularly apt one as regards the next result, which is to the effect that (i) to every truth-value function TV there corresponds a *two-valued* conditional probability function P of Kolmogorov's, and (ii) to every P of that sort there corresponds a truth-value function TV, such that:

$$TV(B \supset A) = P(A,B) \text{ for every } A \text{ and } B \text{ of } L_P.$$

Proof of it is in two parts.

Part One: Let TV be an arbitrary truth-value function, and let P be the binary function defined thus:

$$P(A,B) = TV(B \ldots A).$$[9]

Then:

(1) Since $0 \leq TV(B \supset A)$ and $TV(A \supset A) = 1$, P meets constraints **RP1** and **RP2.**

(2) Suppose B is P-normal. Then $P(C,B) \neq 1$ for some statement C, i.e. $TV(B \supset C) \neq 1$ for some statement C. But if $TV(B \supset C) = 0$, then $TV(B) = 1$, and in that case $TV(B \supset \neg A) = 1$ if and only if $TV(B \supset A) = 0$. So, $TV(B \supset \neg A) + TV(B \supset A) = 1$. So, P meets constraint **RP3.**

(3) Since $TV(C \supset (A \wedge B)) = TV((B \wedge C) \supset A) \times TV(C \supset B)$, $TV(C \supset (A \wedge B)) = TV(C \supset (B \wedge A))$, and $TV((B \wedge C) \supset A) = TV((C \wedge B) \supset A)$, P meets constraints **RR4–RP6.**

(4) $TV((C \vee \neg C) \supset (C \wedge \neg C)) = 0$. So, P meets constraint **RP7.**

(5) If $TV((A \vee \neg A) \supset B) = 0$, then $TV(B) = 0$ and hence $TV(B \supset A) = 1$ for any statement A of L_P. So, P meets constraint **RK.**

So, P constitutes a two-valued *conditional* probability function of Kolmogorov's such that:

$$P(A,B) = TV(B \supset A) \text{ for every } A \text{ and } B \text{ of } L_P.$$

Part Two: Let P be an arbitrary two-valued conditional probability function of Kolmogorov's, and let TV be the unary function defined thus:

$$TV(A) = P_T(A).$$

Then:

(1) $P_T(\neg A) = 1 - P_T(A)$ by **RP3** and Lemma 4(t) of Appendix I. Hence

$$TV(\neg A) = 1 - TV(A).$$

(2) Note that by Lemma 4(s) and IE $P_T(A \wedge B) = P(A,B) \times P_T(B)$ and $P_T(A \wedge B) = P(B,A) \times P_T(A)$. So, $P_T(A \wedge B) = 0$ if either of $P_T(A)$ and $P_T(B)$ equals 0. Suppose, on the other hand, that both equal 1. Then $P_T(\neg B) = 0$ by **RP3** and Lemma 4(t) and, since $P_T(A \wedge \neg B) = P(A,\neg B) \times P_T(\neg B)$ by Lemma 4(s), $P_T(A \wedge \neg B) = 0$. Hence, $P_T(A \wedge B) = 1$ by Lemmas 4(i) and 4(t). Consequently,

$$TV(A \wedge B) = TV(A) \times TV(B).$$

So, TV constitutes a truth-value function. Now

$$P_{\mathrm{T}}(B \supset A) = 1 - P_{\mathrm{T}}(B)) + P_{\mathrm{T}}(B \wedge A)$$

by the definition of '$\supset$', **RP3**, and Lemmas 4(i) and 4(t). Hence

$$P_{\mathrm{T}}(B \supset A) = 1 - P_{\mathrm{T}}(B) + P(A,B) \times P_{\mathrm{T}}(B)$$

by Lemma 4(s) and IE. So, if $P_{\mathrm{T}}(B) = 1$, then $P_{\mathrm{T}}(B \supset A) - P(A,B) = 0$; and, if $P_{\mathrm{T}}(B) = 0$, then $P_{\mathrm{T}}(B \supset A) = 1$, but by **RK** $P(A,B) = 1$ as well. So, $P_{\mathrm{T}}(B \supset A) = P(A,B)$ and hence TV constitutes a truth-value function such that

$$TV(B \supset A) = P(A,B) \textit{ for every } A \textit{ and } B \textit{ of } L_P.$$

Section 3. Probability Functions Defined on Sets of Statements

We have observed that for absolute probability functions $P(A) = 1$ when A is logically true, and hence $P(A) = 0$ when A is logically false. Other semantic notions are similarly reflected in the values of absolute and relative probability functions, e.g. $P(A \supset B) = 1$ and $P(B,A) = 1$ when A logically implies B, and $P(A \wedge B) = 0$ and $P(\neg B,A) = 1$ when A is logically incompatible with B. There is however no simple way in which probability functions reflect semantic notions which involve (possibly infinite) sets of statements, such as the inconsistency of a set of statements or being a logical consequence of a set of statements. This observation motivates investigating probability functions which are defined on sets of statements of L_P.

Absolute Probability Functions

Beginning with absolute probability functions and *finite* sets of statements, we formulate four constraints to characterise the functions. The sets of statements are, as **AS$\wedge$** attests, interpreted *conjunctively* so that for an inconsistent set X of statements $P(X)$ will equal 0.

AS1 $0 \leq P(X)$

AS2 $P(\varnothing) = 1$

AS$\neg$ $P(X \cup \{\neg A\}) = P(X) - P(X \cup \{A\})$

AS$\wedge$ $P(X \cup \{A \wedge B\}) = P(X \cup \{A,B\})$

If P satisfies these constraints, then the function P', defined by

$$P'(A) = P(\{A\}),$$

is sure to meet constraints **AP1–AP6** and hence to constitute an absolute probability function for L_P.

AP1: $0 \leq P(\{A\})$ by **AS1**.

AP2: $P(\{\neg(A \wedge \neg A)\})$

$= 1 - P(\{A \wedge \neg A\})$ (**AS¬**, **AS2**)
$= 1 - P(\{A\} \cup \{\neg A\})$ (**AS∧**)
$= 1 - P(\{A\}) + P(\{A\} \cup \{A\})$ (**AS¬**)
$= 1$

AP3: $P(\{A \wedge B\}) + P(\{A \wedge \neg B\}) = P(\{A,B\}) + P(\{A,\neg B\}) = P(\{A\})$ by **AS∧** and **AS¬**.

AP4: $P(\{A \wedge B\}) \leq P(\{B \wedge A\})$ by **AS∧**.

AP5: $P(\{A \wedge (B \wedge C)\}) \leq P(\{(A \wedge B) \wedge C\})$ by **AS∧**.

AP6: $P(\{A \wedge A\}) \leq P(\{A\})$ by **AS∧**.

Conversely, given any absolute probability function P', one can define a function P that meets constraints **AS1**, **AS2**, **AS¬**, and **AS∧**. When X is a finite non-empty set of statements whose members are, in alphabetical order, $A_1, A_2, \ldots, A_n$, understand by ΛX the conjunction $(\ldots(A_1 \wedge A_2) \wedge \ldots) \wedge A_n$ of those members; and let $\Lambda\varnothing$ be T. The function P defined by

$$P(X) = P'(\Lambda X)$$

is then an absolute probability function of the kind introduced in this section. For:

AS1: $0 \leq P'(\Lambda X)$ by **AP1**.

AS2: $P'(\Lambda\varnothing) = 1$ by **AP2**.

AS¬: $P'(\Lambda(X \cup \{\neg A\}))$
$= P'(\Lambda X \wedge \neg A)$ (IE)
$= P'(\Lambda X) - P'(\Lambda X \wedge A)$ (**AP3**)
$= P'(\Lambda X) - P'(\Lambda(X \cup \{A\}))$ (IE)

AS∧: $P'(\Lambda(X \cup \{A \wedge B\})) = P'(\Lambda X \wedge (A \wedge B)) = P'(\Lambda(X \cup \{A,B\}))$ by IE.

Hence P meets **AS1**, **AS2**, **AS¬**, and **AS∧**.

It is clear then that absolute probability functions defined on statements and those defined on finite sets of statements stand in one-to-one correspondence. In particular, corresponding to Carnap's absolute probability functions, i.e. the functions characterised in Section 1 by the constraint:

AC *If $P'(A) = 0$, then A is logically false,*

one has the absolute probability functions defined on sets of statements which meet the following constraint **ASC**,

ASC *If $P(X) = 0$, then X is logically inconsistent.*

But infinite sets of statements can also be admitted as arguments of absolute probability functions. In this case we add to **AS1**, **AS2**, **AS¬**, and **AS∧** a constraint which relates the probability of any infinite set to that of its finite subsets. If X is infinite, $P(X)$ cannot exceed the value of P for any finite subset X' of X. For the condition that $P(Y) \leq P(Z)$ if $Z \subseteq Y$ which holds for finite sets by Lemma 10(b) should obviously be extended to infinite sets. So for infinite X, $P(X)$ should be smaller than or equal to the *infimum*[10] of the set of reals $\{P(X') : X' \subseteq_f X\}$, where '$X' \subseteq_f X$' abbreviates '$X'$ is a finite subset of X'. Following the common practice of setting $P(X)$ equal to this infimum, we adopt:

ASω $\quad P(X) = \inf_{X'} \{P(X') : X' \subseteq_f X\}$.

Relative Probability Functions

Turning now to relative probability functions on sets of statements, we again first formulate constraints for finite sets and extend consideration to infinite sets later. Suppose P a binary function from the finite sets of statements of L_P to the reals, and suppose it to meet the following six constraints, where X is said to be *P-normal* if $P(Y,X) \neq 1$ for at least one set Y of statements and is said to be *P-abnormal* otherwise:

RS1 *If* $Y \subseteq X$, *then* $P(Y,X) = 1$

RS2 $0 \leq P(Y,X)$

RS3 $P(Z \cup Y,X) = P(Z,Y \cup X) \times P(Y,X)$

RS4 $\varnothing$ *is P-normal*

RS5 *If X is P-normal, then* $P(\{\neg A\},X) = 1 - P(\{A\},X)$

RS6 $P(\{A \wedge B\},X) = P(\{A,B\},X)$[11]

Then, as we go on to show, the function P', defined as

$$P'(A,B) = P(\{A\},\{B\}),$$

is sure to meet constraints **RP1–RP7** and thus to constitute a relative probability function that ranges over the individual statements of L_P.

RP1: $0 \leq P(\{A\},\{B\})$ by **RS2**.

RP2: $P(\{A\},\{A\}) = 1$ by **RS1**.

RP3: If $P(\{B\},\{C\}) \neq 1$ for some B of L_P, then $P(\{A\},\{B\}) + P(\{\neg A\},\{B\}) = 1$ by **RS5**.

RP4: $P(\{A \wedge B\},\{C\})$

$= P(\{A,B\},\{C\})$ (**RS6**)

$= P(\{A\},\{B,C\}) \times P(\{B\},\{C\})$ (**RS3**)

$= P(\{A\},\{B \wedge C\}) \times P(\{B\},\{C\})$ (Lemma 10(f))

RP5: $P(\{A \wedge B\},\{C\}) = P(\{B \wedge A\},\{C\})$ by **RS6**.

RP6: $P(\{A\},\{B \wedge C\}) = P(\{A\},\{C \wedge B\})$ by Lemma 10(f).

RP7: $P(\{A \wedge \neg A\},\varnothing) = 0$ by **RS4** and Lemma 10(h)). So, $P(\{A \wedge \neg A\},\{\neg(A \wedge \neg A)\}) = 0$ by Lemma 10(i). Hence $\neg(A \wedge \neg A)$ is P'-normal.

Conversely, given any relative probability function P' in Popper's sense, one can define a function P by means of

$$P(Y,X) = P'(\Lambda Y,\Lambda X)$$

that meets constraints **RS1–RS6**. For

RS1: If $Y \subseteq X$, ΛX logically implies ΛY; hence $P'(\Lambda X,\Lambda Y) = 1$ by Theorem 3(b) and **RP2**.

RS2: $0 \leq P'(\Lambda X,\Lambda Y)$ by **RP1**.

RS3: $P'(\Lambda(Z \cup Y),\Lambda X)$

$= P'(\Lambda Z \wedge \Lambda Y,\Lambda X)$ (IE)

$= P'(\Lambda Z,\Lambda Y \wedge \Lambda X) \times P'(\Lambda Y,\Lambda X)$ (**RP4**)

$= P'(\Lambda Z,\Lambda(Y \cup X)) \times P'(\Lambda Y,\Lambda X)$ (IE)

RS4: $\Lambda\varnothing = \mathsf{T}$ by stipulation and T is P'-normal by Lemma 4(t).

RS5: If X is P-normal, ΛX is P'-normal. Hence $P'(\neg A,\Lambda X) = 1 - P'(A,\Lambda X)$ by **RP3**.

RS6: $P'(A \wedge B,\Lambda X) = P'(\Lambda\{A,B\},\Lambda X)$ by IE.

Consequently relative probability functions defined on statements and those defined on finite sets of statements stand in one-to-one correspondence. In particular, we have corresponding families of Kolmogorov, Carnap 1952, and Carnap 1950 functions. The counterpart of constraint

RK *If $P(B,\mathsf{T}) = 0$, then B is P-abnormal*

is

RSK *If $P(X,\varnothing) = 0$, then X is P-abnormal,*

Carnap's 1952 functions are characterised by

RC52 *If B is P-abnormal, then B is logically false.*

The corresponding constraint for relative probability functions defined on sets of statements is

RSC52 *If X is P-abnormal, then X is logically inconsistent.*

Finally, the counterpart of

RC50 *If $P(A,\mathsf{T}) = 0$, then A is logically false,*

which is characteristic of Carnap's 1950 functions, is

RSC50 *If $P(X,\varnothing) = 0$, then X is logically inconsistent.*

Infinite sets of statements can also be admitted as arguments of relative probability functions. In this case we add to **RS1–RS6** the constraint

RSω $P(Y,X) = \inf_{Y'} \{P(Y',X) : Y' \subseteq_f Y\}$.

By Lemmas 15(e) and 15(f) the sets of statements the relative functions meeting **RS1–RS6** are defined on are understood conjunctively, as were the sets which serve as arguments of the absolute probability functions considered before. Disjunctively understood sets are, however, also of interest in probability theory. In Chapter 5, and again in Chapter 7, we will make use of binary functions G which are interdefinable with absolute probability functions for the sets of statements of L_P. As long as the sets of statements are finite, the definition of G in terms of P takes this simple form:

$$G(X,Y) = P(\{\Lambda X \supset \mathrm{V}Y\}),$$[12]

which leaves no doubt that the first argument is understood conjunctively and the second disjunctively.

Section 4. Intuitionistic Probability Functions

The development of intuitionistic logic was motivated by epistemic considerations, a fact reflected in the account of the meaning of sentential connectives given by intuitionistic semantics. Informal intuitionistic semantics explains the meaning of a connective in terms of what counts as conclusive evidence for complex statements formed with its help. Now the interpretation of probability that we have primarily in mind is that $P(A)$ measures the evidential support received by A. The evidence for A is inconclusive, except when $P(A)$ equals 1. So, our favoured understanding of probability and probability functions amounts to a generalisation of the feature which sustains intuitionistic semantics. Hence intuitionistic probability functions will be functions which respect intuitionistic semantics in the manner in which the probabilities of different statements are related to one another. This means that the formulation of constraints for intuitionistic probability functions must be directly guided by the intuitionistic account of the meaning of connectives.

Absolute Probability Functions for L_I

The functions presented in this section take as arguments statements and sets of statements of a propositional language L_I which differs from L_P only in that '$\vee$' and '$\supset$' are primitive connectives, as they are not definable in intuitionistic logic in terms of '$\wedge$' and '$\neg$'. By an *absolute probability function for L_I* we shall mean any function on the statements of L_I which meets the following constraints **AI1–AI9**.

AI1 $0 \leq P(A))$

AI2 $\max_{A \in L}\{P(A)\} = 1$

AI3 $\max_{B \in L}\{P(A \wedge B)\} = P(A)$

AI4 $P(A \wedge B) \leq P((A \wedge B) \wedge B)$

AI5 $P(A \wedge (B \wedge C)) = P(A \wedge (C \wedge B))$

AI6 $P(A \wedge (B \wedge (C \wedge D))) = P(A \wedge ((B \wedge C) \wedge D))$

AI7 $P(A \wedge (B \vee C)) = P(A \wedge B) + P(A \wedge C) - P(A \wedge (B \wedge C))$

AI8 $P((A \supset B) \wedge C) = \max_{D \in L}\{P(C \wedge D) : P((D \wedge A) \wedge B) = P(D \wedge A)\}$

AI9 $P(A \wedge \neg B) = \max_{C \in L}\{P(A \wedge C) : P(C \wedge B) = 0\}$

The minimum and maximum of a set of real numbers are respectively the smallest and the largest of those numbers. Of course, there need not be a smallest or largest number in a set of reals. Hence constraints **AI2**, **AI3**, **AI8**, and **AI9** imply the existence of the respective maxima; **AI3** e.g. is to the effect that for any B of L_I $P(A \wedge B) \leq P(A)$ and that there does exist at least one B such that $P(A \wedge B) = P(A)$.

AI1–AI6 are obviously based upon the constraints for absolute probability functions for L_P. Being autonomous constraints, they have to guarantee the substitutivity of equivalences involving just conjunction. **AI4–AI6** achieve this by stipulating that equivalent statements are indiscernible under P. (The notion of indiscernibility was introduced at the end of Section 2.) As far as conjunction and disjunction are concerned, intuitionistic logic does not differ from classical logic. Consequently, the relationship between the probabilities of conjunctions and disjunctions is the same as for absolute probability functions for L_P: **AI7** is equivalent to Lemma 1(v).

Conclusive evidence for a negation $\neg B$ is evidence that rules out B, hence it implies something that is incompatible with B. So, $\neg B$ is a logical consequence of any statement C which is inconsistent with B, i.e. it is the weakest statement inconsistent with B and has therefore the highest degree of support among all those statements that are inconsistent with B. If C is any such statement, there can be no evidential support at all for $C \wedge B$: $P(C \wedge B) = 0$. These considerations explain **AI9**. As to **AI8**, note that conclusive evidence for $A \supset B$ is evidence for something which, together with A, implies B. Hence $A \supset B$ is the weakest statement which in conjunction with A implies B; $A \supset B$ therefore has the highest degree of support among all such statements D. And if $D \wedge A$ implies B, the evidential support for $(D \wedge A) \wedge B$ can be no less than that for $D \wedge A$.

Probability Functions for L_P are a subclass of Probability Functions for L_I

Constraints **AI1–AI9** are implied by **AP1–AP6**, as we go on to show.

AI1 = **AP1**.

AI2: By Lemma 2(b) $P(A) \leq 1$ and by **AP2** $P(\neg(A \wedge \neg A)) = 1$.

AI3: By Lemma 1(a) $P(A \wedge B) \leq P(A)$ and by Lemma 1(c) $P(A \wedge A) = P(A)$.

AI4: $P(A \wedge B) = P((A \wedge B) \wedge B)$ by IE.

AI5–AI6: By IE.

AI7: By Lemma 1(v) and IE.

AI8: $P(((A \supset B) \wedge A) \wedge B) = P((A \supset B) \wedge A)$ by IE. Suppose $P((D \wedge A) \wedge B) = P(D \wedge A)$. Then: $P((A \supset B) \wedge D)$

$= P(D \wedge \neg A) + P(D \wedge B) - P((D \wedge \neg A) \wedge B)$ (IE, Lemma 1(v))

$= P(D) - P(D \wedge A) + P((D \wedge A) \wedge B)$ (**AP3**, IE)

$= P(D)$

Hence $P((A \supset B) \wedge C) \geq P(C \wedge D)$ by Lemmas 1(j) and 1(a).

AI9: $P(\neg B \wedge B) = 0$ by Lemma 1(f). Suppose $P(C \wedge B) = 0$. Then $P(C \wedge \neg B) = P(C)$ by **AP3**. So, $P(A \wedge C) \leq P(A \wedge \neg B)$ by Lemmas 1(j) and 1(a).

So, P is an absolute probability function for intuitionistic propositional logic, if it is one for classical propositional logic. However, the classical functions constitute a proper subset of the intuitionistic ones, as the following exemplifies. Define the unary function P for a propositional language with a single atomic statement as follows:

$\mathrm{P}(E_1) = 1/2$

$\mathrm{P}(\neg E_1) = 1/4$

$\mathrm{P}(E_1 \wedge \neg E_1) = 0$

$\mathrm{P}(E_1 \vee \neg E_1) = 3/4$

$\mathrm{P}(E_1 \supset E_1) = 1$

All other values of P are then to be calculated by means of the constraints, on the assumption that every compound statement is equivalent under P to one of E_1, $\neg E_1$, $E_1 \wedge \neg E_1$, $E_1 \vee \neg E_1$, and $E_1 \supset E_1$. It is easy to verify that P meets **AI1–AI9**; yet P clearly does not constitute an absolute probability function of Popper's, since it accords a value other than 1 to the truth-functional truth $E_1 \vee \neg E_1$, which is of course not intuitionistically valid.

Popper's absolute probability functions can be obtained by adding to **AI1–AI9** this constraint, to the effect that there is always conclusive evidence for $A \vee \neg A$, something intuitionistically unacceptable, of course:

LEM $P(A \vee \neg A) = 1.$

Given this additional constraint, **AP1–AP6** are derivable from **AI1–AI9**. For:

AP1 = **AI1**.

AP2: $P(\neg(A \wedge \neg A)) = 1$ by Lemma 7(t).

AP3: $P(A \wedge B) + P(A \wedge \neg B)$ equals $P(A \wedge (B \vee \neg B)) + P(A \wedge (B \wedge \neg B))$ by **AI7**, hence equals $P(A)$ by **LEM**, Lemmas 7(p), 7(l), and 7(r).

AP4: $P(A \wedge B) = P(B \wedge A)$ by Lemma 7(d).

AP5: $P(A \wedge (B \wedge C)) \leq P((A \wedge B) \wedge C)$ by Lemma 7(e).

AP6: $P(A) = P(A \wedge A)$ by Lemma 7(i).

Intuitionistic Absolute Probability Functions for Sets of Statements

Absolute probability functions defined on finite sets of statements of L_I are characterised by the following constraints:

AS1 $0 \leq P(X)$

AS2 $P(\varnothing) = 1$

ASI3 $P(X \cup Y) \leq P(X)$

AS∧ $P(X \cup \{A \wedge B\}) = P(X \cup \{A,B\})$

AS∨ $P(X \cup \{A \vee B\}) = P(X \cup \{A\}) + P(X \cup \{B\}) - P(X \cup \{A,B\})$

ASI⊃ $P(X \cup \{A \supset B\}) = \max_{Y \subseteq_f L} \{P(X \cup Y) : P(Y \cup \{A\}) = P(Y \cup \{A,B\})\}$

ASI¬ $P(X \cup \{\neg A\}) = \max_{Y \subseteq_f L} \{P(X \cup Y) : P(Y \cup \{A\}) = 0\}$

The expression '$Y \subseteq_f L$' which occurs in **ASI⊃** and **ASI¬** signifies that Y ranges over the finite sets of statements of L_I.

Since the sets of statements intended here are finite and understood conjunctively, they correspond to conjunctions of statements. Absolute probability functions for the finite sets of statements of L_I can be turned into absolute probability functions for the statements of L_I, and conversely, as we will show next. The constraints for the former functions are, however, simpler than those for the latter. **AI4–AI5** can be dispensed with because the concept of set already implies that the order of set members is immaterial.

If P is a function which meets the seven constraints, then the function P', defined by

$$P'(A) = P(\{A\}),$$

is sure to meet constraints **AI1–AI9**.

AI1: $0 \leq P(\{A\})$ by **AS1**.

AI2: $P(\{A\}) \leq 1$ by **AS2** and **ASI3**. And $P(\{A \supset A\}) = 1$ by Lemma 11(e) and **AS2**.

AI3: $\max_{B \in L}\{P(\{A \wedge B\})\} = \max_{B \in L}\{P(\{A,B\})\}$ by **AS∧**. But $P(\{A,B\}) \leq P(\{A\})$ by **ASI3** and $P(\{A,B \supset B\}) = P(\{A\})$ by Lemma 11(e). So, $\max_{B \in L}\{P(\{A \wedge B\})\} = P(\{A\})$.

AI4–AI6: By **AS∧**.

AI7: By **AS∧** and **AS∨**.

AI8: By **AIS⊃** and **AS∧**.

AI9: By **AIS¬** and **AS∧**.

Conversely, given any absolute probability function P' for L_I, one can define a function P on the finite sets of statements of L_I which meets **AS1**, **AS2**, **ASI3**, **AS∧**, **AS∨**, **AIS⊃**, and **AIS¬**. When X is a finite non-empty set of statements whose members are, in alphabetical order, $A_1, A_2, \ldots, A_n$, understand by ΛX the conjunction $(\ldots(A_1 \wedge A_2) \wedge \ldots) \wedge A_n$ of those members, and let $\Lambda\varnothing$ be $A \supset A$, A some arbitrary but fixed statement. The function P defined by

$$P(X) = P'(\Lambda X)$$

is then an absolute probability function for the finite sets of statements of L_I. For

AS1: $0 \leq P'(\Lambda X)$ by **AI1**.

AS2: $P'(\Lambda\varnothing) = P'(A \supset A) = 1$ by Lemma 7(m).

ASI3: $P'(\Lambda(X \cup Y)) = P'(\Lambda X \wedge \Lambda Y)$ by Lemma 7(f) and $P'(\Lambda X \wedge \Lambda Y) \leq P'(\Lambda X)$ by Lemma 7(a).

AS$\wedge$: $P'(\Lambda(X \cup \{A \wedge B\})) = P'(\Lambda X \wedge (A \wedge B)) = P'(\Lambda(X \cup \{A,B\}))$ by Lemma 7(f).

AS$\vee$: $P'(\Lambda(X \cup \{A \vee B\})) = P'(\Lambda X \wedge (A \vee B))$ by Lemma 7(f) and $P'(\Lambda X \wedge (A \vee B)) = P'(\Lambda X \wedge A) + P'(\Lambda X \wedge B) - P'(\Lambda X \wedge (A \wedge B))$ by **AI7**. Hence **AS$\vee$** by Lemma 7(f).

ASI$\supset$: $P'(\Lambda(X \cup \{A \supset B\}))$

$$= P'(\Lambda X \wedge (A \supset B)) \qquad \text{(Lemma 7(f))}$$

$$= \max_{D \in L} \{P'(\Lambda X \wedge D) : P'((D \wedge A) \wedge B) = P'(D \wedge A)\} \qquad \textbf{(AI8)}$$

$$= \max_{Z \subseteq_f L} \{P'(\Lambda X \wedge \Lambda Z) : P'((\Lambda Z \wedge A) \wedge B) = P'(\Lambda Z \wedge A)\}$$

Hence **AIS$\supset$** by Lemma 7(f).

AS$\neg$: Analogously.

Hence the absolute probability functions for L_I and those for the finite sets of statements of L_I stand in one-to-one correspondence.

Relative Probability Functions for L_I

The constraints placed on the relative probability functions for L_I are motivated by the same semantic considerations as those for absolute functions. They are in fact equivalent to constraints proposed by Popper and Miller (1986). By a *P-normal* statement of L_I we mean of course a statement B such that $P(A,B) \neq 1$ for at least one statement A of L_I.

RI1 $0 \leq P(A,B)$

RI2 $P(A,A) = 1$

RI3 $P(A \wedge B,C) = P(A,B \wedge C) \times P(B,C)$

RI4 $P(A \wedge B,C) \leq P(B \wedge A,C)$

RI5 $P(A,B \wedge C) \leq P(A,C \wedge B)$

RI6 *At least one statement of L_I is P-normal*

RI7 $P(A,C) + P(B,C) = P(A \wedge B,C) + P(A \vee B,C)$

RI8 $P(A \supset B,C) = \max_{D \in L}\{P(D,C) : P((D \wedge A) \wedge B,C) = P(D \wedge A,C)\}$

RI9 *If C is P-normal, then* $P(\neg A,C) = \max_{D \in L} \{P(D,C) : P(A \wedge D,C) = 0\}$

The close affinity of these constraints with those for relative probability functions for L_P is even more obvious than in the case of absolute probability functions; the only one of **RP1–RP7** not among **RI1–RI9** is

RP3 *If B is P-normal,* $P(A,B) + P(\neg A,B) = 1$.

The arguments which establish that **RI7–RI9** are implied by **RP1–RP7** also parallel those used in the absolute case. And it is not difficult to show that adding the constraint

LEM $P(A \vee \neg A, B) = 1$

to **RI1–RI9** allows the derivation of **RP3**.

Intuitionistic relative probability functions for sets of statements of L_I will be discussed in Chapter 10, where we will also throw more light on the motivation for **RI1–RI9**.

Chapter 2

The Probabilities of Infinitary Statements and of Quantifications

Probability functions for a quantificational language require constraints which link the probability of quantified statements to the probabilities of their instances. Attention is thereby focussed on substitutional accounts of quantification. Such accounts view universal quantifications as conjunctions of their instances, which means, in case there are infinitely many instances, that they function as if they were statements of an infinitary language, i.e. a language permitting conjunctions of infinitely many conjuncts and disjunctions of infinitely many disjuncts.

The substitutional account of quantification is not the only reason for investigating probability functions for an infinitary language. It is also of interest that such a language allows us to represent an infinite set of statements, when conjunctively understood, by a single statement, namely the conjunction of its members, so that e.g. an infinite set of premisses can be condensed into one statement. An account of probability functions for an infinitary language L_ω is given in Section 1. While absolute probability functions for L_ω are just extensions of those for L_P, only a subclass of the relative probability functions for L_P extend to L_ω functions. This is one of the concerns of Section 2, which also considers additional constraints for relative probability functions for L_ω, constraints concerning infinitary conjunctions when they are the second argument of such a function.

In Section 3 objectual and substitutional accounts of the semantics of quantificational languages are rehearsed and classes of probability functions are defined for languages L_Q^0 and L_Q. The probability functions for L_Q^0 represent the objectual and those for L_Q the substitutional account of quantification.

Section 1. Probability Functions for L_ω

The infinitary language L_ω is an extension of L_P whose conjunctions may have countably many conjuncts rather than just the customary two. L_ω resembles the language $L_{\omega_1\omega}$ in Keisler (1971), but up to a point only: (i) whereas Keisler has just the infinitary conjunction sign 'Λ' and the infinitary disjunction one 'V', we also have the finitary (i.e. standard) '$\wedge$' and '$\vee$', this to underscore the fact that L_ω is an *extension* of L_P, and (ii) whereas Keisler treats 'V' as a primitive sign, we define 'V' in terms of 'Λ'.

Peculiar to L_ω will be the conjunction ΛA, where A (= $\langle A_1, A_2, ..., A_n, ...\rangle$) is an empty, finite, or ω-sequence[1] of statements of L_ω to be known for short as a *sequence*.[2] $\mathrm{V}A$

will be short for $\neg\Lambda A^{\neg}$, where $A^{\neg}$ consists of the negations of the various elements of $\boldsymbol{A}$. Our account of $\Lambda\boldsymbol{A}$ and $\mathrm{V}\boldsymbol{A}$, by the way, also differs from Keisler's in that for him $\boldsymbol{A}$ would be a *set* rather than a *sequence* of statements. Ours, already used in Karp (1964), seems the more natural one; it also underlines the present assumption that infinitary conjunctions have at most denumerably many conjuncts.

Semantics for L_ω simply extend the usual truth-functional semantics for L_P, i.e. truth and falsity of complex statements are explained relative to truth-value assignments to atomic statements. A statement of L_ω is *true on an assignment* α of truth-values to the atomic statements of L_ω if and only if (i) in case the statement is atomic, it is assigned the truth-value truth by α; (ii) in case it is of the sort $\neg A$, A is not true on α; (iii) in case it is of the sort $A \wedge B$, both A and B are true on α; and (iv) in case it is of the sort $\Lambda\boldsymbol{A}$, $A_1, A_2, ..., A_n, ...$ are all true on α. The relation of logical consequence for L_ω is then characterised semantically as follows, where A is any statement of L_ω and X any set of such statements:

$$X \vdash A$$

if and only if

A is true on every truth-value assignment to the atomic statements on which every member of X is true.

Constraints for Infinitary Conjunctions

The constraints placed on the absolute probability functions defined on the language L_ω will be the six constraints **AP1–AP6** of Chapter 1 plus a further constraint that deals with infinitary conjunctions. In Section 3 of that chapter we adopted the view that the probability of an infinite set of statements, conjunctively understood, is the infimum of the probabilities of its finite subsets, and hence of the finite conjunctions of its members. Hence an appropriate constraint would be

$$P(\Lambda\boldsymbol{A} \wedge B) = \inf_{Z'} \{P(B \wedge \Lambda Z') : Z' \subseteq_f \mathrm{S}(\boldsymbol{A})\},$$

where $\mathrm{S}(\boldsymbol{A})$ is the set whose members are the elements of the sequence $\boldsymbol{A}$, '$Z' \subseteq_f \mathrm{S}(\boldsymbol{A})$' abbreviates '$Z'$ is a finite subset of $\mathrm{S}(\boldsymbol{A})$', and $\Lambda Z'$ is an arbitrary conjunction of the (finitely many) members of Z'. But since in L_ω the conjuncts of infinite conjunctions occur in a sequence, we can express the infimum as the limit of a sequence of conjunctions, this thanks to Lemma 1(a). So, we adopt

APΛ $\quad P(\Lambda\boldsymbol{A} \wedge B) = \lim_{n\to\infty} P((A_1 \wedge ... \wedge A_n) \wedge B)$

as the characteristic constraint for infinitary conjunctions.[3] Here (i) $\boldsymbol{A}$ is an empty, finite, or ω–sequence of statements of L_ω, (ii) $A_1, A_2, ..., A_n, ...$ are the elements of the sequence $\boldsymbol{A}$, and (iii) $\Lambda\boldsymbol{A}$ is a conjunction with $A_1, A_2, ..., A_n, ...$ as its conjuncts.[4]

As for the constraints placed on the relative probability functions for L_ω, they will be the seven constraints **RP1–RP7** plus this one, which achieves with the help of the limit of a sequence of numbers what **RSω** achieves using the infimum of a set of numbers:

RPΛ $P(\Lambda A, B) = \underset{n\to\infty}{\text{limit}}\, P(A_1 \wedge \ldots \wedge A_n, B)$

Constraints **APΛ** and **RPΛ** entail the *Infinite Additivity Laws* for the absolute and the relative probability functions of L_ω, to wit:

$$\textit{If } P(A_i \wedge A_j) = 0 \textit{ for every } i \textit{ and } j \textit{ with } i \neq j, \textit{ then } P(\mathrm{V}A) = \sum_{i=1}^{\infty} P(A_i)$$

in the absolute case, and this one in the relative case:

$$\textit{If } P(A_i \wedge A_j, C) = 0 \textit{ for every } i \textit{ and } j \textit{ with } i \neq j, \textit{ then } P(\mathrm{V}A, C) = \sum_{i=1}^{\infty} P(A_i, C).$$

We prove Infinite Additivity first for the absolute probability functions of L_ω and then for the relative ones.

Theorem 1. $P(\Lambda A) = \underset{n\to\infty}{\text{limit}}\, P(A_1 \wedge \ldots \wedge A_n)$.

Proof by **AP3** and **AP4**. ¤

Theorem 2. $P(\mathrm{V}A) = \underset{n\to\infty}{\text{limit}}\, P(A_1 \vee \ldots \vee A_n)$.

Proof: $P(\mathrm{V}A)$

$= P(\neg\Lambda\langle\neg A_1, \neg A_2, \ldots, \neg A_n, \ldots\rangle)$ (Def. 'V')

$= 1 - P(\Lambda\langle\neg A_1, \neg A_2, \ldots, \neg A_n, \ldots\rangle)$ (Lemma 2(a))

$= 1 - \underset{n\to\infty}{\text{limit}}\, P(\neg A_1 \wedge \ldots \wedge \neg A_n)$ (Theorem 1)

$= 1 - \underset{n\to\infty}{\text{limit}}\, P(\neg(A_1 \vee \ldots \vee A_n))$ (IE)

$= 1 - \underset{n\to\infty}{\text{limit}}\, [1 - P(A_1 \vee \ldots \vee A_n)]$ (Lemma 2(a))

$= \underset{n\to\infty}{\text{limit}}\, P(A_1 \vee \ldots \vee A_n)$ ¤

Theorem 3. *If* $P(A_i \wedge A_j) = 0$ *for every* i *and* j *with* $i \neq j$, *then* $P(\mathrm{V}A) = \sum_{i=1}^{\infty} P(A_i)$.

Proof: We show, by induction on n, that $P(A_1 \vee \ldots \vee A_n) = \sum_{i=1}^{n} P(A_i)$. By Lemma 1(y) $P((A_1 \vee \ldots \vee A_n) \wedge A_{n+1}) = 0$. Hence

$$P((A_1 \vee \ldots \vee A_n) \vee A_{n+1}) = P(A_1 \vee \ldots \vee A_n) + P(A_{n+1})$$

by Lemma 1(j). So, $P(A_1 \vee \ldots \vee A_n) \vee A_{n+1}) = \sum_{i=1}^{n+1} P(A_i)$. ¤

Theorem 4. $P(\mathrm{V}A, B) = \underset{n\to\infty}{\text{limit}}\, P(A_1 \vee \ldots \vee A_n, B)$.

Proof: Obvious when B is P-abnormal. So, let B be P-normal. Then

$P(\mathrm{V}A, B)$

$= P(\neg\Lambda\langle\neg A_1, \neg A_2, \ldots, \neg A_n, \ldots\rangle, B)$ (Def. 'V')

$= 1 - P(\Lambda\langle\neg A_1, \neg A_2, \ldots, \neg A_n, \ldots\rangle, B)$ (**RP3**)

$= 1 - \underset{n\to\infty}{\text{limit}}\, P(\neg A_1 \wedge \ldots \wedge \neg A_n, B)$ (**RPΛ**)

$= 1 - \underset{n\to\infty}{\text{limit}}\, P(\neg(A_1 \vee \ldots \vee A_n), B)$ (IE)

$= 1 - \underset{n\to\infty}{\text{limit}}\, [1 - P(A_1 \vee \ldots \vee A_n, B)]$ (**RP3**)

$= \underset{n\to\infty}{\text{limit}}\, P(A_1 \vee \ldots \vee A_n, B)$ ¤

Theorem 5. *If* $P(A_i \wedge A_j, C) = 0$ *for every i and j with* $i \neq j$, *then* $P(\bigvee A, C) = \sum_{i=1}^{\infty} P(A_i, C)$.

Proof: We show, by induction on n, that $P(A_1 \vee ... \vee A_n, B) = \sum_{i=1}^{n} P(A_i, B)$. By Lemma 5(a) $P((A_1 \vee ... \vee A_n) \wedge A_{n+1}, B) = 0$. Hence by IE

$$P(A_1 \vee ... \vee A_n) \vee A_{n+1}, B) = P(A_1 \vee ... \vee A_n, B) + P(A_{n+1}, B),$$

the *Finite Additivity Law*. So, $P((A_1 \vee ... \vee A_n) \vee A_{n+1}, B) = \sum_{i=1}^{n+1} P(A_i, B)$. ¤

Probability Functions on Sets of Statements of L_ω

To conclude this section we now formulate constraints appropriate for absolute and relative probability functions defined on sets of statements of L_ω. To be added to the constraints formulated in Chapter 1, Section 3, is the following for absolute probability functions:

ASΛ $\quad P(X \cup \{\Lambda A\}) = \lim_{n\to\infty} P(X \cup \{A_1, ..., A_n\})$,

which is equivalent by Lemma 10(p) to

ASΛ $\quad P(X \cup \{\Lambda A\}) = \inf_{Z'} \{P(X \cup Z') : Z' \subseteq_f \mathrm{S}(A)\}$,

where $\mathrm{S}(A)$ is, as before, the set whose members are the elements of the sequence A, and, in the presence of **ASω**, equivalent to

ASΛ' $\quad P(X \cup \{\Lambda A\}) = P(X \cup \mathrm{S}(A))$.

For relative probability functions we have:

RSΛ $\quad P(\{\Lambda A\}, X) = \lim_{n\to\infty} P(\{A_1, ..., A_n\}, X)$

which is equivalent by Lemma 15(m) to

RSΛ $\quad P(\{\Lambda A\}, X) = \inf_{Z'} \{P(Z', X) : Z' \subseteq_f \mathrm{S}(A)\}$

with $\mathrm{S}(A)$ as just explained, and, in the presence of **RSω**, equivalent to

RSΛ' $\quad P(\{\Lambda A\}, X) = P(\mathrm{S}(A), X)$.

Section 2. Families of Infinitary Relative Probability Functions

Relative Probability Functions for L_ω *Extend Relative Probability Functions for* L_P

Since the relative probability functions for L_ω satisfy the constraints **RP1–RP7**, their restrictions to L_P are of course relative probability functions for L_P. However, not every relative probability function for L_P is the restriction of a L_ω function, which means that the effect of constraint **RPΛ** is not merely to extend finitary to infinitary functions. Shown first will be that the restrictions to L_P of the functions in question meet the following double-limit formula:

DL1 *For any sequence* $\langle A_1,...,A_n,...\rangle$ *of statements,*

$$\lim_{m\to\infty}\lim_{n\to\infty} P(A_1 \vee ... \vee A_m, A_1 \vee ... \vee A_n) = 1.$$

The following three identities hold for any relative probability function P for L_ω, any statement B, and any sequence $\mathbf{A} = \langle A_1,...,A_n,...\rangle$ of statements of L_ω:

$$P(\Lambda\mathbf{A},B) = \lim_{n\to\infty} P(A_1 \wedge ... \wedge A_n,B),$$

$$P(\mathrm{V}\mathbf{A},B) = \lim_{n\to\infty} P(A_1 \vee ... \vee A_n,B),$$

and

$$P(A,\mathrm{V}\mathbf{B}) = \lim_{n\to\infty} P(A,B_1 \vee ... \vee B_n)$$

(The first is **RPΛ**, the second and third are proved as Theorems 4 and 10.) The fourth identity, which may appear equally plausible, is, however, not generally satisfied by relative probability functions for L_ω; it therefore constitutes an additional constraint,

RPΛ1 $P(A,\Lambda\mathbf{B}) = \lim_{n\to\infty} P(A,B_1 \wedge ... \wedge B_n)$,

met by some but not all relative probability functions for L_ω. As we shall show, the restrictions to L_P of the L_ω functions that do satisfy **RPΛ1** are relative probability functions for L_P which meet this second double-limit formula as well as **DL1**:

DL2 *For any sequence* $\langle A_1,...,A_n,...\rangle$ *of statements,*

$$\lim_{m\to\infty}\lim_{n\to\infty} P(A_1 \wedge ... \wedge A_n, A_1 \wedge ... \wedge A_m) = 1.$$[5]

We shall then show that there is a constraint for relative probability functions for L_ω that is weaker than **RPΛ1**, namely

RPΛ2 $$P(A,\Lambda\mathbf{B}) = \begin{cases} \lim_{n\to\infty} P(A,B_1 \wedge ... \wedge B_n) & \textit{if} \\ \quad \lim_{m\to\infty}\lim_{n\to\infty} P(B_1 \wedge ... \wedge B_n, B_1 \wedge ... \wedge B_m) = 1 & \\ 1 \quad \textit{otherwise} & \end{cases}$$

and that this constraint is tantamount to:

RPΛ2' *If* $P(\Lambda\mathbf{B},(B_1 \wedge ... \wedge B_m)) = 0$ *for every m, then* $P(C,\Lambda\mathbf{B}) = 1$ *for every statement C of* L_ω.

RPΛ2, as opposed to **RPΛ1**, does not further reduce the class of relative probability functions for L_P which are restrictions to L_P of L_ω functions.

We begin by establishing further consequences of the constraints for L_ω.

Theorem 6. $P(\Lambda\mathbf{A} \wedge B,C) = \lim_{n\to\infty} P((A_1 \wedge ... \wedge A_n) \wedge B,C)$.

Proof: $P(\Lambda\mathbf{A} \wedge B,C)$

$= P(\Lambda\mathbf{A},B \wedge C) \times P(B,C)$ **(RP4)**

$= \lim_{n\to\infty} P(A_1 \wedge ... \wedge A_n,B \wedge C) \times P(B,C)$ **(RPΛ)**

$= \lim_{n\to\infty} [P(A_1 \wedge ... \wedge A_n,B \wedge C) \times P(B,C)]$

$= \lim_{n\to\infty} P((A_1 \wedge ... \wedge A_n) \wedge B,C)$ (**RP4**) ¤

Theorem 7. $P(\mathrm{V}\boldsymbol{A} \wedge B,C) = \lim_{n\to\infty} P((A_1 \vee ... \vee A_n) \wedge B,C)$.

Proof like that of Theorem 6, but using Theorem 4 in place of **RP**Λ. ¤

Theorem 8. $\lim_{n\to\infty} P(A_1 \vee ... \vee A_n, \mathrm{V}\boldsymbol{A}) = 1$.

Proof by Theorem 4 and **RP2**. ¤

Theorem 9. $P(\mathrm{V}\boldsymbol{A}, A_1 \vee ... \vee A_n) = 1$.

Proof: $P(A_1 \vee ... \vee A_{n+m}, A_1 \vee ... \vee A_n) = 1$ by Lemma 4(bb). So, $P(\mathrm{V}\boldsymbol{A}, A_1 \vee ... \vee A_n) = 1$ by Theorem 4. ¤

Theorem 10. $P(A,\mathrm{V}\boldsymbol{B}) = \lim_{n\to\infty} P(A,B_1 \vee ... \vee B_n)$.

Proof: $P(A,\mathrm{V}\boldsymbol{B})$

$= P(A \wedge \mathrm{V}\boldsymbol{B},\mathrm{V}\boldsymbol{B})$ (Lemma 4(u))

$= \lim_{n\to\infty} P(A \wedge (B_1 \vee ... \vee B_n),\mathrm{V}\boldsymbol{B})$ (Theorem 7)

$= \lim_{n\to\infty} [P(A,(B_1 \vee ... \vee B_n) \wedge \mathrm{V}\boldsymbol{B}) \times P(B_1 \vee ... \vee B_n,\mathrm{V}\boldsymbol{B})]$ (**RP4**)

$= \lim_{n\to\infty} P(A,(B_1 \vee ... \vee B_n) \wedge \mathrm{V}\boldsymbol{B}) \times \lim_{n\to\infty} P(B_1 \vee ... \vee B_n,\mathrm{V}\boldsymbol{B})$

$= \lim_{n\to\infty} P(A,(B_1 \vee ... \vee B_n) \wedge \mathrm{V}\boldsymbol{B})$ (Theorem 8)

$= \lim_{n\to\infty} P(A,B_1 \vee ... \vee B_n)$ (Lemma 4(x), Theorem 9) ¤

Theorems 8 and 10 deliver **DL1** and hence in particular proof that if a relative probability function P for L_P is to serve as the restriction to L_P of one for L_ω, then P must meet **DL1** as a constraint. Proof of the converse result, that any relative probability function for L_P meeting **DL1** is the restriction of a L_ω function, will be found in Chapter 5, which is concerned with the extension of functions from one language to another, more comprehensive one.

Theorem 11. *Let* $\boldsymbol{A} = \langle A_1,...,A_n,...\rangle$ *be any sequence of statements of* L_ω. *Then* $\lim_{m\to\infty} \lim_{n\to\infty} P(A_1 \vee ... \vee A_m, A_1 \vee ... \vee A_n) = 1$.

Proof: $\lim_{m\to\infty} \lim_{n\to\infty} P(A_1 \vee ... \vee A_m, A_1 \vee ... \vee A_n)$

$= \lim_{m\to\infty} P(A_1 \vee ... \vee A_m, \mathrm{V}\boldsymbol{A})$ (Theorem 10)

$= 1$ (Theorem 8) ¤

A Counterexample to ***DL1***

Proof that some relative probability functions for L_P do not meet the constraint

$$\lim_{m\to\infty} \lim_{n\to\infty} P(A_1 \vee ... \vee A_m, A_1 \vee ... \vee A_n) = 1$$

is simple enough. *First*, let P_0 be the total function for L_P such that

$$\mathrm{P}_0(E_i) = 0 \textit{ for each atomic statement } E_i \textit{ of } L_P,$$

a function easily verified to constitute an absolute probability function for L_P. *Second*, E_1, E_2, ..., and E_k being for any k from 1 on the alphabetically first k atomic statements of L_P, let P_k be this function on the statements of L_P compounded from E_1, E_2, ..., and E_k alone:

(i) for each state-description $\pm E_1 \wedge \dots \wedge \pm E_k$,

$$P_k(\pm E_1 \wedge \dots \wedge \pm E_k) = \begin{cases} 1/k & \textit{if } \pm E_i = E_i \textit{ for exactly one } i, \; 1 \leq i \leq k \\ 0 & \textit{otherwise,} \end{cases}$$

and

(ii) for every other statement A of L_P compounded from E_1, E_2, ..., and E_k alone and not logically false,

$$P_k(A) = P_k(W_1) + P_k(W_2) + \dots + P_k(W_h),$$

where W_1, W_2, ..., and W_h are the h ($h \geq 1$) state-descriptions of L_P in E_1, E_2, ..., and E_k to whose disjunction A is logically equivalent, a function easily verified to constitute an absolute probability function on the statements of L_P compounded from E_1, E_2, ..., and E_k alone. And, *third*, let $P(A,B)$ be this total binary function on the statements of L_P:

$$P(A,B) = \begin{cases} P_0(A \wedge B) / P_0(B) & \textit{if } P_0(B) \neq 0 \\ \lim\limits_{k \to \infty} [P_k(A \wedge B) / P_k(B)] & \textit{if } P_0(B) = 0 \textit{ but } P_k(B) \neq 0 \textit{ for some } k \textit{ or other} \\ 1 & \textit{otherwise,} \end{cases}$$

a function which can easily be shown to be a relative probability function for L_P.

These preliminaries attended to, note that, where $h = (2^k - 1) \times 2^{k-n}$ and W_1, W_2, ..., and W_h are the state-descriptions in E_1, E_2, ..., E_k to whose disjunction $E_1 \vee \dots \vee E_n$ ($n \leq k$) is logically equivalent, exactly n of W_1, W_2, ..., and W_h have exactly one atomic statement as an unnegated conjunct. Hence, for $n \leq k$ again,

$$P_k(E_1 \vee \dots \vee E_n) = n/k > 0.$$

But, since for $m \leq n$, the conjunction of $E_1 \vee \dots \vee E_m$ and $E_1 \vee \dots \vee E_n$ is logically equivalent to $E_1 \vee \dots \vee E_m$,

$$P(E_1 \vee \dots \vee E_m, E_1 \vee \dots \vee E_n) = \lim_{k \to \infty} [P_k(E_1 \vee \dots \vee E_m) / P_k(E_1 \vee \dots \vee E_n)].$$

So,

$$\lim_{m \to \infty} \lim_{n \to \infty} P(E_1 \vee \dots \vee E_m, E_1 \vee \dots \vee E_n)$$
$$= \lim_{m \to \infty} \lim_{n \to \infty} \lim_{k \to \infty} (m/k)/(n/k) = \lim_{m \to \infty} \lim_{n \to \infty} m/n = 0 \neq 1.$$

I.e. the function P is a relative probability function for L_P which does not meet constraint **DL1**.

*Counterexamples to **RP∧1***

As reported at the beginning of this section, the counterpart

RP∧1 $P(A, \wedge B) = \lim\limits_{n \to \infty} P(A, B_1 \wedge \dots \wedge B_n)$

of the constraint

$$\textbf{RP}\Lambda \qquad P(\Lambda\boldsymbol{A},B) = \lim_{n\to\infty} P(A_1 \wedge \ldots \wedge A_n,B),$$

does not generally hold true. For proof of this we construct a first relative probability function P for L_ω such that, for a certain statement A and sequence $\boldsymbol{B}$, the limit

$$\lim_{n\to\infty} P(A,B_1 \wedge \ldots \wedge B_n)$$

does not exist, and then construct a second one such that the limit in question does exist but is not equal to $P(A,\Lambda\boldsymbol{B})$.

The first function, call it P_1, is arrived at as follows. (i) With $E_1, E_2, \ldots, E_n, \ldots$ the atomic statements of L_P, let the absolute probability function P_0 for L_P have these values for the state-descriptions in $E_1, E_2, \ldots, E_n$:

$$\mathrm{P}_0(\pm E_1 \wedge \ldots \wedge \pm E_n) = \begin{cases} 2/3 \times 1/2^n & \textit{if } \pm E_1 = \neg E_1 \\ 2/3 \times 1/2^n & \textit{if } \pm E_1 = E_1, \textit{ n is even, and } \pm E_i = E_i \textit{ for all } i, \\ & \qquad 2 \le i \le n \\ 1/3 \times 1/2^n & \textit{if } \pm E_1 = E_1, \textit{ n is odd, and } \pm E_i = E_i \textit{ for all } i, \\ & \qquad 2 \le i \le n \\ 0 & \textit{if } \pm E_1 = E_1, \pm E_i = \neg E_i \textit{ for at least one } i \textit{ from } 2 \\ & \qquad \textit{through } n, \textit{ and the smallest such } i \textit{ is even} \\ 1/2^n & \textit{if } \pm E_1 = E_1, \pm E_i = \neg E_i \textit{ for at least one } i \textit{ from } 2 \\ & \qquad \textit{through } n, \textit{ and the smallest such } i \textit{ is odd,} \end{cases}$$

(ii) for every other statement A of L_P compounded from $E_1, E_2, \ldots,$ and E_n alone and not logically false, let P_0 have this value:

$$\mathrm{P}_0(A) = \mathrm{P}_0(W_1) + \mathrm{P}_0(W_2) + \ldots + \mathrm{P}_0(W_h),$$

where $W_1, W_2, \ldots,$ and W_h are the h ($h \ge 1$) state-descriptions of L_P in $E_1, E_2, \ldots,$ and E_n to whose disjunction A is logically equivalent,

(iii) recursively extend P_0 to an absolute probability function P_ω for L_ω by means of these two clauses, stated here for arbitrary P:

$$P(A \wedge \neg B) = P(A) - P(A \wedge B)$$

and

$$P(A \wedge \Lambda\boldsymbol{B}) = \lim_{n\to\infty} P(A \wedge (B_1 \wedge \ldots \wedge B_n)),$$

and

(iv) define the function P_1 thus:

$$\mathrm{P}_1(A,B) = \begin{cases} \mathrm{P}_\omega(A \wedge B) / \mathrm{P}_\omega(B) & \textit{if } \mathrm{P}_\omega(B) \neq 0 \\ 1 & \textit{otherwise.} \end{cases}$$

P_1 can be shown to meet constraints **RP1–RP7** and **RPΛ**, a task we leave to the reader.

Consider now this value of P_1:

$$\mathrm{P}_1(E_1,E_2 \wedge \ldots \wedge E_n).$$

It equals 1/2 when n is even, and 1/3 when n is odd, this for every n from 2 on. So,

$$\lim_{n\to\infty} P_1(E_1, E_2 \wedge \ldots \wedge E_n)$$

does not exist.

As for the second function, call it P_2, it is generated as follows. (i) With $E_1, E_2, \ldots,$ $E_n, \ldots$ the atomic statements of L_P as before, let the absolute probability function P_0 for L_P have these values for the state-descriptions in $E_1, E_2, \ldots, E_n$:

$$P_0(\pm E_1 \wedge \ldots \wedge \pm E_n) = 1/2^{n+1},$$

(ii) for every other statement A of L_P compounded from $E_1, E_2, \ldots,$ and E_n alone and not logically false, let P_0 have this value:

$$P_0(A) = P_0(W_1) + P_0(W_2) + \ldots + P_0(W_h),$$

where $W_1, W_2, \ldots,$ and W_h are the h ($h \geq 1$) state-descriptions of L_P in $E_1, E_2, \ldots,$ and E_n to whose disjunction A is logically equivalent,

(iii) recursively extend P_0 to an absolute probability function P^1_ω for L_ω by the same clauses as before,

(iv) let P^2_ω be the absolute probability function for L_ω such that $P^2_\omega(E_i) = 1$ for every $i = 1, 2,$ $\ldots$, and

(v) define the function P_2 thus:

$$P_2(A,B) = \begin{cases} P^1_\omega(A \wedge B) / P^1_\omega(B) & \textit{if } P^1_\omega(B) \neq 0 \\ P^2_\omega(A \wedge B) / P^2_\omega(B) & \textit{if } P^1_\omega(B) = 0 \textit{ but } P^2_\omega(B) \neq 0 \\ 1 \quad \textit{otherwise.} \end{cases}$$

This done, let $\boldsymbol{A}$ be the infinite sequence $\langle E_2, \ldots, E_n, \ldots \rangle$. Then by definition $P_2(\neg E_1, E_2 \wedge \ldots \wedge E_n) = 1/2$ for every $n = 1, 2, \ldots$. So, $\lim_{n\to\infty} P_2(\neg E_1, E_2 \wedge \ldots \wedge E_n) = 1/2$. But, since (a) $P^1_\omega(\Lambda\boldsymbol{A})$ equals $\lim_{n\to\infty} P_0(E_2 \wedge \ldots \wedge E_n)$, hence $\lim_{n\to\infty} 1/2^n$, and hence 0, (b) $P^2_\omega(\Lambda\boldsymbol{A}) = 1$, and (c) $P^2_\omega(\neg E_1 \wedge \Lambda\boldsymbol{A}) = 0$, $P_2(\neg E_1, \Lambda\boldsymbol{A})$ has to equal 0. So, $\lim_{n\to\infty} P_2(\neg E_1, E_2 \wedge \ldots \wedge E_n)$ is not equal to $P_2(\neg E_1, \Lambda\boldsymbol{A})$.

Constraints ***RPΛ1*** *and* ***RPΛ2***

Now consider the second double-limit constraint mentioned at the beginning of this section:

DL2 *For any sequence* $\langle A_1, \ldots, A_n, \ldots \rangle$ *of statements,*

$$\lim_{m\to\infty} \lim_{n\to\infty} P(A_1 \wedge \ldots \wedge A_n, A_1 \wedge \ldots \wedge A_m) = 1.$$

Intriguingly enough, for a given sequence $\boldsymbol{B}$,

For every statement A, $P(A, \Lambda\boldsymbol{B}) = \lim_{n\to\infty} P(A, B_1 \wedge \ldots \wedge B_n)$

is equivalent to this case of **DL2**,

$$\lim_{m\to\infty} \lim_{n\to\infty} P(B_1 \wedge \ldots \wedge B_n, B_1 \wedge \ldots \wedge B_m) = 1,$$

which itself is equivalent to

$$P(\Lambda\boldsymbol{B},B_1 \wedge ... \wedge B_n) > 0 \textit{ for at least one } n.$$

It will follow from the first of these two results that the restrictions to L_P of the relative probability functions for L_ω meeting all nine of constraints **RP1–RP7**, **RPΛ**, and **RPΛ1** meet both **DL1** and **DL2**.

Theorem 12. *Let* $\boldsymbol{B} = \langle B_1,...,B_n, ...\rangle$ *be a sequence of statements of* L_ω. *Then the following are equivalent to one another:*

(a) *For every statement A of* L_ω, $P(A,\Lambda\boldsymbol{B}) = \lim_{n\to\infty} P(A,B_1 \wedge ... \wedge B_n)$;

(b) $\lim_{m\to\infty} \lim_{n\to\infty} P(B_1 \wedge ... \wedge B_n,B_1 \wedge ... \wedge B_m) = 1$;

(c) $P(\Lambda\boldsymbol{B},B_1 \wedge ... \wedge B_n) > 0$ *for at least one n.*

Proof: (i) Suppose (a). Then $P(\Lambda\boldsymbol{B},\Lambda\boldsymbol{B}) = \lim_{m\to\infty} \lim_{n\to\infty} P(B_1 \wedge ... \wedge B_n,B_1 \wedge ... \wedge B_m)$ by **RPΛ**. Hence (b) by **RP2**.

(ii) Suppose (b). Then $\lim_{m\to\infty} P(\Lambda\boldsymbol{B},B_1 \wedge ... \wedge B_m) = 1$ by **RPΛ**. But, if $P(\Lambda\boldsymbol{B},B_1 \wedge ... \wedge B_n)$ equalled 0 for every n, then $\lim_{m\to\infty} P(\Lambda\boldsymbol{B},B_1 \wedge ... \wedge B_m)$ would likewise equal 0. Hence (c).

(iii) Suppose (c), and in particular $P(\Lambda\boldsymbol{B},B_1 \wedge ... \wedge B_m) > 0$. Then, by **RP4**,

$$P(A \wedge \Lambda\boldsymbol{B},B_1 \wedge ... \wedge B_m) = P(A,\Lambda\boldsymbol{B} \wedge (B_1 \wedge ... \wedge B_m)) \times P(\Lambda\boldsymbol{B},B_1 \wedge ... \wedge B_m).$$

But, since $P(B_1 \wedge ... \wedge B_m,\Lambda\boldsymbol{B}) = 1$ by **RP2**, **RPΛ**, and Lemma 4(d),

$$P(A,\Lambda\boldsymbol{B} \wedge (B_1 \wedge ... \wedge B_m)) = P(A,\Lambda\boldsymbol{B})$$

by Lemma 4(x). So,

$$\begin{aligned}
P(A,\Lambda\boldsymbol{B})
&= P(A \wedge \Lambda\boldsymbol{B},B_1 \wedge ... \wedge B_m) \,/\, P(\Lambda\boldsymbol{B},B_1 \wedge ... \wedge B_m)\\
&= \lim_{n\to\infty} P(A \wedge (B_1 \wedge ... \wedge B_n),B_1 \wedge ... \wedge B_m) \,/\\
&\qquad \lim_{n\to\infty} P(B_1 \wedge ... \wedge B_n,B_1 \wedge ... \wedge B_m) \qquad \text{(Theorem 6, } \mathbf{RP\Lambda}\text{)}\\
&= \lim_{n\to\infty} [P(A \wedge (B_1 \wedge ... \wedge B_n),B_1 \wedge ... \wedge B_m) \,/\, P(B_1 \wedge ... \wedge B_n,B_1 \wedge ... \wedge B_m)]\\
&= \lim_{n\to\infty} P(A,B_1 \wedge ... \wedge B_n) \qquad (\mathbf{RP4}\text{, IE})
\end{aligned}$$

Hence (a). ¤

So, the relative probability functions for L_ω which meet **RP1–RP7**, **RPΛ** plus **RPΛ1** constitute a subclass of those meeting **RP1–RP7** and **RPΛ**. It follows from Theorem 12 that the restrictions to L_P of the former functions meet the restriction to L_P of the functions meeting the double-limit constraint **DL2**, and, as the two examples illustrate, they constitute a proper subclass of the restrictions to L_P of the relative probability functions for L_ω meeting **RP1–RP7** and **RPΛ**.

Mentioned at the beginning of this section was this weaker limit formula for $P(A,\Lambda\boldsymbol{B})$:

$$\textbf{RP}\Lambda\textbf{2} \qquad P(A,\Lambda\boldsymbol{B}) = \begin{cases} \lim\limits_{n\to\infty} P(A,B_1 \wedge \ldots \wedge B_n) \;\; \textit{if} \\ \qquad \lim\limits_{m\to\infty} \lim\limits_{n\to\infty} P(B_1 \wedge \ldots \wedge B_n, B_1 \wedge \ldots \wedge B_m) = 1 \\ 1 \;\; \textit{otherwise.} \end{cases}$$

The constraint declares $\Lambda\boldsymbol{B}$ P-abnormal if $\lim_{m\to\infty} \lim_{n\to\infty} P(B_1 \wedge \ldots \wedge B_n, B_1 \wedge \ldots \wedge B_m) \neq 1$. Thanks to **RP1**, Lemmas 4(a) and 4(d), and Lemma 5(b), this double limit always exists. It follows from Theorem 12 that **RPΛ2** is a weaker constraint than **RPΛ1**. And, since the function P_2 in the second example does not meet **RPΛ2**, the relative probability functions for L_ω which meet **RP1–RP7** and **RPΛ** plus **RPΛ2** constitute a proper subclass of those meeting **RP1–RP7** and **RPΛ**.

Adopting constraint **RPΛ2** in addition to **RP1–RP7** and **RPΛ** does not affect the class of restrictions to L_P. Indeed any restriction to L_P of some relative probability function for L_ω that meets constraints **RP1–RP7** and **RPΛ** is also the restriction to L_P of a relative probability function for L_ω meeting constraints **RP1–RP7** and **RPΛ** plus **RPΛ2**. As already indicated, this result will not be proved here, but in Chapter 5.[6] However, our second example can be used to illustrate the situation. Let P^1_ω and P^2_ω be as before and define P_3 as follows:

$$P_3(A,B) = \begin{cases} P^1_\omega(A \wedge B) \,/\, P^1_\omega(B) \;\; \textit{if}\; P^1_\omega(B) \neq 0 \\ 1 \;\; \textit{otherwise.} \end{cases}$$

The function P_3 has the same restriction to L_P as the function P_2: indeed, $P^1_\omega(C)$ being distinct from 0 for every statement C of L_P that is not logically false, the function P^2_ω is never invoked in the definition of P_2 when the second argument B is a statement of L_P. But P_3 meets **RPΛ2** (though not **RPΛ1**), whereas P_2 does not.

Finally, we establish that **RPΛ2** is equivalent to this simpler constraint:

RPΛ2' *If* $P(\Lambda\boldsymbol{B}, B_1 \wedge \ldots \wedge B_n) = 0$ *for every n, then* $\Lambda\boldsymbol{B}$ *is P-abnormal.*

Theorem 13. *Let* $\boldsymbol{B} = \langle B_1, \ldots, B_n, \ldots\rangle$ *be a sequence of statements of* L_ω. *Then the following two are equivalent:*

(a) *For every statement A of* L_ω,

$$P(A,\Lambda\boldsymbol{B}) = \begin{cases} \lim\limits_{n\to\infty} P(A,B_1 \wedge \ldots \wedge B_n) \\ \qquad \textit{if} \lim\limits_{m\to\infty} \lim\limits_{n\to\infty} P(B_1 \wedge \ldots \wedge B_n, B_1 \wedge \ldots \wedge B_m) = 1 \\ 1 \;\; \textit{otherwise.} \end{cases}$$

(b) *If* $P(\Lambda\boldsymbol{B}, B_1 \wedge \ldots \wedge B_n) = 0$ *for every n, then* $P(A,\Lambda\boldsymbol{B}) = 1$ *for every statement A of* L_ω.

Proof: (i) Assume (a), let A be any statement of L_ω, and let $P(\Lambda\boldsymbol{B}, B_1 \wedge \ldots \wedge B_n) = 0$ for every n. Then $\lim_{m\to\infty} \lim_{n\to\infty} P(B_1 \wedge \ldots \wedge B_n, B_1 \wedge \ldots \wedge B_m) \neq 1$ by Theorem 12, and hence $P(A,\Lambda\boldsymbol{B}) = 1$ by the assumption. Hence (b).

(ii) Assume (b). If, on the one hand,

$$\lim_{m\to\infty} \lim_{n\to\infty} P(B_1 \wedge ... \wedge B_n, B_1 \wedge ... \wedge B_m) = 1,$$

then

$$P(A,\wedge \boldsymbol{B}) = \lim_{n\to\infty} P(A,B_1 \wedge ... \wedge B_n)$$

by Theorem 12. If, on the other hand,

$$\lim_{m\to\infty} \lim_{n\to\infty} P(B_1 \wedge ... \wedge B_n, B_1 \wedge ... \wedge B_m) \neq 1,$$

then, again by Theorem 12, $P(\wedge\boldsymbol{B},B_1 \wedge ... \wedge B_n) = 0$ for every n, and hence $P(A,\wedge\boldsymbol{B}) = 1$. Hence (a). ¤

Section 3. Probability Functions for L_Q^0 and L_Q

The Language L_Q^0

The language L_Q^0 is an extension of the propositional language L_P. The primitive signs of L_Q^0 are (i) a countable number of predicates, one of them possibly the identity sign '=', (ii) a denumerable number of individual variables, say, $x_1, x_2, ..., x_n, ...$, (iii) at least one and at most denumerably many individual terms, (iv) the two connectives '¬' and '∧', (v) the quantifier letter '∀', and (vi) the two parentheses '(' and ')'. Its atomic statements will be statements of the sort $R(T_1,T_2,...,T_n)$, where R is an n-adic predicate and $T_1, T_2, ..., T_n$ are n not necessarily distinct terms. Its compound statements will be the customary negations, conjunctions, and universal quantifications that can be compounded from these atomic statements. However, to avoid identical quantifiers from overlapping and thereby dispense with the irksome distinction between free and bound variables, we shall count $(\forall x)A$ a statement of L_Q^0 only if the result $A(T/x)$ of replacing the variable x everywhere in A by the term T is itself a statement. As for the substitution instances of a universal quantification $(\forall x)A$, they will of course be the countably many results of replacing the variable x in A by a term. Acknowledged thereby as universal quantifications are also statements of the sort $(\forall x)A$ where x does not occur in A. Each of these quantifications, known as *vacuous quantifications*, has just A as its substitution instance. The set of substitution instances of a quantification $(\forall x)A$, with terms drawn from the set $\mathcal{D}$, will be referred to as $\mathrm{I}((\forall x)A,\mathcal{D})$. If we want to emphasise that '=' is one of the predicates, we refer to the language as $L_{Q=}^0$.

L_Q^0 differs from the usual first-order languages in that it is not a requirement that there be infinitely many individual terms; the set $\mathcal{D}_0$ of these terms is assumed to be non-empty and countable. We will shortly introduce further first-order languages, distinguished from L_Q^0 by the number of terms available. It becomes clear that and why terms additional to those in $\mathcal{D}_0$ are needed, when one considers first-order semantics. Two types of semantics are commonly employed, substitutional and objectual.

Substitutional semantics treat a universal quantification as true when all its substitution instances are, a condition which is reasonable only on the assumption that there exists a term for every individual in the domain of quantification. But the countable set $\mathcal{D}_0$ of individual terms of L^0_Q may of course not contain a name for every individual in the domain. So, in order to ensure that all finite domains and, among infinite domains, at least the denumerable ones can be treated properly, L^0_Q needs to be enlarged by a denumerably infinite set $\mathcal{D}_\omega$ of additional terms.

Objectual semantics let the substitution instance $A(T/x)$ represent all cases of predicating the predicate A of individuals in the domain by letting the reference of T vary over the whole domain of quantification. The definition (see clause (iv) below) requires that the term T does not occur in A. Hence, since a statement may contain any number of nested quantifiers, an unlimited supply of terms is required. And since this may not be forthcoming within L^0_Q, denumerably many additional terms are needed.

So, both semantic approaches require, though for different reasons, a denumerable infinity of additional terms. Adding the denumerably many terms in $\mathcal{D}_\omega$ to L^0_Q results in the language $L^0_Q(\mathcal{D}_\omega)$. The set $\mathcal{D}_0 \cup \mathcal{D}_\omega$, containing all individual and additional terms, is itself denumerably infinite. And since neither of the two semantic approaches distinguishes formally between individual and additional terms, it is convenient to think of $\mathcal{D}_0 \cup \mathcal{D}_\omega$ as an undifferentiated set of terms, whose members are, say, t_1, t_2, ..., t_n, ..., and which the pertinent syntactic and semantic definitional clauses make reference to. Since $L^0_Q(\mathcal{D}_\omega)$, so understood, is indeed the most widely used version of a quantificational language, we refer to it simply as L_Q.

A distinction between individual terms and additional terms can however be made in modal contexts. An individual term, intended to pick out a specific individual, can be used to make modal claims about that individual and should therefore have one and the same individual as referent in all possible worlds. Hence identities $T = T'$, with T and T' individual terms, are determinately true or false, independently of possible worlds. The additional terms, on the other hand, serve to refer collectively to all members of the domain. The domain may however vary in extension from possible world to possible world. Questions of identity across possible worlds should therefore not arise for the additional terms, and identities $T = T'$, with at least one of T and T' an additional term, can be true at some possible world and false at others. These considerations will of course also affect probability statements.

We turn now to the semantics for L_Q, i.e. both substitutional and objectual semantics, and give the respective semantic accounts of logical consequence. '$\vdash_o$' will stand for the consequence relation in the objectual sense, while '$\vdash_s$' will be used for logical consequence in the substitutional sense.[7]

Objectual Semantics for L_Q

D being a domain, we understand by a *D-interpretation of L_Q* any function that pairs each term of L_Q with a member of D and each *n-adic predicate of L_Q* with a subset of the nth power

D^n of D. And, $\mathcal{I}_D$ being a D-interpretation of L_Q and T a term of L_Q, we understand by a *T-variant* of $\mathcal{I}_D$ any D-interpretation of L_Q that is like $\mathcal{I}_D$ except for possibly pairing with T a member of D other than $\mathcal{I}_D(T)$.

Supposing now that D is a domain, A is a statement of L_Q, and $\mathcal{I}_D$ is a D-interpretation of the terms of L_Q, we say that *A is true on* $\mathcal{I}_D$ if and only if:

(i) in the case that A is an atomic statement $R(T_1,T_2,\ldots,T_n)$, the n-tuple $\langle\mathcal{I}_D(T_1),\mathcal{I}_D(T_2),\ldots,\mathcal{I}_D(T_n)\rangle$ belongs to $\mathcal{I}_D(R)$, and in the case that A is an atomic statement of the form $T = T'$, $\mathcal{I}_D(T) = \mathcal{I}_D(T')$;

(ii) in the case that A is a negation $\neg B$, B is not true on $\mathcal{I}_D$,

(iii) in the case that A is a conjunction $B \wedge C$, B and C are both true on $\mathcal{I}_D$, and

(iv) in the case that A is a quantification $(\forall x)B$ and T is any term of L_Q foreign to A, $B(T/x)$ is true on every T-variant of $\mathcal{I}_D$.

The semantic characterisation of logical consequence *in the objectual sense* is then:

$X \vdash_o A$
if and only if
A is true on every D-interpretation
for any domain D on which every statement in S is true.

Substitutional Semantics for L_Q

Substitutional semantics for L_Q ($L_{Q=}$) simply extend the usual truth-functional semantics for L_P, where truth and falsity of complex statements are explained relative to truth-value assignments to the atomic statements of the language. We say of an assignment α of truth-values to the atomic statements of $L_{Q=}$ that it is *identity-normal*, if it heeds self-identity and the substitutivity of identicals, i.e. if $\alpha(T = T) = 1$ for every term T and if, for every atomic statement A, $\alpha(A(T'//T)) = \alpha(A)$, when $\alpha(T = T') = 1$. For conjunctions and negations the usual truth-functional conditions stipulate what it is to be true on a truth-value assignment α, while a quantification $(\forall x)B$ is said to be true on a truth-value assignment α if and only if $B(T/x)$ is true on α for every term T.

Logical consequence *in the substitutional sense* is then characterised semantically as follows:

$X \vdash_s A$
if and only if
A is true on every truth-value assignment (or every identity-normal truth-value assignment,
in the case of $L_{Q=}$*) to the atomic statements on which every member of S is true.*

$\vdash_o$ and $\vdash_s$ are different relations, but the following two results are worth noting.[8]

(1) *If* $X \vdash_o A$, *then* $X \vdash_s A$,

(2) *If A and the statements in X belong to* L_Q^0 *(to* $L_{Q=}^0$*), then* $X \vdash_0 A$ *if and only if* $X \vdash_s A$.

Constraints for Universal Quantifications and Identities

When it comes to characterising probability functions for L_Q, the substitutional account is well-suited to yield appropriate constraints. Since substitutional semantics interprets a universal quantification as an infinite conjunction of its instances, the constraints needed are just those for absolute and relative probability functions for L_ω, suitably adapted, i.e.

AP∀ $P((\forall x)A \wedge B) = \lim_{n\to\infty} P((A(t_1/x) \wedge \ldots \wedge A(t_n/x)) \wedge B)$[9]

and

RP∀ $P((\forall x)A,B) = \lim_{n\to\infty} P(A(t_1/x) \wedge \ldots \wedge A(t_n/x),B)$.

$t_1, t_2, \ldots, t_n, \ldots$ are of course the terms in $\mathcal{D}_0 \cup \mathcal{D}_\omega$. We also note the infimum versions of these constraints which will prove useful later on. For absolute probability functions:

AP∀' $P((\forall x)A \wedge B) = \inf_{X'} \{P(\Lambda X' \wedge B) : X' \subseteq_f \mathrm{I}((\forall x)A, \mathcal{D}_0 \cup \mathcal{D}_\omega)\}$,

where $\mathrm{I}((\forall x)A, \mathcal{D}_0 \cup \mathcal{D}_\alpha)$ is the set of substitution instances $A(T/x)$ obtained with terms from $\mathcal{D}_0 \cup \mathcal{D}_\alpha$, i.e. $\mathrm{I}((\forall x)A, \mathcal{D}_0 \cup \mathcal{D}_\alpha) = \{A(T/x) : T \in \mathcal{D}_0 \cup \mathcal{D}_\alpha\}$. And for relative probability functions:

RP∀' $P((\forall x)A,B) = \inf_{X'} \{P(\Lambda X',B) : X' \subseteq_f \mathrm{I}((\forall x)A, \mathcal{D}_0 \cup \mathcal{D}_\omega)\}$.

When dealing with $L_{Q=}$, we have to add constraints for identity. Not surprisingly, they are, for absolute probability functions,

AP=1 $P(T = T) = 1$

and

AP=2 $P(T = T' \supset (A \equiv A(T'//T))) = 1$,

where $A(T'//T)$ is the result of replacing T by T' at zero or more places in A. For relative probability functions the two constraints are

RP=1 $P(T = T,B) = 1$

and

RP=2 $P(T = T' \supset (A \equiv A(T'//T)),B) = 1$.

Finally, we have to ensure that for individual terms, i.e. the terms in $\mathcal{D}_0$, identities determinately hold or fail to hold. I.e.

AP=3 *If T and T' are terms of* L_Q^0*, then* $P(T = T') = 1$ *or* $P(T = T') = 0$,

and

RP=3 *If T and T' are terms of* L_Q^0*, then* $P(T = T',B) = 1$ *or* $P(T = T',B) = 0$.

Probability Functions for L_Q and L_Q^0

All in all, by an *absolute probability function for L_Q* (i.e. $L_Q^0(\mathcal{D}_\omega)$) we will understand any unary function on the statements of L_Q which meets constraints **AP1–AP6, AP∀, AP=1, AP=2**, and **AP=3**; and by a *relative probability function for L_Q* any binary function on the statements of L_Q which meets constraints **RP1–RP7, RP∀, RP=1, RP=2**, and **RP=3**.

This account of probability functions for L_Q can be labelled substitutional since it, paralleling substitutional semantics, determines the probability of a quantification in terms of the probabilities of its substitution instances. In order to obtain an account that parallels objectual semantics, we say that *P is an absolute (relative) probability function for L_Q^0*, if P is the restriction to the statements of L_Q^0 of an absolute (relative) probability function for L_Q (i.e. $L_Q^0(\mathcal{D}_\omega)$). Calling this account objectual can be justified by noting that the probability of a quantification is not determined merely by the probabilities of its instances in L_Q^0, but with reference to a domain whose membership may well outrun the supply $\mathcal{D}_0$ of individual terms in L_Q^0.

This justification immediately leads to the question whether the supply of terms in L_Q is really sufficient. There being just denumerably many terms in $\mathcal{D}_0 \cup \mathcal{D}_\omega$, the probability functions for L_Q can take account of domains with as many as $\aleph_0$, but not more objects, while objectual semantics envisages domains of any cardinality. In order to overcome this limitation we should bring into the picture sets $\mathcal{D}_\alpha$ of terms whose cardinality α is larger than $\aleph_0$. Any such set $\mathcal{D}_\alpha$ allows one to deal with any number of objects smaller than or equal to α, since assigning a probability value of 1 to identities is a means of ensuring that the number of distinct objects referred to by the terms in $\mathcal{D}_\alpha$ is as small as one likes. Of course, no $\mathcal{D}_\alpha$ can be large enough to cover every possible number of objects. So, in order to take account of totalities of any magnitude we need to consider sets $\mathcal{D}_\alpha$ of every infinite cardinality and the corresponding extensions $L_Q^0(\mathcal{D}_\alpha)$ of L_Q^0.

Non-Denumerable Sets of Terms

$L_Q^0(\mathcal{D}_\alpha)$ is the extension of L_Q^0 which has, over and above the (individual) terms in $\mathcal{D}_0$, the (additional) terms in $\mathcal{D}_\alpha$. More exactly, the statements of $L_Q^0(\mathcal{D}_\alpha)$ are the atomic statements involving the terms in $\mathcal{D}_0 \cup \mathcal{D}_\alpha$ plus all the negations, conjunctions and universal quantifications compounded from these, i.e. the statements as specified in the previous section, except that the terms are those in $\mathcal{D}_0 \cup \mathcal{D}_\alpha$. Absolute probability functions for $L_Q^0(\mathcal{D}_\alpha)$ have to meet **AP1–AP6, AP=1, AP=2, AP=3**, and this generalisation of **AP∀'**, appropriate for sets of terms of any cardinality,

$$\textbf{AP}\forall' \quad P((\forall x)A \wedge B) = \inf_{X'} \{P(\Lambda X' \wedge B) : X' \subseteq_f \mathrm{I}((\forall x)A, \mathcal{D}_0 \cup \mathcal{D}_\alpha)\},$$

where $\mathrm{I}((\forall x)A, \mathcal{D}_0 \cup \mathcal{D}_\alpha)$ is the set of substitution instances $A(T/x)$ obtained with terms from $\mathcal{D}_0 \cup \mathcal{D}_\alpha$, i.e. $\mathrm{I}((\forall x)A, \mathcal{D}_0 \cup \mathcal{D}_\alpha) = \{A(T/x) : T \in \mathcal{D}_0 \cup \mathcal{D}_\alpha\}$. And relative probability functions for $L_Q^0(\mathcal{D}_\alpha)$ meet **RP1–RP7, RP=1, RP=2, RP=3**, and

RP∀' $P((\forall x)A,B) = \inf_{X'} \{P(\Lambda X',B) : X' \subseteq_f \mathrm{I}((\forall x)A,\mathcal{D}_0 \cup \mathcal{D}_\alpha)\}$.

An absolute (relative) probability function for L_Q^0 is then the restriction to L_Q^0 of an absolute (relative) probability function for $L_Q^0(\mathcal{D}_\alpha)$ for some $\mathcal{D}_\alpha$.[10]

Interestingly, even when there are non-denumerably many terms, at most a denumerable number of them play a role in determining the infimum in any particular case:

Theorem 14. *Let P be an absolute probability function for* $L_Q^0(\mathcal{D}_\alpha)$ *and let* $(\forall x)A$ *and B be statements of* $L_Q^0(\mathcal{D}_\alpha)$. *Then there exists a countable subset* $\mathcal{D}$ *of* $\mathcal{D}_0 \cup \mathcal{D}_\alpha$ *such that*
(a) $P((\forall x)A \wedge B) = \inf_{X'} \{P(\Lambda X' \wedge B) : X' \subseteq_f \mathrm{I}((\forall x)A,\mathcal{D})\}$;
(b) *If* $\mathcal{D}'$ *is any countable set of terms with* $\mathcal{D} \subseteq \mathcal{D}' \subseteq \mathcal{D}_0 \cup \mathcal{D}_\alpha$, *then*

$$P((\forall x)A \wedge B) = \inf_{X'} \{P(\Lambda X' \wedge B) : X' \subseteq_f \mathrm{I}((\forall x)A,\mathcal{D}')\}.$$

Proof like that of the next theorem. ¤

Theorem 15. *Let P be a relative probability function for* $L_Q^0(\mathcal{D}_\alpha)$, *and let* $(\forall x)A$ *and B be statements of* $L_Q^0(\mathcal{D}_\alpha)$. *Then there exists a countable subset* $\mathcal{D}$ *of* $\mathcal{D}_0 \cup \mathcal{D}_\alpha$ *such that*
(a) $P((\forall x)A,B) = \inf_{X'} \{P(\Lambda X',B) : X' \subseteq_f \mathrm{I}((\forall x)A,\mathcal{D})\}$;
(b) *If* $\mathcal{D}'$ *is any countable set of terms with* $\mathcal{D} \subseteq \mathcal{D}' \subseteq \mathcal{D}_0 \cup \mathcal{D}_\alpha$, *then*

$$P((\forall x)A,B) = \inf_{X'} \{P(\Lambda X',B) : X' \subseteq_f \mathrm{I}((\forall x)A,\mathcal{D}')\}.$$

Proof: (a) Let $r = \inf_{X'} \{P(\Lambda X',B) : X' \subseteq_f \mathrm{I}((\forall x)A,\mathcal{D}_0 \cup \mathcal{D}_\alpha)\}$. Then there exist terms $T_{n_1}, \ldots, T_{n_{m(n)}}$ such that

$$P(A(T_{n_1}/x) \wedge \ldots \wedge A(T_{n_{m(n)}}/x),B) < r + 1/2^n,$$

this for every n. Let $\mathcal{D}$ be $\bigcup_{n\geq 1}\{T_{n_1},\ldots,T_{n_{m(n)}}\}$. Then $\mathcal{D}$ is countable and

$$\inf_{X'} \{P(\Lambda X',B) : X' \subseteq_f \mathrm{I}((\forall x)A,\mathcal{D})\} = r$$

by the construction of $\mathcal{D}$.

(b) Suppose $\mathcal{D}'$ a countable set of terms with $\mathcal{D} \subseteq \mathcal{D}' \subseteq \mathcal{D}_0 \cup \mathcal{D}_\alpha$. Then, $\mathrm{I}((\forall x)A,\mathcal{D}) \subseteq \mathrm{I}((\forall x)A,\mathcal{D}')$ and

$$\inf_{X'} \{P(\Lambda X',B) : X' \subseteq_f \mathrm{I}((\forall x)A,\mathcal{D}')\} \leq \inf_{X'} \{P(\Lambda X',B) : X' \subseteq_f \mathrm{I}((\forall x)A,\mathcal{D})\}$$

by Lemma 4(d), and since $\mathcal{D}' \subseteq \mathcal{D}_0 \cup \mathcal{D}_\alpha$, $\mathrm{I}((\forall x)A,\mathcal{D}') \subseteq \mathrm{I}((\forall x)A,\mathcal{D}_0 \cup \mathcal{D}_\alpha)$ and

$$P((\forall x)A,B) \leq \inf_{X'} \{P(\Lambda X',B) : X' \subseteq_f \mathrm{I}((\forall x)A,\mathcal{D}')\}$$

by Lemma 4(d) again. So,

$$P((\forall x)A,B) = \inf_{X'} \{P(\Lambda X',B) : X' \subseteq_f \mathrm{I}((\forall x)A,\mathcal{D}')\}$$

by (a). ¤

Much less obviously, this result can be extended to the probability function as a whole, i.e. it can be shown that the values of the probability function depend on at most denumerably many of the terms in $\mathcal{D}_0 \cup \mathcal{D}_\alpha$.

Theorem 16. *Let $\mathcal{D}_\alpha$ be a non-denumerable set of terms. Let P_α be an absolute probability function for $L_Q^0(\mathcal{D}_\alpha)$. There exists then a countable subset $\mathcal{D}^*$ of $\mathcal{D}_0 \cup \mathcal{D}_\alpha$ such that the restriction of P_α to $L_Q^0(\mathcal{D}^*)$ is an absolute probability function for $L_Q^0(\mathcal{D}^*)$.*[11]

Proof like that of the next theorem.

Theorem 17. *Let $\mathcal{D}_\alpha$ be a non-denumerable set of terms and let P_α be a relative probability function for $L_Q^0(\mathcal{D}_\alpha)$. Then there exists a countable subset $\mathcal{D}^*$ of $\mathcal{D}_0 \cup \mathcal{D}_\alpha$ such that the restriction of P_α to $L_Q^0(\mathcal{D}^*)$ is a relative probability function for $L_Q^0(\mathcal{D}^*)$.*

Proof: Let $\mathcal{D}^+ = \{s_{i,j}\}_{i=1,2,\ldots;j=1,2,\ldots}$ be a denumerable set of new terms. Let $\langle A,B\rangle_1$, $\langle A,B\rangle_2$, ..., $\langle A,B\rangle_m$, ... be the denumerably many pairs of statements consisting of a universally quantified statement of $L_Q^0(\mathcal{D}^+)$ and a statement of $L_Q^0(\mathcal{D}^+)$, ordered in such a way that if $\langle A,B\rangle_m$ is the first pair in the ordering to contain the term $s_{i,j}$, then $m > i$. Now define, by recursion on this ordering of pairs of statements, a function * which associates with every new term $s_{i,j} \in \mathcal{D}^+$ a term $s_{i,j}{}^* \in \mathcal{D}_0 \cup \mathcal{D}_\alpha$.
Case 1: $m = 1$. By the hypothesis on the ordering, neither A (= $(\forall x)C$) nor B contains occurrences of any terms in $\mathcal{D}^+$, i.e. A and B contain only terms in $\mathcal{D}_0$; hence A and B are statements of L_Q^0 and therefore of $L_Q^0(\mathcal{D}_\alpha)$. By Theorem 15(a) there exists a countable subset $\mathcal{D}_1 = \{T_{1_1},\ldots,T_{1_n},\ldots\}$ of $\mathcal{D}_0 \cup \mathcal{D}_\alpha$ such that

$$P_\alpha(A,B) = \inf_{X'} \{P_\alpha(\Lambda X',B) : X' \subseteq_f \mathrm{I}((\forall x)C,\mathcal{D}_1)\}$$

So, we stipulate: $s_{11}{}^* = T_{1_1}$, ..., $s_{1j}{}^* = T_{1_j}$, ...
Case 2: $m > 1$. Then A (= $(\forall x)C$) and B are statements of $L_Q^0(\mathcal{D}^+)$ and, in general, both A and B will contain terms in $\mathcal{D}^+$. But if $s_{i,j}$ occurs in either A or B then $m > i$, so that $s_{i,j}{}^*$ has already been determined. Let A^*, C^*, and B^* be the results of replacing, for every term $s_{i,j}$ that occurs in A, C, or B , all occurrences of $s_{i,j}$ in A, C, and B by $s_{i,j}{}^*$, which is of course a term in $\mathcal{D}_0 \cup \mathcal{D}_\alpha$. Then A^*, B^* and C^* are statements of $L_Q^0(\mathcal{D}_\alpha)$ and A^* is $(\forall x)C^*$. By Theorem 15(a) there exists a countable subset $\mathcal{D}_m = \{T_{m_1},\ldots,T_{m_n},\ldots\}$ of $\mathcal{D}_0 \cup \mathcal{D}_\alpha$ such that

$$P_\alpha(A^*,B^*) = \inf_{X'} \{P_\alpha(\Lambda X',B) : X' \subseteq_f \mathrm{I}((\forall x)C^*,\mathcal{D}_m)\}.$$

So, we stipulate: $s_{m1}{}^* = T_{m_1}$, ..., $s_{mj}{}^* = T_{m_j}$, ...

Let $\mathcal{D}^*$ be the countable set of terms $\bigcup_{m=1,\ldots} \mathcal{D}_m$ and let $T_1, T_2, \ldots, T_n, \ldots$ be the terms in $\mathcal{D}^*$. Of course, $\mathcal{D}^* \subseteq \mathcal{D}_0 \cup \mathcal{D}_\alpha$. Further, let P^* be the restriction of P_α to $L_Q^0(\mathcal{D}^*)$. In order to establish that P^* is a relative probability function for $L_Q^0(\mathcal{D}^*)$ all we need to show is that P^* meets **RP∀'**. So let A' be any universally quantified statement $(\forall x)C'$ of $L_Q^0(\mathcal{D}^*)$ and B' any statement of $L_Q^0(\mathcal{D}^*)$. Since $\mathcal{D}^*$ consists of the terms in $\mathcal{D}_0 \cup \mathcal{D}_\alpha$ that are associated with the terms in $\mathcal{D}^+$ by the function *, there must be statements A (= $(\forall x)C$) and B of $L_Q^0(\mathcal{D}^+)$ such that A' is A^* (= $(\forall x)C^*$) and B' is B^*. As A and B are statements of $L_Q^0(\mathcal{D}^+)$, the pair $\langle A,B\rangle$ must occur in the list $\langle A,B\rangle_1$, $\langle A,B\rangle_2$, ..., $\langle A,B\rangle_m$, ... , say as the mth entry. Therefore

$$P_\alpha(A',B') = \inf_{X'} \{P_\alpha(\Lambda X',B') : X' \subseteq_f \mathrm{I}((\forall x)C',\mathcal{D}_m)\}$$

and, since $\mathcal{D}_m \subseteq \mathcal{D}^* \subseteq \mathcal{D}_0 \cup \mathcal{D}_\alpha$,

$$P_\alpha(A',B') = \inf_{X'} \{P_\alpha(\Lambda X',B') : X' \subseteq_f \mathrm{I}((\forall x)C',\mathcal{D}^*)\}$$

by Theorem 15(b). But, since A', B', and C' are statements of $L_Q^0(\mathcal{D}^+)$,

$$P^*(A',B') = P_\alpha(A',B')$$

and

$$P_\alpha(C'(T_1/x) \wedge \ldots \wedge C'(T_n/x),B) = P^*(C'(T_1/x) \wedge \ldots \wedge C'(T_n/x),B).$$

Therefore P^* meets **RP∀'** and consequently is a relative probability function for $L_Q^0(\mathcal{D}^*)$. ¤

As a consequence of this result we have

Theorem 18. *Let P_α be an absolute (relative) probability function for $L_Q^0(\mathcal{D}_\alpha)$ and P_0 its restriction to L_Q^0. There exists then an absolute (relative) probability function P for $L_Q^0(\mathcal{D}_\omega)$ whose restriction to L_Q^0* is *also P_0.*

So, no new probability functions for L_Q^0 can be obtained by considering in addition to $\mathcal{D}_\omega$ further infinite sets of terms. The restrictions to L_Q^0 of probability functions for $L_Q^0(\mathcal{D}_\omega)$ are all the probability functions for L_Q^0 there are. Therefore, the account above of absolute and relative probability functions for L_Q^0 captures all there is to capture and is justly called objectual.

Probability Functions on Sets of Statements of L_Q

To conclude this section we now formulate constraints appropriate for absolute and relative probability functions defined on sets of statements of L_Q. To be added to the constraints formulated in Chapter 1, Section 3, is the following for absolute probability functions:

AS∀ $P(X \cup \{(\forall x)A\}) = \lim_{n\to\infty} P(X \cup \{A(\mathrm{t}_1/x),\ldots,A(\mathrm{t}_n/x)\})$,[12]

which is equivalent by Lemma 10(q) to

AS∀ $P(X \cup \{(\forall x)A\}) = \inf_{Z'} \{P(X \cup Z') : Z' \subseteq_f \mathrm{I}((\forall x)A,\mathcal{D}_0 \cup \mathcal{D}_\omega)\}$,

where $\mathrm{I}((\forall x)A,\mathcal{D}_0 \cup \mathcal{D}_\omega)$ is the set of substitution instances $A(T/x)$ of $(\forall x)A$, and, in the presence of **ASω**, equivalent to

AS∀' $P(X \cup \{(\forall x)A\}) = P(X \cup \mathrm{I}((\forall x)A,\mathcal{D}_0 \cup \mathcal{D}_\omega))$.

For relative probability functions we have:

RS∀ $P(\{(\forall x)A\}, X) = \lim_{n\to\infty} P(\{A(\mathrm{t}_1/x),\ldots,A(\mathrm{t}_n/x)\},X)$,

which is equivalent by Lemma 15(n) to

RS∀ $P(\{(\forall x)A\}, X) = \inf_{Z'} \{P(Z',X) : Z' \subseteq_f \mathrm{I}((\forall x)A,\mathcal{D}_0 \cup \mathcal{D}_\omega)\}$,

with $\mathrm{I}((\forall x)A,\mathcal{D}_0 \cup \mathcal{D}_\omega)$ as just explained, and, in the presence of **RSω**, equivalent to

RS∀' $P(\{(\forall x)A\}, X) = P(\mathrm{I}((\forall x)A,\mathcal{D}_0 \cup \mathcal{D}_\omega),X)$.

Chapter 3

Relative Probability Functions and Their T–Restrictions

A one-to-one correspondence between absolute and relative probability functions exists in the case of Kolmogorov functions. We noted in Section 2 of Chapter 1 that if P is a relative probability function for L_P which meets constraint

RK *If* $P(B,\mathrm{T}) = 0$, *then* $P(A,B) = 1$ *for every statement* A *of* L_P,

and if P_T (the T-restriction of P) is the absolute probability function such that

$$P_\mathrm{T}(A) = P(A,\mathrm{T}),^1$$

then

$$P(A,B) = \begin{cases} P_\mathrm{T}(A \wedge B) \,/\, P_\mathrm{T}(B) & \text{if } P_\mathrm{T}(B) \neq 0 \\ 1 & \text{otherwise.} \end{cases}$$

Kolmogorov functions constitute only a subclass of the relative probability functions for L_P, but for any relative probability function P, the T-restriction P_T stands out as a counterpart of P. The present chapter is concerned with the relationship between relative probability functions and their T-restrictions.

The question considered in the first section is whether relative (or conditional) probabilities are probabilities of conditionals, i.e. whether these identities obtain:

$$P(A,B) = P_\mathrm{T}(B \supset A).$$

It will emerge that this relationship between a relative probability function and its T-restriction holds only for a very limited range of functions.

In the second section we identify classes of absolute probability functions which are T-restrictions of Carnap 1950, Carnap 1952, and Kolmogorov functions. The third section deals with probability functions defined on sets of statements and in particular the correspondence between relative functions and the absolute ones which are their ∅-restrictions.

Section 1. Conditional Probabilities and the Probabilities of Conditionals

A conjecture which has had considerable appeal is that

$$P(A,B) = P_\mathrm{T}(B \supset A),$$

where P_T is the T-restriction of P. A more general conjecture has it that the probability of A relative to B plus a further statement equals the probability of 'If B, then A' relative to that further statement. Some writers sought in this way to obtain an analysis of certain conditional statements in terms of conditional or relative probability.

We begin our discussion of the conjecture with a result about truth-functional conditionals, known as the *Excess Law*. The proof is a simplification by Dorn in (1992–93) of a result of Popper's in (1963). Contrary to the expectations of many in the past and possibly still some today,

$$P(B \supset A, C) > P(A, B \wedge C) \textit{ for every } A, B, \textit{ and } C \textit{ of } L_P,$$

provided that

$$\text{(i) } P(B,C) < 1$$

and

$$\text{(ii) } P(A,B \wedge C) < 1.$$

Proof of the law is as follows. By Lemma 4(y), **RP2**, and **RP3**

$$\begin{aligned} &P(B \supset A,C) - P(A,B \wedge C) \\ &\quad = 1 - P(B,C) + P(A,B \wedge C) \times P(B,C) - P(A,B \wedge C) \\ &\quad = (1 - P(B,C)) \times (1 - P(A,B \wedge C)). \end{aligned}$$

But by hypotheses (i) and (ii)

$$(1 - P(B,C)) \times (1 - P(A,B \wedge C)) > 0.$$

So,

$$P(B \supset A,C) > P(A,B \wedge C).$$

In particular, when C is T, we have

$$\textit{If } P(B,\mathrm{T}) < 1 \textit{ and } P(A,B) < 1, \textit{ then } P_{\mathrm{T}}(B \supset A) > P(A,B).$$

Now assume that a relative probability function P does satisfy

$$P(B \supset A,\mathrm{T}) = P(A,B).$$

Then either $P(B,\mathrm{T}) = 1$ or $P(A,B) = 1$ for every statement A, i.e. B is P-abnormal, and hence $P(B,\mathrm{T}) = 0$ by Lemmas 4(t) and 4(w). So, $P(B,\mathrm{T})$ equals either 0 or 1, and when $P(B,\mathrm{T}) = 0$ B is P-abnormal, i.e. P is two-valued and meets constraint **RK**. So,

Theorem 1. *Any relative probability function P for L_P that meets*

$$P(B \supset A,\mathrm{T}) = P(A,B)$$

is a two-valued Kolmogorov function.

But two-valued Kolmogorov functions are truth-value functions, as we saw in Section 2 of Chapter 1, and so the conditional probability $P(A,B)$ is but the truth-value of the conditional $B \supset A$.

Advocates of the idea that certain conditional statements can be analysed in terms of conditional probabilities did of course not have material conditionals in mind. Assume therefore that L_P has been enriched by the addition of the connective '$\rightarrow$', which is intended to be a non-truth-functional connective yielding conditional statements. We shall refer to the resulting language as $L_{\rightarrow}$. Furthermore, assume that '$\rightarrow$' is characterised by this constraint on relative probability functions for $L_{\rightarrow}$:

RP$\rightarrow$ $P(B \rightarrow A,C) = P(A,B \wedge C)$.

Theorem 2. *Let P be binary function defined on the sets of statements of* $L_{\rightarrow}$ *which meets* **RP1–RP7**, *and* **RP$\rightarrow$**. *Then:*
(a) *For every statement B, either* $P(B,\text{T}) = 1$ *or B is P-abnormal;*
(b) *P is a Kolmogorov function;*
(c) *P is 2-valued;*
(d) $P(B \rightarrow A,C) = P(B \supset A,C)$.

Proof: (a) By Lemma 4(cc) and IE,
$P((B \rightarrow A) \wedge A,\text{T}) + P((B \rightarrow A) \wedge B,\text{T}) = P((B \rightarrow A) \wedge (A \wedge B),\text{T}) + P((B \rightarrow A) \wedge (A \vee B),\text{T})$
Hence by **RP4** and **RP$\rightarrow$**, $P(A,A \wedge B) \times P(A,\text{T}) + P(A,B) \times P(B,\text{T})$
$= P(A,A \wedge B) \times P(A \wedge B,\text{T}) + P(A,B) \times P(A \vee B,\text{T})$.
So by Lemma 4(b) and **RP4**, $P(A,\text{T}) = P(A,B) \times P(A \vee B,\text{T})$, hence by Lemma 4(a), $P(A,\text{T}) \leq P(A,B)$. Putting '$\neg B$' for '$A$' yields $P(\neg B,\text{T}) \leq P(\neg B,B)$. But since by **RP2** and **RP3** $P(\neg B,B) = 0$ unless B is P-abnormal, and since by **RP3** and Lemma 4(t) $P(\neg B,\text{T}) = 0$ if and only if $P(B,\text{T}) = 1$, either $P(B,\text{T}) = 1$ or B is P-abnormal.

(b) By (a) P meets **RK**.

(c) If $P(B,\text{T}) = 1$, then, by Lemma 4(x), $P(A,B)$ equals $P(A,\text{T})$, which equals either 1 or 0 by (a) and Lemma 4(w); and if B is P-abnormal, $P(A,B) = 1$.

(d) $P(B \rightarrow A,C) = P(A,B \wedge C)$ by **RP$\rightarrow$**. But $P(A,B \wedge C) = P(B \supset A,C)$ by (b), (c), and Theorem 1. ¤

Hence any connective $\rightarrow$ that is characterised by constraint **RP$\rightarrow$** behaves exactly like the material conditional. Theorem 2 amounts to Lewis's celebrated triviality result in (1976) except that Lewis on the one hand considers in effect only Kolmogorov functions, on the other hand admits partial functions, namely functions for which $P(A,B)$ is not defined, when $P(B,\text{T}) = 0$.

Section 2. Relativising Probability Functions Defined on Statements

Families of Absolute and Relative Probability Functions

As explained before, we define as the T-restriction of a relative probability function P the absolute probability function P_{T} such that

$$P_{\text{T}}(A) = P(A,\text{T}).$$

And, given a set Π of relative probability functions and a set Π' of absolute ones, we shall say that *the functions in* Π *relativise those in* Π' if

(i) each function in Π has one in Π' as its T-restriction, and

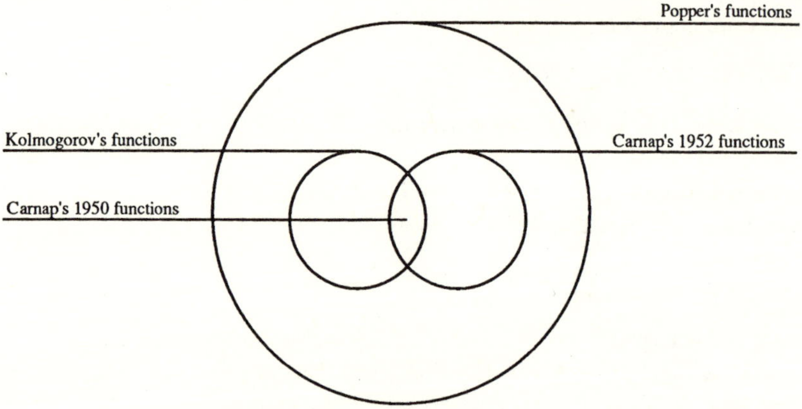

Figure 1

(ii) each function in Π' is the T-restriction of one in Π.

We shall consider sets of probability functions for the three kinds of languages which have been introduced in Chapters 1 and 2: L_P, L_Q, and L_ω; but since many of the results in this section and the next hold for probability functions defined for any of these languages, we shall usually talk of probability functions for a language L in order to achieve greater generality.

Two families of absolute probability functions for L have been introduced in Chapter 1, namely those of Popper and those of Carnap, the latter being a subfamily of the former, characterised by the additional constraint

AC *If $P(A) = 0$, then A is logically false.*

Of relative probability functions for L four families have been distinguished in Chapter 1, namely Popper's functions, Kolmogorov's functions, which meet the additional constraint

RK *If $P(B,\top) = 0$, then $P(A,B) = 1$ for every statement A of L,*

Carnap's 1952 functions, which meet the additional constraint

RC52 *If $P(A,B) = 1$ for every statement A of L, then B is logically false,*

and Carnap's 1950 functions, which meet the additional constraints **RK** and **RC52**, but can also be characterised by the single constraint

RC50 *If $P(A,\top) = 0$, then A is logically false.*

The relationship of the four families of relative probability functions is shown in Figure 1.

We now offer as examples some very simple functions that may serve to illustrate various families of relative probability functions. Table 1 displays a Carnap 1950 function for a

	$P(A,B)$	B: ⊤	E	$\neg E$	⊥
	⊤	1	1	1	1
A	E	1/2	1	0	1
	$\neg E$	1/2	0	1	1
	⊥	0	0	0	1

Table 1

	$P(A,B)$	B: ⊤	E	$\neg E$	⊥
	⊤	1	1	1	1
A	E	0	1	0	1
	$\neg E$	1	0	1	1
	⊥	0	0	0	1

Table 2

language with a single atomic statement E. Values for P are given for the statements E, $\neg E$, ⊤ (i.e. $E \vee \neg E$), and ⊥ (i.e. $E \wedge \neg E$) as arguments; any other compound statement is logically equivalent to one of these and the value of P for such a statement as argument is therefore the same as its value for the statement among E, $\neg E$, ⊤, and ⊥ to which it is equivalent. Table 2 displays a Carnap 1952 function that is not a Carnap 1950 function and Table 3 a Kolmogorov function that is not a Carnap 1950 function.

Carnap Functions

It is important to note that the constraints **AC**, **RC50**, and **RC52**, making use as they do of the semantic notions of logical truth and falsity, are not autonomous constraints. Hence none of the various types of Carnap functions is characterised autonomously. Autonomous characterisations will become available in Chapters 7 and 8 of Part Two, where different sets of constraints for probability functions will be employed. Needed here, for the discussion of relativisation, is a fact already proved in Chapter 1, namely that any absolute probability function of Popper's assigns the value 0 to any logical falsehood and the value 1 to any logical truth. We also draw on the result that if P is a relative probability function, then any logical falsehood A is P-abnormal (this follows from Lemma 4(f) by IE), and consequently $P(A,⊤) = 0$ (by Lemmas 4(w) and 4(t)).

Theorem 3. *Let P be a relative probability function for L, i.e. a function meeting constraints* **RP1–RP7***; and let* $P_⊤$ *be its* ⊤*-restriction. Then:*

(a) $P_⊤$ *is an absolute probability function, i.e. a function meeting* **AP1–AP6***;*

(b) *If P also meets* **RP∀***, then* $P_⊤$ *meets* **AP∀***;*

		B			
	P(*A*,*B*)	T	*E*	¬*E*	⊥
	T	1	1	1	1
A	*E*	0	1	0	1
	¬*E*	1	1	1	1
	⊥	0	1	0	1

Table 3

(c) *If P also meets* **RP**Λ, *then* $P_{\top}$ *meets* **AP**Λ*;*

(d) *P meets* **RC50** *if and only if* $P_{\top}$ *meets* **AC**.

Proof: (a) By Theorem 2 in Chapter 1 and the fact that T is *P*-normal, i.e. Lemma 4(t).

(b) By **RP**$\forall$ and **RP4** $P((\forall x)A \wedge B,\top) = \lim_{n\to\infty} P((A(t_1/x), \wedge \ldots \wedge A(t_n/x)) \wedge B,\top)$. So, $P_{\top}$ meets **AP**$\forall$.

(c) $P(\Lambda A \wedge B,\top) = \lim_{n\to\infty} P((A_1 \wedge \ldots \wedge A_n) \wedge B,\top)$ by Theorem 6 of Chapter 2. So, $P_{\top}$ meets **AP**Λ.

(d) Immediate. ¤

Thus every relative probability function of Popper's – and hence every Kolmogorov as well as every Carnap 1952 one – has an absolute probability function of Popper's as its T-restriction. Moreover, every Carnap 1950 relative probability function has an absolute probability function of Carnap's as its T-restriction.

Theorem 4. *Let* P_1 *be a Popper absolute probability function for L, let* P_2 *be an absolute Carnap function for L, and let P be this binary function:*

$$P(A,B) = \begin{cases} P_1(A \wedge B) / P_1(B) & \textit{if } P_1(B) \neq 0 \\ P_2(A \wedge B) / P_2(B) & \textit{if } P_1(B) = 0 \textit{ but } P_2(B) \neq 0 \\ 1 \quad \textit{if } P_2(B) = 0. \end{cases}$$ [2]

Then:

(a) *P is a Carnap 1952 relative probability function for L;*

(b) *If* P_1 *and* P_2 *meet* **AP**$\forall$, *then P meets* **RP**$\forall$*;*

(c) *If* P_1 *and* P_2 *meet* **AP**Λ, *then P meets* **RP**Λ*;*

(d) *P has* P_1 *as its* T*-restriction.*

Proof: (a) That *P* meets each of **RP1–RP7** and **RC52** is established by cases.

RP1: By **AP1**.

RP2: $P(A,A) = 1$ by definition when $P_2(A) = 0$. So, suppose $P_2(A) \neq 0$. Since $P_1(A \wedge A) = P_1(A)$ and $P_2(A \wedge A) = P_2(A)$ by Lemma 1(c), *P* meets **RP2** whether or not $P_1(A) \neq 0$.

RP3: Suppose *B* is *P*-normal and hence, by the definition of *P*, $P_2(B) \neq 0$. Since $P_1(B) = P_1(A \wedge B) + P_1(\neg A \wedge B)$ and $P_2(B) = P_2(A \wedge B) + P_2(\neg A \wedge B)$ by **AP3** and **AP4**, *P* meets

RP3 whether or not $P_1(B) \neq 0$.

RP4: P meets **RP4** by definition when $P_2(C) = 0$, and hence by Lemma 1(b) $P_2(B \wedge C) = 0$. So, suppose *first* that $P_2(C) \neq 0$, but $P_2(B \wedge C) = 0$. Then $P_1(B \wedge C) = 0$ by **AC**, and hence $P_1((A \wedge B) \wedge C) = 0$ by Lemmas 1(b) and 1(e). So, P meets **RP4** whether or not $P_1(C) \neq 0$. Suppose *next* that both $P_2(C) \neq 0$ and $P_2(B \wedge C) \neq 0$ but $P_1(B \wedge C) = 0$, in which case $P_1(A \wedge (B \wedge C)) = P_1((A \wedge B) \wedge C) = 0$ by Lemmas 1(b) and 1(e). Then P meets **RP4** whether or not $P_1(C) = 0$. Suppose *finally* that both $P_2(C) \neq 0$ and $P_1(B \wedge C) \neq 0$. Then P meets **RP4** by Lemma 1(e).

RP5: By **AP4**.

RP6: By **AP4** and Lemma 1(n).

RP7: $P_1(\mathsf{T}) = 1$ by **AP2**, hence $P(A \wedge \neg A, \mathsf{T}) = 0$ by Lemma 1(f). So, T is P-normal and P meets **RP7**.

RC52: Suppose A is P-abnormal. Then $P_2(A) = 0$ by the definition of P and A is logically false by **AC**, since P_2 is a Carnap function. So, P meets **RC52**.

(b) and (c) Immediate from the definition of P.

(d) Let A be an arbitrary statement of L. Then, since $P_1(\mathsf{T}) = 1$ by **AP2** and $P_1(A \wedge \mathsf{T}) = P_1(A)$ by Lemma 2(d), $P(A,\mathsf{T}) = P_1(A)$. So, P_1 is the T-restriction of P. ¤

Thus each one of Popper's absolute probability function is the T-restriction of a Carnap 1952 relative probability function. So,

Theorem 5. *Carnap's 1952 relative probability functions for L relativise Popper's absolute probability functions for L.*

But each one of Popper's absolute probability function, being the T-restriction of a Carnap 1952 relative probability function, is that of a Popper one. So:

Theorem 6. *Popper's relative probability functions for L relativise Popper's absolute probability functions for L.*

Unlike Carnap's 1952 relative probability functions, which relativise *all* of Popper's absolute probability functions, the rest of Popper's relative probability functions relativise *those only that are not Carnap ones.* In virtue of Theorem 3, any relative probability function P of Popper's that is not a Carnap 1952 one has an absolute probability function P' of Popper's as its T-restriction. But that P' cannot be a Carnap function. Indeed, since P is not a Carnap 1952 function, P cannot be a Carnap 1950 one either. So, P does not meet **RC50**, and as a result P' does not meet **AC**. Thus, those among Popper's relative probability functions that are not Carnap 1952 ones have as their T-restrictions those, *but only those*, among Popper's absolute probability functions that are *not* Carnap ones.

Our next theorem exploits the fact that if a Carnap 1952 relative probability function P is not a Carnap 1950 one, then there is at least one statement B in L that is not logically false and yet such that $P(B,\mathrm{T}) = 0$. It delivers, we shall see, our second result on the relativisation of Popper's absolute probability functions.

Theorem 7. *Let P be a Carnap 1952 relative probability function for L that is not a Carnap 1950 one, and let $P^{\#}$ be this function for L:*

$$P^{\#}(A,B) = \begin{cases} P(A,B) & \textit{if } P(B,\mathrm{T}) \neq 0 \\ 1 & \textit{otherwise.} \end{cases}$$

Then:

(a) *$P^{\#}$ is a Popper relative probability function for L;*

(b) *$P^{\#}$ is not a Carnap 1952 one;*

(c) *$P^{\#}$ has the same* T*-restriction as P.*

Proof: (a) That $P^{\#}$ meets each of the constraints **RP1–RP7** is trivial in most cases and is proved in the others as follows:

RP3: Suppose B is $P^{\#}$-normal. Then $P(B,\mathrm{T}) \neq 0$, hence B is P-normal by Lemma 4(w), and hence $P^{\#}$ automatically meets **RP3**.

RP4: That $P^{\#}$ meets **RP4** is trivially true when $P(C,\mathrm{T}) = 0$ and hence $P(B \wedge C,\mathrm{T}) = 0$ by Lemma 4(e). So, suppose $P(C,\mathrm{T}) \neq 0$ *but* $P(B \wedge C,\mathrm{T}) = 0$, in which case $P((A \wedge B) \wedge C,\mathrm{T}) = 0$ by Lemmas 4(e) and 4(q). Since $P(C,\mathrm{T}) \neq 0$, $P(A \wedge B,C) = P((A \wedge B) \wedge C,\mathrm{T}) / P(C,\mathrm{T})$ and $P(B,C) = P(B \wedge C,\mathrm{T}) / P(C,\mathrm{T})$ by **RP4**, hence $P^{\#}(A \wedge B,C) = P^{\#}(B,C) = 0$, and hence $P^{\#}$ again meets **RP4**. Suppose, on the other hand, that $P(C,\mathrm{T}) \neq 0$ and $P(B \wedge C,\mathrm{T}) \neq 0$. Then $P^{\#}$ automatically meets **RP4**.

RP7: T is $P^{\#}$-normal by Lemma 4(t) and the definition of $P^{\#}$.

(b) Since P is not a Carnap 1950 function, there is a B in L that is not logically false but such that $P(B,\mathrm{T}) = 0$, hence there is a B in L that is not logically false but such that $P^{\#}(A,B) = 1$ for every A in L. Hence $P^{\#}$ does not meet **RC52**, and hence is not a Carnap 1952 function.

(c) Since $P(\mathrm{T},\mathrm{T}) \neq 0$ by **RP2**, $P^{\#}(A,\mathrm{T}) = P(A,\mathrm{T})$ for every A in L, and hence $P^{\#}$ and P have the same T-restriction. ¤

Now let P' be any Popper absolute probability function that is not a Carnap one. By virtue of Theorem 4 there is a Carnap 1952 relative probability function P that has P' as its T-restriction. But P cannot be a Carnap 1950 function: if it met **RC50**, then P' would meet **AC**, which by hypothesis it cannot do. So, P' is the T-restriction of a Carnap 1952 relative probability function that is not a Carnap 1950 one. Thus, by virtue of Theorem 7, P' is also the T-restriction of a Popper relative probability function that is not a Carnap 1952 one. So,

Theorem 8. *Those among Popper's relative probability functions for L that are not Carnap 1952 ones relativise those, but only those, among Popper's absolute probability functions for L that are not Carnap ones.*

Kolmogorov Functions

Turning now to Kolmogorov's relative probability functions, we first recall that the T-restriction of such a function is just the absolute probability function from which it is defined.

Theorem 9. *Let P be a Kolmogorov relative probability function for L. Then:*

$$P(A,B) = \begin{cases} P_{\mathrm{T}}(A \wedge B) \,/\, P_{\mathrm{T}}(B) & \textit{if } P_{\mathrm{T}}(B) \neq 0 \\ 1 \textit{ otherwise.} \end{cases}$$

Proof: If $P(B,\mathrm{T}) \neq 0$, then $P(A \wedge B,\mathrm{T}) = P(A,B) \times P(B,\mathrm{T})$ by **RP4** and IE, and hence $P(A,B) = P_{\mathrm{T}}(A \wedge B) \,/\, P_{\mathrm{T}}(B)$. And if $P(B,\mathrm{T}) = 0$, $P(A,B) = 1$ by **RK**. ¤

Since every Kolmogorov relative probability function P is a Popper one, its T-restriction is by virtue of Theorem 3 sure to be an absolute probability function of Popper's, and, more specifically, a Carnap one if P is a Carnap 1950 function, a non-Carnap one otherwise. And it is easily verified that, if P_1 is any Popper absolute probability function, the function

$$P(A,B) = \begin{cases} P_1(A \wedge B) \,/\, P_1(B) & \textit{if } P_1(B) \neq 0 \\ 1 \quad \textit{otherwise} \end{cases}$$

is a Kolmogorov relative probability function, and that P is a Carnap 1950 function if and only if P_1 is a Carnap absolute function. So,

Theorem 10. (a) *Kolmogorov's relative probability functions for L relativise Popper's absolute probability functions for L;*
(b) *Carnap's 1950 relative probability functions for L relativise Carnap absolute ones for L;*
(c) *Those of Kolmogorov's relative probability functions that are not Carnap 1950 ones relativise those of Popper's absolute probability functions that are not Carnap ones.*

The result can obviously be strengthened thus:

Theorem 11. (a) *Kolmogorov's relative probability functions for L match one-to-one Popper's absolute probability functions for L;*
(b) *Carnap 1950 relative probability functions for L and Carnap absolute ones for L match one-to-one;*
(c) *Those of Kolmogorov's relative probability functions that are not Carnap 1950 ones match one-to-one those of Popper's absolute probability functions that are not Carnap ones.*

We now turn to those of Popper's relative probability functions that are neither Carnap 1952 functions nor Kolmogorov ones. Suppose P such a function. By virtue of Theorem 3 P_{T} is an absolute probability function of Popper's for L. Furthermore, since P does not meet **RC52**, there is a statement of L, say D_1, that is not logically false but P-abnormal, so that $P_{\mathrm{T}}(D_1) = 0$ by Lemmas 4(w) and 4(t). And since P does not meet **RK** either, there is a statement of L, say D_2, such that D_2 is P-normal and $P_{\mathrm{T}}(D_2) = 0$. As a result the statement $\neg D_1 \wedge D_2$ is not logically false. For suppose $\neg D_1 \wedge D_2$ is logically false. Then $\neg D_1 \wedge D_2$ is

P-abnormal and hence $P(\neg D_1 \wedge D_2, D_2) = 0$ by Lemma 4(w), $P(\neg D_1, D_2) = 0$ by Lemma 4(u), and, since D_2 is P-normal, $P(D_1, D_2) = 1$ by **RP3**. However, $P(D_1, D_2) = 0$ by Lemma 4(w). So, as claimed, $\neg D_1 \wedge D_2$ is not logically false, i.e. D_2 does not logically imply D_1. So:

Theorem 12. *Let P be a relative probability function of Popper's for L that is neither a Carnap 1952 function nor a Kolmogorov one. Then $P_\top$ is an absolute probability function of Popper's which is not a Carnap one and there are statements D_1 and D_2 of L such that*

(i) *D_1 and D_2 are not logically false,*
(ii) *D_2 does not logically imply D_1,*
(iii) $P_\top(D_1) = P_\top(D_2) = 0$.[3]

Theorem 13. *Let D_1 and D_2 be statements of L and assume that neither D_1, nor D_2, nor $\neg D_1 \wedge D_2$ is logically false. Let P_1 be an absolute probability function of Popper's such that $P_1(D_1) = P_1(D_2) = 0$, let P_2 be a Carnap absolute probability function, and let P be this binary function:*

$$P(A,B) = \begin{cases} P_1(A \wedge B) \,/\, P_1(B) & \text{if } P_1(B) \neq 0 \\ P_2((A \wedge B) \wedge \neg D_1) \,/\, P_2(B \wedge \neg D_1) & \text{if } P_1(B) = 0 \text{ but } P_2(B \wedge \neg D_1) \neq 0 \\ 1 & \text{if } P_2(B \wedge \neg D_1) = 0 \end{cases}$$ [4]

Then:

(a) *P is a Popper relative probability function;*
(b) *P is not a Carnap 1952 relative probability function;*
(c) *P is not a Kolmogorov relative probability function;*
(d) *P has P_1 as its $\top$-restriction;*
(e) *If P_1 and P_2 meet* **AP∀**, *then P meets* **RP∀**;
(f) *If P_1 and P_2 meet* **APΛ**, *then P meets* **RPΛ**.

Proof: (a) That P meets each of **RP1–RP7** is established by cases.

RP1: By **AP1**;

RP2: $P(A,A) = 1$ by definition when $P_2(A \wedge \neg D_1) = 0$. So, suppose $P_2(A \wedge \neg D_1) \neq 0$. Since by IE $P_1(A \wedge A) = P_1(A)$ and $P_2((A \wedge A) \wedge \neg D_1) = P_2(A \wedge \neg D_1)$, P meets **RP2** whether or not $P_1(A) \neq 0$.

RP3: Suppose B is P-normal and hence, by the definition of P, that $P_2(B \wedge \neg D_1) \neq 0$. Since $P_1(B) = P_1(A \wedge B) + P_1(\neg A \wedge B)$ and $P_2(B \wedge \neg D_1) = P_2(A \wedge (B \wedge \neg D_1)) + P_2(\neg A \wedge (B \wedge \neg D_1))$ by **AP3** and **AP4**, P meets **RP3** whether or not $P_1(B) \neq 0$.

RP4: P meets **RP4** by definition when $P_2(C \wedge \neg D_1) = 0$, and hence $P_2((B \wedge C) \wedge \neg D_1) = 0$. So, suppose *first* that $P_2(C \wedge \neg D_1) \neq 0$, but $P_2((B \wedge C) \wedge \neg D_1) = 0$, and hence $P_2(((A \wedge B) \wedge C) \wedge \neg D_1) = 0$. Then $P_1(B \wedge C) = P_1((A \wedge B) \wedge C) = 0$ by the hypothesis on P_1, and hence P meets **RP4** whether or not $P_1(C) \neq 0$. Suppose *next* that both $P_2(C \wedge \neg D_1) \neq 0$ and $P_2((B \wedge C) \wedge \neg D_1) \neq 0$, but $P_1(B \wedge C) = 0$, in which case

Relative Probability Functions		Absolute Probability Functions
Carnap 1950	match 1–1	Carnap
Kolmogorov that are not Carnap 1950	match 1–1	Popper that are not Carnap
Carnap 1952 that are not Carnap 1950	relativise	Popper that are not Carnap
Popper that are neither Carnap 1952 nor Kolmogorov	relativise	Popper that evaluate to 0 for at least 2 non-equivalent statements that are not logically false

Table 4

$P_1(A \wedge (B \wedge C)) = P_1((A \wedge B) \wedge C) = 0$ by Lemmas 1(b) and IE. Hence P meets **RP4** whether or not $P_1(C) = 0$. Suppose *finally* that $P_2(C \wedge \neg D_1) \neq 0$, $P_2((B \wedge C) \wedge \neg D_1) \neq 0$, and $P_1(B \wedge C) \neq 0$, in which case $P_1(C) \neq 0$ by Lemma 1(b). Then P meets **RP4** by Lemma 1(e).

RP5: By **AP4**.

RP6: By **AP4** and IE.

RP7: $P_1(\mathsf{T}) = 1 \neq 0$ by **AP2** and $P_1((A \wedge \neg A) \wedge \mathsf{T}) = 0$ by Lemma 1(g), hence $P(A \wedge \neg A,\mathsf{T}) = 0$. So, T is P-normal and P meets **RP7**.

(b) Since $P_2(D_1 \wedge \neg D_1) = 0$ by Lemma 1(f), $P(A,D_1) = 1$ for every A in L. But by hypothesis D_1 is not logically false. So, P does not meet **RC52**.

(c) Since $P_1(\mathsf{T}) = 1$, and $P_1(D_2 \wedge \mathsf{T}) = 0$ by Lemma 1(b) and the hypothesis on P_1, $P(D_2,\mathsf{T}) = P_1(D_2,\mathsf{T}) / P_1(\mathsf{T}) = 0$. But since by hypothesis $P_1(D_2) = 0$ and $D_2 \wedge \neg D_1$ is not logically false, $P_2(D_2 \wedge \neg D_1) \neq 0$ by **AC** and $P(A \wedge \neg A,D_2) = P_2(((A \wedge \neg A) \wedge D_2) \wedge \neg D_1) / P_1(D_2 \wedge \neg D_1)$, and since $P_2(((A \wedge \neg A) \wedge D_2) \wedge \neg D) = 0$ by Lemma 1(g) and IE, $P(A \wedge \neg A,D_2) = 0$ and D_2 is P-normal. Thus **RK** is not met and P is not a Kolmogorov function.

(d) Let A be an arbitrary statement in L. Since $P_1(\mathsf{T}) = 1 \neq 0$, $P(A,\mathsf{T}) = P_1(A \wedge \mathsf{T})$. But $P_1(A \wedge \mathsf{T}) = P_1(A)$ by IE. Hence $P(A,\mathsf{T}) = P_1(A)$. So, P_1 is the T-restriction of P.

(e) and (f) Immediate from the definition of P. ¤

So, our final relativisation theorem:

Theorem 14. *Those among Popper's relative probability functions that are neither Carnap 1952 functions nor Kolmogorov ones relativise those among Popper's absolute probability functions on L that are not Carnap ones and evaluate to 0 for at least two statements in L that are neither logically false and nor logically equivalent.*[5]

We conclude this section with tables that summarise the foregoing relativisation theorems and corollaries thereof. Table 4 covers the four types of relative probability functions of Popper's that are mutually exclusive; Table 5 covers the remaining 6 types. The results summarised in Table 5 are of course already implicit in Table 4.

Relative Probability Functions		Absolute Probability Functions
Kolmogorov	match 1–1	Popper
Carnap 1952	relativise	Popper
Popper that are not Kolmogorov	relativise	Popper that are not Carnap
Popper that are not Carnap 1952	relativise	Popper that are not Carnap
Popper that are not Carnap 1950	relativise	Popper that are not Carnap
Popper	relativise	Popper

Table 5

Section 3. Relativising Probability Functions Defined on Sets of Statements

In Chapter 1, Section 3, we introduced probability functions that are defined on sets of statements, rather than statements. Absolute functions were characterised by the constraints

AS1 $0 \leq P(X)$

AS2 $P(\varnothing) = 1$

AS¬ $P(X \cup \{\neg A\}) = P(X) - P(X \cup \{A\})$

AS∧ $P(X \cup \{A \wedge B\}) = P(X \cup \{A,B\})$

plus, in case infinite sets of statements are admitted as arguments,

ASω $P(X) = \inf_{X'} \{P(X') : X' \subseteq_f X\}$.

The relative probability functions on sets of statements were characterised by

RS1 *If* $Y \subseteq X$, *then* $P(Y,X) = 1$

RS2 $0 \leq P(Y,X)$

RS3 $P(Z \cup Y,X) = P(Z,Y \cup X) \times P(Y,X)$

RS4 $\varnothing$ *is P-normal*

RS5 *If X is P-normal, then* $P(\{\neg A\},X) = 1 - P(\{A\},X)$

RS6 $P(\{A \wedge B\},X) = P(\{A,B\},X)$

and, when infinite sets of statements can be arguments of P,

RSω $P(Y,X) = \inf_{Y'} \{P(Y',X) : Y' \subseteq_f Y\}$.

We also formulated constraints corresponding to the various subfamilies of relative probability functions. Among Popper's absolute probability functions that are defined on sets of statements the Carnap ones are singled out by the constraint

ASC *If $P(X) = 0$, then X is logically inconsistent,*

while among Popper's relative functions the Kolmogorov functions meet

RSK *If $P(X,\varnothing) = 0$, then $P(Y,X) = 1$ for every Y,*

the Carnap 1952 functions meet

RSC52 *If $P(Y,X) = 1$ for every Y, then X is logically inconsistent,*

and the Carnap 1950 functions are singled out by

RSC50 *If $P(X,\varnothing) = 0$, then X is logically inconsistent.*

We will now offer proof that Popper's relative probability functions defined on sets of statements relativise Popper's absolute ones. By $P_\varnothing$ we mean of course the $\varnothing$-restriction of P, i.e. the function such that

$$P_\varnothing(S) = P(S,\varnothing).$$

Theorem 15. *Let P be a relative probability function of Popper's defined on sets of statements, i.e. a function meeting constraints* **RS1–RS6**. *Then:*

(a) *$P_\varnothing$ is an absolute probability function defined on sets of statements, i.e. a function meeting* **AS1**, **AS2**, **AS¬**, *and* **AS∧**;

(b) *If P also meets* **RSω**, *then $P_\varnothing$ meets* **ASω**;

(c) *If P also meets* **RS∀**, *then $P_\varnothing$ meets* **AS∀**;

(d) *If P also meets* **RSΛ**, *then $P_\varnothing$ meets* **ASΛ**;

(e) *P meets* **RSC50**, *if and only if $P_\varnothing$ meets* **ASC**.

Proof: (a) **AS1**: $0 \le P(Y,X)$ by **RS2**.

AS2: $P(\varnothing,\varnothing) = 1$ by **RS1**.

AS¬: $P(X \cup \{\neg A\},\varnothing) = P(X,\varnothing) - P(X \cup \{A\},\varnothing)$ by Lemma 15(d) and **RS4**.

AS∧: $P(X \cup \{A \wedge B\},\varnothing) = P(X \cup \{A,B\},\varnothing)$ by Lemma 15(e).

(b) **ASω**: $P(X,\varnothing) = \inf_{X'} \{P(X',\varnothing) : X' \subseteq_f X\}$ by **RSω**.

(c) $P(\{X \cup (\forall x)A\},\varnothing)$

$$= P(\{(\forall x)A\},X) \times P(X,\varnothing) \qquad \textbf{(RS3)}$$
$$= \lim_{n\to\infty} P(\{A(t_1/x),\ldots,A(t_n/x)\},X) \times P(X,\varnothing) \qquad \textbf{(RS}\forall\textbf{)}$$
$$= \lim_{n\to\infty} P(X \cup \{A(t_1/x),\ldots,A(t_n/x)\}) \qquad \textbf{(RS3)}$$

(d) Like (c).

(e) Immediate. ¤

Conversely, one can obtain a relative probability function from an absolute one by Kolmogorov's device.

Theorem 16. *Let P_1 be a Popper absolute probability function for sets of statements, and let P be this binary function:*

$$P(Y,X) = \begin{cases} P_1(Y \cup X) \,/\, P_1(X) & \text{if } P_1(X) \neq 0 \\ 1 & \text{otherwise.} \end{cases}$$

Then:

(a) *P is a Kolmogorov relative probability function for sets of statements;*

(b) *If P_1 meets* **AS**ω*, then P meets* **RS**ω*;*

(c) *If P_1 meets* **AS**$\forall$*, then P meets* **RS**$\forall$*;*

(d) *If P_1 meets* **AS**Λ*, then P meets* **RS**Λ*;*

(e) *P has P_1 as its $\varnothing$-restriction.*

Proof: (a) That P meets each of **RS1–RS6** and **RSK** is immediate in most cases; so only some of them are proved here:

RS3: $P(Z \cup Y,X) = P(Z,Y \cup X) \times P(Y,X)$ by definition when $P_1(X) = 0$, and hence, by Lemma 10(c), $P_1(Y \cup X) = 0$. So, suppose *first* that $P_1(X) \neq 0$ but $P_1(Y \cup X) = 0$ and therefore $P_1(Z \cup Y \cup X) = 0$, in which case **RS3** holds. Suppose *next* that $P_1(X) \neq 0$ and $P_1(Y \cup X) \neq 0$, in which case **RS3** holds, too. So P meets **RS3**.

RS4: $P_1(\{A \wedge \neg A\} \cup \varnothing) = 0$ by Lemma 10(f). So, since $P_1(\varnothing) = 1$ by **AS2**, $P(\{A \wedge \neg A\},\varnothing) = 0$ and hence $\varnothing$ is P-normal.

RS5: Suppose X is P-normal. Then $P_1(X) \neq 0$ by the definition of P and hence $P(\{\neg A\},X) = 1 - P(\{A\},X)$ by **AS3**.

RSK: Suppose $P(X,\varnothing) = 0$. Then $P_1(X) = 0$ by the definition of P and **AS2**. But then $P(Y,X) = 1$ for every Y by the definition of P. So, P meets **RSK**.

(b), (c), and (d) Immediate from the definition of P.

(e) Let X be an arbitrary set of statements of L. Then, since $P_1(\varnothing) = 1$ by **AS2**, $P(X,\varnothing) = P_1(X)$. So, P_1 is the $\varnothing$-restriction of P. ¤

So, Kolmogorov's, and hence Popper's, relative probability functions that are defined on sets of statements relativise Popper's absolute functions. All other relativisation results that were obtained in Section 1 and summarised in Tables 4 and 5 can also be duplicated for functions defined on sets of statements. Since the proofs do not contain anything new, we omit them.

Chapter 4

Representing Relative Probability Functions by Means of Classes of Measure Functions

Any Kolmogorov relative probability function P can be represented by its T-restriction P_{T}, which is an absolute probability function, in the sense that P and P_{T} determine one another:

$$P_{\mathrm{T}}(A) = P(A,\mathrm{T}),$$

and

$$P(A,B) = \begin{cases} P_{\mathrm{T}}(A \wedge B) \,/\, P_{\mathrm{T}}(B) & \textit{if } P_{\mathrm{T}}(B) \neq 0 \\ 1 \quad \textit{otherwise.} \end{cases}$$

We also found in Chapter 3, Section 2, that any absolute probability function represents in this sense a Kolmogorov relative function.

Here we deal with the question whether relative probability functions generally can be represented by absolute ones. It is clear that only in the case of Kolmogorov functions will a single absolute probability function be sufficient. In the other cases a multiplicity of unary functions is required and these are generally measure functions (encountered in Chapter 1, Section 1) and not specifically absolute probability functions. Moreover, the class of representing measure functions is ordered by a relation which parallels an ordering of statements by Császár's relation $\geq_P$, defined thus:

$$A \geq_P B \ \textit{ if and only if } \ P(A, A \vee B) > 0.$$

The relative probability function is then determined by this ordered class of measure functions by the definition

$$P(A,B) = \begin{cases} M(A \wedge B) \,/\, M(B) & \textit{if } M \textit{ is the first function in the ordered class with } M(B) > 0 \\ 1 \quad \textit{if there is no function with } M(B) > 0. \end{cases}$$

Császár's relation will be explored in Section 1. In Section 2 so-called Type I-ordered classes of measure functions are identified and it is proved that every relative probability function for L_P can be represented by means of a Type I-ordered class of measure functions and every Type I-ordered class of measure functions on L_P represents a relative probability function. Section 3 focuses on relative probability functions for L_P which meet the constraint **DL1** of Chapter 2. It is found that each of these functions can be represented by a well-ordered class of absolute probability functions for L_P. And Section 4 establishes that, similarly, relative probability functions for L_ω can be represented by a well-ordered class of absolute probability functions for L_ω. Definitions and results which hold for both L_P and L_ω will be phrased in terms of a language L.

Section 1. Császár's Relations

Three relations, which originated in Császár (1955), will play a critical role in what follows. They are defined thus for arbitrary statements A and B of L (which may be L_P or L_ω) and relative probability function P for L:

$$A \geq_P B \text{ if and only if } P(A,A \vee B) > 0,$$

$$A >_P B \text{ if and only if } A \geq_P B \text{ but not } B \geq_P A,$$

$$A \sim_P B \text{ if and only if } A \geq_P B \text{ and } B \geq_P A.$$

Following general practice we shall say that A and B are *P-commensurable* if $A \sim_P B$; and, as the relation will prove to be an equivalence relation, we shall talk of *P-commensurability classes*. As we know of no label for the relation $\geq_P$, we shall refer to it as *Császár's relation.* Incidentally, it follows from the foregoing definition that A and B are P-commensurable if and only if $P(A,A \vee B) \times P(B,B \vee A) > 0$, a point noted by Császár (1955) and Rényi (1964).

We now offer proof that Császár's relation $\geq_P$ is reflexive, serial, and transitive, and that his commensurability relation $\sim_P$, being reflexive, symmetrical and transitive, constitutes an equivalence relation.

Theorem 1. $A \geq_P A$.

Proof: $P(A,A \vee A) = 1$ by **RP2** and IE.

Theorem 2. *If not* $A \geq_P B$, *then* $P(A,A \vee B) = 0$.

Proof by **RP1**.

Theorem 3. *Either* $A \geq_P B$ *or* $B \geq_P A$.

Proof: If not $A \geq_P B$, then $P(A,A \vee B) = 0$ by Theorem 2, hence $P(B,B \vee A) = 1$ by Lemma 4(hh) and IE, and hence $B \geq_P A$. ¤

Theorem 4. $A >_P B$ *if and only if* $P(B,B \vee A) = 0$.

Proof: If $A >_P B$, then not $B \geq_P A$, and hence $P(B,B \vee A) = 0$ by Theorem 2; and if $P(B,B \vee A) = 0$, then not $B \geq_P A$, hence $A \geq_P B$ by Theorem 3, and hence $A >_P B$. ¤

Theorem 5. (a) *If* $A \geq_P B$ *and* $B \geq_P C$, *then* $A \geq_P C$;
(b) *If* $A >_P B$ *and* $B \geq_P C$, *then* $A >_P C$;
(c) *If* $A \geq_P B$ *and* $B >_P C$, *then* $A >_P C$.

Proof: (a) Assume $P(A,A \vee B) > 0$ and $P(B,B \vee C) > 0$. Then $P(B \vee A,(B \vee C) \vee A) > 0$ by Lemma 4(ii). So $P(A \vee B,A \vee (B \vee C)) > 0$ by IE and $P(A,A \vee (B \vee C)) > 0$ by Lemma 4(jj). But

$$P(A,A \vee (B \vee C)) = P(A,A \vee C) \times P(A \vee C,A \vee (C \vee B))$$

by Lemma 4(jj). So $P(A, A \vee C) > 0$ by **RP1**, i.e. $A \geq_P C$.

(b) Suppose $A >_P B$ and $B \geq_P C$. Then $A \geq_P C$ by (a). But if $C \geq_P A$, then $B \geq_P A$ by (a), contrary to the supposition. Hence $A >_P C$.

(c) By a similar reasoning. ¤

Theorem 6. (a) *If* $P(A,B) > 0$, *then* $A \geq_P B$;
(b) *If* $P(A,B) > 0$ *and* $P(B,A) > 0$, *then* $A \sim_P B$.

Proof: $P(A, A \vee B) + (1 - P(A,B)) \times P(B, A \vee B) = 1$ by Lemma 4(kk), and $P(B, B \vee A) \leq 1$ by Lemma 4(a). So, if $P(A,B) > 0$, then $P(A, A \vee B) > 0$. ¤

Theorem 7.[1] (a) $\top \geq_P A$;
(b) *If* $P(A,\top) > 0$, *then* $A \geq_P B$;
(c) $A \geq_P \bot$;
(d) *If A is P-abnormal, then* $B \geq_P A$.

Proof: (a) By IE and **RP2**. (b) By (a), Theorem 6(a), and Theorem 5(a). (c) By IE and **RP2**. (d) If A is P-abnormal, then $P(B,A) = 1$ and hence $B \geq_P A$ by Theorem 6(a). ¤

Theorem 8. (a) $A \sim_P A$;
(b) *If* $A \sim_P B$, *then* $B \sim_P A$;
(c) *If* $A \sim_P B$ *and* $B \sim_P C$, *then* $A \sim_P C$.

Proof of (a) is by Theorem 1, proof of (b) is trivial, and proof of (c) is by Theorem 5(a). ¤

So, as claimed above, $\sim_P$ is reflexive, symmetrical, and transitive, and hence constitutes an equivalence relation. Császár's relation $\geq_P$ can be so extended as to hold between P-commensurability classes of statements of L as well as between statements of L:

$$[A]_{\sim P} \geq_P [B]_{\sim P} \quad \textit{if and only if} \quad A \geq_P B.$$

And since

$$[A]_{\sim P} \geq_P [B]_{\sim P} \ \textit{and} \ [B]_{\sim P} \geq_P [A]_{\sim P}$$

amounts to

$$[A]_{\sim P} = [B]_{\sim P},$$

the P-commensurability classes are linearly ordered by $\geq_P$. By Theorem 7 the statements that are P-commensurable with $\top$ (i.e. the statements A such that $P(A,\top) > 0$) form the $\geq_P$-first equivalence class, while the statements P-commensurable with $\bot$ (i.e. the P-abnormal statements) form the $\geq_P$-last one.

The next two theorems will be put to use in Section 2.

Theorem 9. (a) $A \geq_P A \wedge B$;
(b) $A \vee B \geq_P A$;
(c) *If* $A \geq_P B$, *then* $A \sim_P A \vee B$.

Proof: (a) By IE and **RP2**. (b) By IE and by **RP2**. (c) Suppose $A \geq_P B$. Then $A \geq_P A \vee B$ by IE. But $A \vee B \geq_P A$ by (b). So $A \sim_P A \vee B$. ¤

Theorem 10. *If $A \sim_P B$ and $B \geq_P C$, then $A \sim_P B \vee C$.*

Proof by Theorem 9(c) and Theorem 8(c). ¤

Section 2. Representation by Type I-Ordered Classes of Measures

Given the results in Section 1, we may now embark upon proof of the *Representation Theorem for L* (Theorem 24). The theorem, according to which every binary function for L (hence, for L_P or for L_ω) meeting **RP1–RP7** can be represented by means of a so-called *Type I-ordered class of measure functions on L*, is a generalisation of a result in Császár (1955) for the relative probability functions in Rényi (1955). Our results differ from those in van Fraassen (1976) in that (i) for every relative probability function for L our representing class of unary functions is unique (modulo multiplicative constants) and (ii) these functions are total rather than partial ones. Proof of a Representation Theorem for L_ω alone is in Spohn (1986), with the relative probability functions of Popper's for L_ω represented there by means of classes of absolute probability functions for L_ω. Our own Representation Theorem has Spohn's as a special case which we attend to in Section 4.

The quotient function Q_P

Given a binary function P for L meeting **RP1–RP7**, we can define a function, the *probability quotient of P*, which, in a sense, compares the absolute probabilities of A and B, even in the case where $P(A,\top) = P(B,\top) = 0$:

$$Q_P(B,A) = P(B,A \vee B) \,/\, P(A,A \vee B).$$

Q_P is a partial function: though $Q_P(B,A)$ is defined whenever $A \geq_P B$, it is not when $B >_P A$.

Theorem 11. (a) $0 \leq Q_P(B,A)$;
(b) $0 < Q_P(B,A)$ *only if* $B \geq_P A$;
(c) $Q_P(B,A) = 0$ *if and only if* $A >_P B$.

Consequently, $Q_P(B,A)$ has a positive value when A and B are P-commensurable; $Q_P(B,A) = 0$ when $A >_P B$; and $Q_P(B,A)$ is not defined when $B >_P A$.

Theorem 12. *If $A \geq_P B$ and $A \geq_P C$, then $Q_P(B,A) \leq Q_P(B \vee C,A)$.*

Proof: Suppose $A \geq_P B$ *and* $A \geq_P C$.
Case 1: $A >_P B$, i.e. $P(B,B \vee A) = 0$. Then $Q_P(B,A) = 0$ and hence $Q_P(B,A) \leq Q_P(B \vee C,A)$ by Theorem 11(a).
Case 2: $A \sim_P B$. Then $A, B, A \vee B, A \vee C,$ and $B \vee C$ are all P-commensurable by Theorem 10. And all the probabilities in the following three consequences by IE of Lemma 4(jj) are

larger than 0:

$$P(A,A \vee (B \vee C)) = P(A,A \vee B) \times P(A \vee B,A \vee (B \vee C)),$$
$$P(B,A \vee (B \vee C)) = P(B,A \vee B) \times P(A \vee B, A \vee (B \vee C)), \text{ and}$$
$$P(B,A \vee (B \vee C)) = P(B,B \vee C) \times P(B \vee C,A \vee (B \vee C)).$$

Hence $P(B \vee C,A \vee (B \vee C)) \,/\, P(A,A \vee (B \vee C)) = P(B,A \vee B) \,/\, [P(A,A \vee B) \times P(B,B \vee C)]$. But $P(B,B \vee C) \leq 1$ by Lemma 4(a). So, $Q_P(B,A) \leq Q_P(B \vee C,A)$. ¤

Theorem 13. *If $A \geq_P B$, then $Q_P(B \wedge C,A) \leq Q_P(B,A)$.*

Proof: Suppose $A \geq_P B$. Then $A \geq_P B \wedge C$ and $A \geq_P B \wedge \neg C$ by Theorems 9(a) and 5(a). Hence $Q_P(B \wedge C,A) \leq Q_P((B \wedge C) \vee (B \wedge \neg C),A)$ by Theorem 12. So Theorem 13 by IE. ¤

The measure functions $M_{P,A}$

We next define in terms of $Q_P(B,A)$ a function $M_{P,A}(B)$ which extends $Q_P(B,A)$ to statements B for which it is undefined, namely those such that $B >_P A$:

$$M_{P,A}(B) = \sup_{C \in L} \{Q_P(B \wedge C,A) : C \sim_P A\},$$

where the supremum of a set of real numbers is the smallest upper bound of those numbers and where $M_{P,A}(B)$ is set equal to ∞, when there is no finite upper bound. Since $A \geq_P B \wedge C$ when $A \sim_P C$, $Q_P(B \wedge C,A)$ is always defined. $M_{P,A}(B)$ equals $Q_P(B,A)$ when either A and B are P-commensurable or $A >_P B$. And, for fixed A, $M_{P,A}(B)$ turns out to be a measure function for L.

Theorem 14. (a) *If $A \geq_P B$, then $M_{P,A}(B) = Q_P(B,A)$;*
(b) *If $A >_P B$, then $M_{P,A}(B) = 0$;*
(c) *If $A \sim_P B$, then $0 < M_{P,A}(B) < \infty$;*
(d) *$M_{P,A}(A) = 1$;*
(e) *If A is P-abnormal, then $M_{P,A}(B) = 1$ for every B;*
(f) *If A is P-abnormal, then $M_{P,B}(A) = 0$ for every P-normal B;*
(g) *If A is P-normal, then $M_{P,B}(A) > 0$ for at least one P-normal B.*

Proof: (a) By Theorem 13, if $A \sim_P B$; and by Theorems 11 and 13 if $A >_P B$. (b) By (a) and Theorem 11. (c) By (a) and Theorem 11. (d) By (a). (e) Let A be P-abnormal and $A \sim_P C$. Then C is P-abnormal by Lemma 4(ad), and $B \wedge C$ is P-abnormal by Lemma 4(v). Hence $Q(B \wedge C,A) = 1$ by Lemma 4(ll). So (e). (f) Suppose A P-abnormal and B P-normal. Then $B >_P A$ by Lemma 4(ad). Hence $M_{P,B}(A) = 0$ by (b). (g) Suppose A P-normal. Then $M_{P,A}(A) = 1$ by (d) and so $M_{P,B}(A) \neq 0$ for at least one P-normal statement B of L. ¤

Theorem 15. *Let A be P-normal, $A \geq_P B$, and C be any statement of L. Then $Q_P(B,A) = Q_P(B \wedge C,A) + Q_P(B \wedge \neg C,A)$.*

Proof: $A \vee B$ is P-normal by Lemma 4(ll). Hence

$$P(B,A \vee B) = P(B \wedge C,A \vee B) + P(B \wedge \neg C,A \vee B)$$

by Lemma 4(i).

$$P(B \wedge C, A \vee B) = P(B \wedge C, A \vee (B \wedge C)) \times P(A \vee (B \wedge C), A \vee B)$$

by Lemma 4(jj) and IE. Similarly,

$$P(B \wedge \neg C, A \vee B) = P(B \wedge \neg C, A \vee (B \wedge \neg C)) \times P(A \vee (B \wedge \neg C), A \vee B),$$

$$P(A, A \vee B) = P(A, A \vee (B \wedge C)) \times P(A \vee (B \wedge C), A \vee B), \text{ and}$$

$$P(A, A \vee B) = P(A, A \vee (B \wedge \neg C)) \times P(A \vee (B \wedge \neg C), A \vee B).$$

Hence Theorem 15 by the definition of Q_P. ¤

Theorem 16. *Let A be P-normal, and B and C be any statements of L. Then* $M_{P,A}(B) = M_{P,A}(B \wedge C) + M_{P,A}(B \wedge \neg C)$.

Proof: $\sup_{D \in L} \{Q_P(B \wedge D, A) : D \sim_P A\}$

$= \sup_{D \in L} \{Q_P((B \wedge C) \wedge D, A) + Q_P((B \wedge \neg C) \wedge D, A) : D \sim_P A\}$ (Theorems 15, 9(a))

$\leq \sup_{D \in L} \{Q_P((B \wedge C) \wedge D, A) : D \sim_P A\} + \sup_{D \in L} \{Q_P((B \wedge \neg C) \wedge D, A) : D \sim_P A\}$

Let D_1 and D_2 be any statements of L which are P-commensurable with A. Then

$Q_P((B \wedge C) \wedge D_1, A) + Q_P((B \wedge \neg C) \wedge D_2, A)$

$\leq Q_P(((B \wedge C) \wedge D_1) \vee ((B \wedge C) \wedge D_2), A) + Q_P(((B \wedge \neg C) \wedge D_1) \vee ((B \wedge \neg C) \wedge D_2), A)$ (Theorem 12)

$= Q_P((B \wedge C) \wedge (D_1 \vee D_2), A) + Q_P((B \wedge \neg C) \wedge (D_1 \vee D_2), A)$ (IE)

Hence by Theorem 10

$\sup_{D \in L} \{Q_P((B \wedge C) \wedge D, A) : D \sim_P A\} + \sup_{D \in L} \{Q_P((B \wedge \neg C) \wedge D, A) : D \sim_P A\}$

$\leq \sup_{D \in L} \{Q_P((B \wedge C) \wedge D, A) + Q_P((B \wedge \neg C) \wedge D, A) : D \sim_P A\}$.

So, $M_{P,A}(B) = M_{P,A}(B \wedge C) + M_{P,A}(B \wedge \neg C)$. ¤

Putting together the results obtained about $M_{P,A}$ we can now establish that, as long as A is P-normal, $M_{P,A}$ is a measure function in the sense of Chapter 1, Section 1. On the other hand, we have already seen that if A is P-abnormal then $M_{P,A}(B) = 1$ no matter the statement B.

Theorem 17. *If A is P-normal, then* $M_{P,A}$ *is a measure function, i.e.* $M_{P,A}$ *satisfies constraints* **AP1** *and* **AP3–AP6**.

Proof: (i) $0 \leq M_{P,A}(B)$ by **RP2** and the definition of $M_{P,A}$. So, $M_{P,A}$ meets **AP1**. (ii) $M_{P,A}$ meets **AP3** by Theorem 16. (iii) $M_{P,A}$ meets **AP4–AP6** by IE. ¤

Next we investigate how $M_{P,A}$ and $M_{P,A'}$ are related when A and A' are P-commensurable.

Theorem 18. *Let* $A \geq_P A'$ *and* $A' \geq_P B$. *Then* $Q_P(B,A) = Q_P(B,A') \times Q_P(A',A)$.

Proof by Lemma 4(jj) and Theorem 9(c). ¤

Theorem 19. (a) *If* $A \sim_P A'$, *then* $M_{P,A}(B) = M_{P,A}(A') \times M_{P,A'}(B)$;

(b) $A \sim_P A'$ *if and only if* $M_{P,A}(B) = k \times M_{P,A'}(B)$ *for every statement B of L, where* $0 < k < \infty$.

Proof: (a) Assume $A \sim_P A'$. Then

$$
\begin{aligned}
&M_{P,A}(B)\\
&\quad = \sup_{C \in L} \{Q_P(B \wedge C, A) : C \sim_P A\}\\
&\quad = \sup_{C \in L} \{Q_P(B \wedge C, A') \times Q_P(A', A) : C \sim_P A\} && \text{(Theorem 18)}\\
&\quad = Q_P(A', A) \times \sup_{C \in L} \{Q_P(B \wedge C, A') : C \sim_P A'\}\\
&\quad = M_{P,A}(A') \times M_{P,A'}(B) && \text{(Theorem 14(a))}
\end{aligned}
$$

(b) (i) Assume $A \sim_P A'$. Then $M_{P,A}(B) = k \times M_{P,A'}(B)$ by (a) and $0 < k < \infty$ by Theorem 14(c). (ii) Assume $A >_P A'$. Then $M_{P,A}(A') = 0$, while $M_{P,A'}(A') = 1$ by Theorems 14(b) and 14(d). So there exists no finite, positive number k such that $M_{P,A}(A') = k \times M_{P,A'}(A')$. ¤

Assuming that A is P-normal, $M_{P,A}(\top)$ is the highest value that $M_{P,A}$ attains. The next theorems show that when $M_{P,A}(\top) < \infty$ $M_{P,A}$ is an absolute probability function, multiplied by a constant, while when $M_{P,A}(\top) = \infty$, $M_{P,A}$ is a measure function which is not a multiple of an absolute probability function.

Theorem 20. (a) $M_{P,A}(B) \leq M_{P,A}(\top)$;
(b) $1 \leq M_{P,A}(\top)$.

Proof: (a) $M_{P,A}(\top) = \sup_{C \in L} \{Q_P(C, A) : C \sim_P A\}$ by IE; and

$$M_{P,A}(B) = \sup_{C \in L} \{Q_P(B \wedge C, A) : C \sim_P A\}.$$

But $Q_P(B \wedge C, A) \leq Q_P(C, A)$ by Theorem 13. So (a). (b) By (a) and Theorem 14(d). ¤

Theorem 21. *Let* $M_{P,A}(\top) < \infty$. *Then*
(a) *If* $A' \sim_P A$, *then* $M_{P,A'}(B) / M_{P,A'}(\top) = M_{P,A}(B) / M_{P,A}(\top)$;
(b) *If* A *is* P*-normal, the function* P', *defined thus*

$$P'(B) = M_{P,A}(B) / M_{P,A}(\top)$$

meets **AP1–AP6**;
(c) $M_{P,\top}$ *is an absolute probability function;*
(d) *If* $A \geq_P B$, *then* $P'(B) = \inf_{C \in L} \{P(B, B \vee C) : C \sim_P A\}$.

Proof: (a) By Theorems 19 and 20(b). (b) Suppose A P-normal. Since $M_{P,A}$ meets **AP1** and **AP3–AP6** by Theorem 17, so does P'. Moreover, $P'(\top) = 1$. So P' meets **AP2**. (c) By (b) and Theorem 14(d). (d) Suppose $A \geq_P B$. Then $M_{P,A}(B) = Q_P(B, A)$ by Theorem 14(a), and hence

$$
\begin{aligned}
&P'(B)\\
&\quad = Q_P(B, A) / M_{P,A}(\top)\\
&\quad = Q_P(B, A) / \sup_{C \in L} \{Q_P(C, A) : C \sim_P A\}\\
&\quad = \inf_{C \in L} \{Q_P(B, A) / Q_P(C, A) : C \sim_P A\}\\
&\quad = \inf_{C \in L} \{Q_P(B, C) : C \sim_P A\} && \text{(Theorem 18)}
\end{aligned}
$$

But, since (i) $B \vee C \sim_P A$ by Theorem 10, (ii) $P(C, C \vee B) \leq 1$ by Lemma 4(a), and (iii)

$P(B,B \vee (B \vee C)) = P(B,B \vee C)$ by IE, $P'(B)$ has to equal $\inf_{C \in L} \{P(B,B \vee C) : C \sim_P A\}$. ¤

Theorem 22. *Let* $P(A,\mathrm{T}) > 0$. *Then*
(a) $M_{P,A}(\mathrm{T}) < \infty$;
(b) $M_{P,A}(B) = M_{P,A}(\mathrm{T}) \times P(B,\mathrm{T})$.

Proof: (a) $A \sim_P \mathrm{T}$ by Theorem 7, and $M_{P,A}(\mathrm{T}) < \infty$ by Theorem 14(c). (b) $M_{P,A}(B) = M_{P,A}(\mathrm{T}) \times M_{P,\mathrm{T}}(B)$ by Theorem 19(a). But $M_{P,\mathrm{T}}(B) = P(B,\mathrm{T})$ by the definition of $M_{P,\mathrm{T}}(B)$. ¤

So, to each P-commensurability class of statements of L there corresponds a class of measure functions for L, any two of which are identical modulo a multiplicative factor (Theorem 19). We conceive of the classes of measure functions as being linearly ordered, namely by an ordering relation that parallels the linear ordering of P-commensurability classes in terms of Császár's relation. The functions $M_{P,A}$ for statements A with $P(A,\mathrm{T}) > 0$ come first in the linear ordering and are multiples of $M_{P,\mathrm{T}}$, which is an absolute probability function, namely the T-restriction of P (Theorem 22). Further, if $M_{P,A}(\mathrm{T})$ is finite, then so is $M_{P,B}(\mathrm{T})$ for every statement B of L that is P-commensurable with A (Theorem 19). The function $M_{P,A}(C) / M_{P,A}(\mathrm{T})$ thus meets **AP1–AP6** (Theorem 21).

The Spectrum Class $\mathcal{M}_P$

Still supposing P to be a binary function on the statements of L that meets **RP1–RP7**, let $\mathcal{M}_P$ be a class of measure functions for L which contains, for every P-commensurability class of P-normal statements of L, exactly one measure function $M_{P,A}$ with A drawn from that class. The supposition is tantamount to there existing, *for every P-normal statement B of L*, a statement A of L such that $A \sim_P B$ and $M_{P,A}$ belongs to $\mathcal{M}_P$, and there existing, for every member M of $\mathcal{M}_P$, a P-normal statement A of L such that $M = M_{P,A}$. Regard the members of $\mathcal{M}_P$ as ordered by Császár's relation, as before: for any M and M' in $\mathcal{M}_P$, $M \geq M'$ when M is $M_{P,A}$, M' is $M_{P,A'}$ and $A \geq_P A'$. Although $\mathcal{M}_P$ is not a unique class of functions, alternative choices differ only inessentially from one another. Thanks to Theorem 19, any class of functions that plays the role of $\mathcal{M}_P$ can be obtained from any other such class by multiplying each function in the class by an appropriate factor. So, we take the liberty of talking as if $\mathcal{M}_P$ was a unique class, to be called the *spectrum class* of P.

The interest of the spectrum class $\mathcal{M}_P$ lies in the fact that it contains the same information as P itself. P can be recovered from $\mathcal{M}_P$ by virtue of this equation:

$$P(A,B) = \begin{cases} 1 \quad \textit{if } M(B) = 0 \textit{ for every member } M \textit{ of } \mathcal{M}_P \\ M(A \wedge B) / M(B) \quad \textit{where } M \\ \qquad \textit{is the } \geq_P\textit{-first member of } \mathcal{M}_P \textit{ with } M(B) > 0. \end{cases}$$

Note that by virtue of Theorem 14 all possible cases are covered on the right-hand side. The equation is a generalisation of the definition of a Kolmogorov function by means of its

T-restriction. Indeed, if P is a Kolmogorov function then every P-normal statement is P-commensurable with T, which means that there is only one P-commensurability class of P-normal statements.

For proof of the equation assume first that B is P-abnormal. Then, on the one hand, $P(A,B) = 1$, while, on the other hand, $M_{P,C}(B) = 0$ for every P-normal C, and hence, in particular, $M(B) = 0$ for every member M of $\mathcal{M}_P$. Assume secondly that B is P-normal and that M is the $\gtrsim_P$-first member of $\mathcal{M}_P$ with $M(B) > 0$. By Theorems 14(b) and 14(c) M is then $M_{P,C}$ for some statement C with $C \sim_P B$. But by Theorem 19(b) $M_{P,C} = k \times M_{P,B}$ for some real number k. $B \gtrsim_P A \wedge B$ by Theorem 9(a). Hence $M_{P,B}(A \wedge B) = Q_P(A \wedge B,B)$ by Theorem 14(a). But $Q_P(A \wedge B,B) = P(A \wedge B,B) / P(B,B)$ by the definition of Q_P and IE. So $M_{P,B}(A \wedge B) = P(A \wedge B,B) = P(A,B)$ by **RP2** and Lemma 4(u). And since $M_{P,B}(B) = 1$ by Theorem 14(d), $P(A,B) = M_{P,C}(A \wedge B) / M_{P,C}(B)$.

Type I-ordered classes of measure functions

Since P and $\mathcal{M}_P$ determine one another, we say that P is *represented* by its spectrum class $\mathcal{M}_P$. The representation result naturally leads to the question what is required of a class $\mathcal{M}$ of measure functions if it is to represent a function meeting **RP1–RP7**. Can the requirements be formulated without reference to an already given such function P? The answer is that $\mathcal{M}$ has to be *Type I-ordered*. The definition of this type of ordered class is this:

> A class $\mathcal{M}$ of measure functions is *Type I-ordered* by $\geq$ if it meets the conditions **O1– O4**, which are:

O1 *$\mathcal{M}$ is linearly ordered by $\geq$;*

O2 *$\mathcal{M}$ has a first member M_0 under $\geq$, i.e. there exists a member M_0 of $\mathcal{M}$ such that $M_0 \geq M$ for every member M of $\mathcal{M}$, and $0 < M_0(\mathrm{T}) < \infty$;*

O3 *For every statement A of L, if $M(A) > 0$ for some member M of $\mathcal{M}$, then there exists a member M^A of $\mathcal{M}$ such that*
(a) $0 < M^A(A) < \infty$, *and*
(b) $M(A) = 0$ *for every member M of $\mathcal{M}$ such that $M > M^A$*

O4 *For every member M of $\mathcal{M}$*
(a) *there exists a statement A of L such that $M = M^A$, and*
(b) *for every statement B,* $M(B) = \sup_{C \in L} \{M(B \wedge C) : M = M^C\}$.

For given P, the spectrum class $\mathcal{M}_P$ itself is indeed Type I-ordered by $\gtrsim_P$. **O1** holds, since the P-commensurability classes are linearly ordered by $\gtrsim_P$. **O2**: The first member is $M_{P,\mathrm{T}}$, which is an absolute probability function by Theorem 21(c). **O3**: Suppose $M(A) > 0$ for some M in $\mathcal{M}_P$. M is $M_{P,C}$ for some P-normal C by the definition of $\mathcal{M}_P$. Hence A is P-normal by Theorem 14(f). But by the characterisation of $\mathcal{M}_P$ there is a statement B of L such that $A \sim_P B$ and $M_{P,B} \in \mathcal{M}_P$. Hence $0 < M_{P,B}(A) < \infty$ by Theorem 14(c). Suppose M' in $\mathcal{M}_P$

and $M' > M$. Then $M' = M_{P,C}$ for some statement C of L such that $C >_P B$ and hence $C >_P A$. Hence $M_{P,C}(A) = 0$ by Theorem 14(b), i.e. $M'(A) = 0$. Hence $M^A = M_{P,B}$. **O4**: (a) Let M be in $\mathcal{M}_P$. By the characterisation of $\mathcal{M}_P$, M is $M_{P,A}$ for some P-normal statement A of L. Hence $0 < M(A) < \infty$ by Theorem 14(d). If $M' > M$, then $M'(A) = 0$ by Theorem 14(b). Hence $M = M^A$. (b) $M_{P,A}(B) = \sup_{C \in L} \{M_{P,A}(B \wedge C) : C \sim_P A\}$ by Theorems 9(a) and 14(a). But, by Theorem 19 and the characterisation of $\mathcal{M}_P$, $C \sim_P A$ if and only if $M^C = M^A$. Hence $M(B) = \sup_{C \in L} \{M(B \wedge C) : M = M^C\}$.

The Representation of Relative Probability Functions

Given any class $\mathcal{M}$ of measure functions, Type I-ordered by $\geq$, define a binary function $P_{\mathcal{M}}$, to be called *the function generated by* $\mathcal{M}$, thus:

$$\textbf{GEN} \qquad P_{\mathcal{M}}(A,B) = \begin{cases} 1 & \textit{if } M(B) = 0 \textit{ for every member } M \textit{ of } \mathcal{M} \\ M^B(A \wedge B) \,/\, M^B(B) & \textit{otherwise.} \end{cases}$$

Here M^B is the $\geq$-first measure function M in $\mathcal{M}$ with $M(B) > 0$, i.e. M^B has the same significance as in **O3** and **O4**. We establish that $P_{\mathcal{M}}$ is a relative probability function, i.e. meets constraints **RP1–RP7**.

Theorem 23. *Let* $\mathcal{M}$ *be a class of measure functions for L that is Type I-ordered by* $\geq$, *and let* $P_{\mathcal{M}}$ *be defined by* **GEN**. *Then* $P_{\mathcal{M}}$ *meets* **RP1–RP7**.

Proof:

RP1: By **AP1**.

RP2: $M^A(A \wedge A) = M^A(A)$ by Lemma 1(c).

RP3: Suppose B $P_{\mathcal{M}}$-normal. Then $M(B) \neq 0$ for some M in $\mathcal{M}$ and hence by **O3** there exists a $\geq$-first member M^B in $\mathcal{M}$. So, **RP3** by **AP3**.

RP4: Assume *first* that $M(C) = 0$ for every M in $\mathcal{M}$. Then $M(B \wedge C) = 0$ for every M in $\mathcal{M}$ as well, by Lemma 1(b). So

$$P_{\mathcal{M}}(A \wedge B,C) = P_{\mathcal{M}}(A,B \wedge C) \times P_{\mathcal{M}}(B,C),$$

since all these values of $P_{\mathcal{M}}$ equal 1 by **GEN**. Assume *next* that $M = M^C$ for some M in $\mathcal{M}$ and hence

$$P_{\mathcal{M}}(A \wedge B,C) = M^C((A \wedge B) \wedge C) \,/\, M^C(C),$$

but that $M^C(B \wedge C) = 0$. Then $M^C(A \wedge (B \wedge C)) = 0$ by Lemma 1(b), and

$$P_{\mathcal{M}}(A \wedge B,C) = P_{\mathcal{M}}(A,B \wedge C) \times P_{\mathcal{M}}(B,C),$$

since $P_{\mathcal{M}}(A \wedge B,C) = 0$ and $P_{\mathcal{M}}(B,C) = 0$. Assume *finally* that $M^C(B \wedge C) \neq 0$. Then $M(B \wedge C)$ must equal 0 for all $M > M^C$ by **O3** and Lemma 1(b). So $M^C = M^{B \wedge C}$, and

$$P_{\mathcal{M}}(A,B \wedge C) = M^C(A \wedge (B \wedge C)) \,/\, M^C(B \wedge C)$$

by **GEN**. Hence

$$P_{\mathcal{M}}(A \wedge B,C) = P_{\mathcal{M}}(A,B \wedge C) \times P_{\mathcal{M}}(B,C)$$

by **AP5**.

RP5: By **AP4**.

RP6: By **AP5**.

RP7: $0 < M_0(\top) < \infty$ by **O2**. Hence $M^{\top}$ is M_0 and $P_{\mathcal{M}}(\bot,\top) = 0$ by Lemma 1(g). ¤

So, given a class $\mathcal{M}$ of measure functions which is Type I-ordered by a relation $\geq$, the binary function $P_{\mathcal{M}}$ defined by **GEN** meets **RP1–RP7** and hence, in case L is L_P, is a relative probability function for L_P. Furthermore, $\mathcal{M}$ is the spectrum class $\mathcal{M}_{P_{\mathcal{M}}}$ of $P_{\mathcal{M}}$, i.e. for every $P_{\mathcal{M}}$-normal statement A of L there exists a measure function M in $\mathcal{M}$ and for every M in $\mathcal{M}$ there exists a $P_{\mathcal{M}}$-normal statement A of L such that $M_{P_{\mathcal{M}},A} = k \times M$ *for some real number k*, $0 < k < \infty$. We omit the proof of this result, which makes use of theorems already established in this section.

So, Type I-ordered classes $\mathcal{M}$ of measure functions correspond to relative probability functions P in the sense that $\mathcal{M}$ is the spectrum class $\mathcal{M}_P$ of P and P is the function $P_{\mathcal{M}}$ generated by $\mathcal{M}$; or simply, $\mathcal{M}$ *represents* P. Hence, summing up our results, we state the *Representation Theorem.*

Theorem 24. *Every binary function P on L meeting* **RP1–RP7** *can be represented by means of a Type I-ordered class* $\mathcal{M}$ *of measure functions on L; and every Type I-ordered class* $\mathcal{M}$ *of measure functions on L represents a binary function P on L meeting* **RP1–RP7**.

It is worth noting that, as long as we are only interested in the conditions under which a binary function on L meeting **RP1–RP7** can be generated from a class of measure functions, not all the conditions **O1–O4** are required. In the proof of Theorem 23 condition **O4** has not been invoked. Hence any class of measure functions satisfying **O1–O3** is capable of generating a function meeting **RP1–RP7**, i.e., in case L is L_P, a relative probability function for L_P. There is then no guarantee of course that the generating class is the spectrum class of the generated function.

Section 3. Representation by Type II-, Type III-, and Type IV-Ordered Classes of Measure Functions

Well-Ordered Classes of Measure Functions

The definition of a Type I order *explicitly* requires that the measure functions in $\mathcal{M}$ form a linear order, but not that they be well-ordered. Our next task is to show that well-orderedness is not *implicit* in the further clauses of the definition, by constructing an example of a Type I-order that is not well-ordered.

Example 1. $E_1, E_2, \ldots$ being the atomic statements of L_P,
(a) define the sets $S_0, S_1, \ldots, S_n, \ldots$, which correspond in effect to some of the state-

descriptions of L_P in $E_1, E_2, \ldots$:

$$S_0 = \{\neg E_i : i \in \mathbf{N}\}$$
$$S_1 = \{E_1\} \cup \{\neg E_i : i \in \mathbf{N}, i \neq 1\}$$
$$\vdots$$
$$S_n = \{E_n\} \cup \{\neg E_i : i \in \mathbf{N}, i \neq n\}$$
$$\vdots$$

(b) given the sets $S_0, S_1, \ldots$, define for each n from 0 on and every statement B of L_P this function which will prove to constitute an absolute probability function for L_P:

$$M_n(B) \;=\; \begin{cases} 1 & \textit{if } S_n \vdash B \\ 0 & \textit{otherwise,} \end{cases}$$

and

(c) $\mathcal{M}_1$ consisting of all the measure functions defined in (b), define this ordering on $\mathcal{M}_1$:

(i) $M_0 \geq M$ *for every* M *in* $\mathcal{M}_1$,

(ii) $M_i > M_j$ *if* i *and* j *are distinct from* 0 *and* $j < i$.

Then $\mathcal{M}_1$ clearly is linearly ordered by > but not well-ordered since the set $\mathcal{M}_1 - \{M_0\}$ does not have a ≥-first element.

It remains to be shown that $\mathcal{M}_1$ is Type I-ordered by ≥. Conditions **O1** and **O2** are obviously met. As for condition **O3**, suppose that $M_i(B) > 0$ for some M_i, in which case $S_i \vdash B$. If $M_0(B) > 0$, **O3** is met. Suppose then that $M_0(B) = 0$, hence that $S_0 \nvdash B$. Then B is logically equivalent to a statement in disjunctive normal form, i.e. to a finite disjunction of finite conjunctions of unnegated and negated atomic statements, i.e. B is logically equivalent to a statement B' of the form $C_1 \vee \ldots \vee C_m$, where none of $C_1, \ldots, C_m$ is a conjunction of only negated atomic statements of L, this since $S_0 \nvdash B$. But $S_i \vdash B$ if and only if $S_i \vdash C_k$ for at least one k. Hence the set consisting of the conjuncts of C_k is a subset of S_i. Hence $S_i \vdash C_k$ if and only if C_k has E_k as one of its conjuncts and E_k is the only atomic statement that is a conjunct of C_k. Further, given that $S_i \vdash B$ for at least one S_i other than S_0, there is sure to exist at least one disjunct of B' that contains exactly one atomic statement as a conjunct. And since B' is a finite disjunction, there exists a largest number i such that E_i occurs as the only unnegated atomic statement in one or more of the disjuncts of B'. So $M^B = M_i$, $0 < M_i(B) < \infty$ and $M_j(B) = 0$ for $j = 0$ and for $j > i$. As for condition **O4**(a), it is also met since $M_0(\neg E_1) > 0$, and, for $i \neq 0$, $M_i(E_i) > 0$, but $M_j(E_i) = 0$ for $j > i$ and for $j = 0$. And as for **O4**(b), let B be any statement of L and C any statement of L such that $S_i \vdash C$ and $S_j \nvdash C$ for $j = 0$ and $j > i$. Then $S_i \vdash B$ if and only if $S_i \vdash B \wedge C$. Hence $M_i(B) = \sup_{C \in L} \{M_i(B \wedge C) : M_i = M^C\}$.

So $\mathcal{M}_1$ is a Type I-ordered class of measure functions which is not well-ordered. And – owing to Theorem 24 – the $P_{\mathcal{M}_1}$-commensurability classes are not well-ordered under

$\geq_{P_{\mathcal{M}_1}}$. Hence the definition of a Type I-order does not imply well-orderedness.

The following constraint, to be known as **RPW**, will be shown to ensure that the commensurability classes under P are well-ordered by $\geq_P$, and hence that the members of any class $\mathcal{M}$ of measure functions representing P are well-ordered:

RPW *For any* ω*-sequence* $\langle A_1, A_2, \ldots \rangle$ *of statements of* L, $P(A_i, A_i \vee A_{i+1}) > 0$ (*i.e.* $A_i \geq_P A_{i+1}$) *for at least one* i.

A Type I-ordered class of measure functions for L will be declared *Type II-ordered* if it meets this further condition:

O5 $\mathcal{M}$ *is well-ordered by* $\geq$.

As a result a Type II-ordered class of measure functions is one that meets conditions **O2, O4,** and **O5, O1** and **O3** then being entailed by them.

Theorem 25. *Let* P *be a binary function for* L *meeting* **RP1–RP7**. *Then* P *meets* **RPW** *if and only if the commensurability classes under* P *are well-ordered by* $\geq_P$.

Proof: (i) Suppose P does not meet **RPW**. Then there exists an ω-sequence of statements $\langle A_1, A_2, \ldots \rangle$ such that $A_2 >_P A_1$, $A_3 >_P A_2$, ..., $A_{i+1} >_P A_i$, ... Hence the commensurability classes under P are not well-ordered. (ii) Suppose the commensurability classes under P are not well-ordered, i.e. there exists an infinitely ascending sequence of such commensurability classes. Selecting a representative statement from each such class yields a sequence of statements $\langle A_1, A_2, \ldots \rangle$ such that $P(A_1, A_1 \vee A_2) = 0$, $P(A_2, A_2 \vee A_3) = 0$, ..., $P(A_i, A_i \vee A_{i+1}) = 0$, ... So P does not meet **RPW**. ¤

Hence, this second Representation Theorem:

Theorem 26. *Every binary function* P *on* L *meeting* **RP1–RP7** *and* **RPW** *can be represented by means of a Type II-ordered class* $\mathcal{M}$ *of measure functions on* L*; and every Type II-ordered class* $\mathcal{M}$ *of measure functions on* L *represents a binary function* P *on* L *meeting* **RP1–RP7** *and* **RPW**.

Proof by Theorems 24 and 25. ¤

Representation by Absolute Probability Functions

The definition of a Type I-ordered class of measure functions allows for the members of the class to take ∞ as a value, in particular $M(\mathrm{T}) = \infty$ is possible. This is demonstrated by the following example:

Example 2. Let $S_0, S_1, \ldots$ be as in Example 1, let the measure functions M_0 and M_1 be defined as follows:

$$M_0(B) = \begin{cases} 1 & \textit{if } S_0 \vdash B \\ 0 & \textit{otherwise} \end{cases}$$

$M_1(B) = \text{card}\{i : i \neq 0 \text{ and } S_i \vdash B\}$,

and let $\mathcal{M}_2 = \{M_0, M_1\}$ and $M_0 > M_1$. Only condition **O4**(b) of a Type I-order needs to be attended to. So let $S_0 \vdash B$ and, as in Example 1, B' be in disjunctive normal form and logically equivalent to B. Let $C_1, \ldots, C_k$ be the disjuncts of B' that are conjunctions of only negated atomic statements. Then $S_0 \nvdash B \wedge \neg(C_1 \vee \ldots \vee C_k)$ and either $S_i \nvdash B \wedge \neg(C_1 \vee \ldots \vee C_k)$ for every $i \neq 0$ and hence $M_1(B) = 0$, or $M_1 = M^{B \wedge \neg(C_1 \vee \ldots C_k)}$ and $M_1(B \wedge \neg(C_1 \vee \ldots \vee C_k)) = M_1(B)$. In either case $M_1(B) = \sup_{C \in L}\{M_1(B \wedge C) : M_1 = M^C\}$ by Lemma 1(a). So $\mathcal{M}_2$ is Type I-ordered and $M_1(\top) = \infty$.

But we can achieve a representation by means of bounded measure functions, and hence by means of absolute probability functions when L is L_P, by imposing this constraint on the binary functions for L that meet **RP1–RP7**:

RPL *For any ω-sequence $\langle A_1, A_2, \ldots \rangle$ of statements of L, if* $\lim_{n \to \infty} P(A_1, A_1 \vee \ldots \vee A_n) = 0$, *then* $P(A_1, A_1 \vee A_n) = 0$ (i.e. $A_n >_P A_1$) *for some n.*

With a Type I-ordered class of measure functions for L declared *Type III-ordered* if it meets this further condition:

O6 *Every member M of $\mathcal{M}$ meets* **AP2**,

we conclude that

Theorem 27. *Every binary function P on L meeting* **RP1–RP7** *and* **RPL** *can be represented by means of a Type III-ordered class $\mathcal{M}$ of measure functions on L; and every Type III-ordered class $\mathcal{M}$ of measure functions on L represents a binary function P on L meeting* **RP1–RP7** *and* **RPL**.

Proof: (a) Suppose P does not meet **RPL**. Then there exists an ω-sequence $\langle A_1, A_2, \ldots \rangle$ such that

$$\lim_{n \to \infty} P(A_1, A_1 \vee \ldots \vee A_n) = 0$$

but

$$P(A_1, A_1 \vee A_n) > 0 \text{ for every } n.$$

Hence $A_1 \sim_P A_1 \vee \ldots \vee A_n$ by Theorem 9(c). Hence

$$M_{P,A_1}(\top) = \sup_{C \in L}\{P(C, C \vee A_1) / P(A_1, A_1 \vee C) : C \sim_P A_1\} = \infty,$$

since $P(A_1, A_1 \vee C)$ can be arbitrarily close to 0, by taking C to be $A_1 \vee \ldots \vee A_n$ for appropriately large n. So, M_{P,A_1} does not meet **AP2** and $\mathcal{M}$ is not Type II-ordered.

(b) Suppose $\mathcal{M}$ is Type I-ordered, but not Type III-ordered. Then $M(\top) = \infty$ for some member M of $\mathcal{M}$. Hence $M_{P,A}(\top) = \infty$ for some statement A of L by Theorem 24, i.e.

$$\sup_{C \in L}\{P(C, C \vee A) / P(A, A \vee C) : C \sim_P A\} = \infty.$$

Since $P(C, C \vee A) \leq 1$ by Lemma 4(a), there must be a sequence $\langle C_1, C_2, \ldots \rangle$ of statements C_i

that are P-commensurable with A such that $\lim_{n\to\infty} P(A,A \vee C_n) = 0$. So,

$$\lim_{n\to\infty} P(A,A \vee C_1 \vee ... \vee C_n) = 0$$

by Lemmas 4(tt) and 4(bb). But $P(A,A \vee C_i) > 0$ for every i. So, P does not meet **RPL**. ¤

So, when L is L_P:

Theorem 28. *A relative probability function P for L_P is represented by means of a class of absolute probability functions for L_P if and only if P meets* **RPL**.

Representation by Well-Ordered Classes of Absolute Probability Functions

As Examples 1 and 2 show, the two constraints **RPW** and **RPL** are independent of one another. They can be combined into one constraint, called in Chapter 2, Section 2, *the Double-Limit Constraint* **DL1**:

DL1 *For any ω-sequence $\langle A_1,A_2, ... \rangle$ of statements of L,*

$$\lim_{m\to\infty} \lim_{n\to\infty} P(A_1 \vee ... \vee A_m,A_1 \vee ... \vee A_n) = 1.$$

Theorem 29. *A binary function for L satisfying* **RP1–RP7** *meets* **DL1** *if and only if it meets* **RPW** *and* **RPL**.

Proof: Note that

$$P(A_1 \vee ... \vee A_m,A_1 \vee ... \vee A_{n+1}) = \prod_{i=m}^{n} P(A_1 \vee ... \vee A_i,A_1 \vee ... \vee A_{i+1})$$

by Lemma 4(jj) and that if

$$\lim_{m\to\infty} \lim_{n\to\infty} P(A_1 \vee ... \vee A_m,A_1 \vee ... \vee A_n) \neq 1,$$

$$\text{i.e. } \lim_{m\to\infty} \prod_{i=m}^{\infty} P(A_1 \vee ... \vee A_i,A_1 \vee ... \vee A_{i+1}) \neq 1,$$

then $\prod_{i=m}^{\infty} P(A_1 \vee ... \vee A_i,A_1 \vee ... \vee A_{i+1}) = 0$ for every m, i.e.

$$\lim_{n\to\infty} P(A_1 \vee ... \vee A_m,A_1 \vee ... \vee A_n) = 0 \text{ for every } m.$$

(a) Suppose P meets **RPL** but not **DL1**. Then there exists a sequence $\langle A_1,A_2,...\rangle$ such that $\lim_{n\to\infty} P(A_1,A_1 \vee ... \vee A_n) = 0$ and so $A_{i_1} >_P A_1$ for some i_1 by **RPL**. But

$$\lim_{n\to\infty} P(A_{i_1},A_1 \vee ... \vee A_n) = 0$$

by Lemma 4(aa), and so $A_{i_2} >_P A_{i_1}$ for some i_2. Hence, repeating this reasoning, there is an infinite sequence $A_1, A_{i_1}, A_{i_2}, ..., A_{i_k}, ...$ such that

$$A_{i_1} >_P A_1, A_{i_2} >_P A_{i_1}, ..., A_{i_{k+1}} >_P A_{i_k}, ...$$

and P fails to meet **RPW**.

(b) Suppose P meets **DL1**. Then

$$\prod_{i=m}^{\infty} P(A_1 \vee ... \vee A_i,A_1 \vee ... \vee A_{i+1}) > 0 \text{ for some } m,$$

and hence $P(A_1 \vee ... \vee A_m,A_1 \vee ... \vee A_{m+1}) > 0$. However, if $A_{i+1} >_P A_i$ for every $i = 1, ..., m$, then

$$A_1 \vee ... \vee A_{m+1} >_P A_1 \vee ... \vee A_m$$

by Theorems 5 and 9(c). So $A_i \geq_P A_{i+1}$ for at least one i and P meets **RPW**. And if

$$\lim_{n\to\infty} P(A_1, A_1 \vee \ldots \vee A_n) = 0,$$

then $P(A_1 \vee \ldots \vee A_i, A_1 \vee \ldots \vee A_{i+1}) = 0$ for at least one $i = 1, 2, \ldots$ and then

$$P(A_1, A_1 \vee A_{i+1}) = 0$$

by Lemma 4(ii). So P meets **RPL.** ¤

With a Type I-ordered class of measure functions for L declared *Type IV-ordered* if it meets both conditions **O5** and **O6** (so that a Type IV-ordered class of measure functions is one that meets conditions **O4, O5**, and **O6**), we conclude that

Theorem 30. *Every binary function P on L meeting* **RP1–RP7** *and* **DL1** *can be represented by means of a Type IV-ordered class* $\mathcal{M}$ *of measure functions on L; and every Type IV-ordered class* $\mathcal{M}$ *of measure functions on L represents a binary function P on L meeting* **RP1–RP7** *and* **DL1**.

So, when L is L_P:

Theorem 31. *Every relative probability function P for* L_P *meeting* **DL1** *can be represented by means of a well-ordered class of absolute probability functions for* L_P*; and every well-ordered class of absolute probability functions for* L_P *represents a relative probability function P for* L_P *which meets* **DL1** .

Section 4. The Representation Theorem for L_ω

We deal in this closing section with the relative probability functions for L_ω. It was shown in Chapter 2, Section 2, that those functions meet constraint **DL1**. So by Theorem 30 every relative probability function P for L_ω is represented by means of a Type IV-ordered class $\mathcal{M}_P$ of measure functions for L_ω, indeed, as we will show, by means of a certain well-ordered class of absolute probability functions for L_ω. In the light of Theorem 21 we have it that – for any given P-normal statement A of L_ω – $M_{P,A}(B) / M_{P,A}(\top)$ meets **AP1–AP6**. We establish that $M_{P,A}(B) / M_{P,A}(\top)$ also meets **AP∧** and is therefore an absolute probability function for L_ω. We proceed to show that there exists a statement D_A of L_ω, P-commensurable with A, such that this function is the same as $M_{P,D_A}(B)$, and the same as $P(B,D_A)$, i.e. the same as $P_{D_A}(B)$, the D_A-restriction of P. Required for the proof of these results are the following theorems which have been proved in Chapter 2 as Theorems 6 and 4:

Theorem 32. (a) $P(\Lambda A \wedge B, C) = \lim_{n\to\infty} P((A_1 \wedge \ldots \wedge A_n) \wedge B, C)$;
(b) $P(\mathrm{V}A, B) = \lim_{n\to\infty} P(A_1 \vee \ldots \vee A_n, B)$.

In the following theorems P is assumed to be a relative probability function for L_ω.

Theorem 33. *For every statement A of L_ω there exists a statement A′ of L_ω such that*
(a) $P(A',A) = 1$;
(b) $P(A,A') = \inf_{C \in L} \{P(A,A \vee C) : C \sim_P A\}$;
(c) $P(A,A') > 0$;
(d) $A' \sim_P A$.

Proof:[2] Let $\inf_{C \in L} \{P(A,A \vee C) : C \sim_P A\}$ equal r. Then $r > 0$ by Theorems 21 and 14(d), and there exists an infinite sequence $\langle C_1,...,C_n,...\rangle$ of statements of L_ω such that $C_n \sim_P A$ for each n from 1 on and $\lim_{n\to\infty} P(A,A \vee C_n) = r$. Now let A' be short for $\mathrm{V}\langle A,C_1,...,C_n,...\rangle$.

(a) By Theorem 32(a) and Lemma 4(bb).

(b) $P(A,A \vee C_1 \vee ... \vee C_n) \leq P(A,A \vee C_n)$ by Lemmas 4(tt) and 4(bb). Hence

$$\lim_{n\to\infty} P(A,A \vee C_1 \vee ... \vee C_n) \leq r.$$

But since $C_1 \vee ... \vee C_n \sim_P A$ by Theorem 9(c),

$$\lim_{n\to\infty} P(A,A \vee C_1 \vee ... \vee C_n) \geq r$$

by the hypothesis on

$$\inf_{C \in L} \{P(A,A \vee C) : C \sim_P A\}.$$

So, $P(A,A') = r$ by Theorem 32(b).

(c) and (d) $P(A,A') > 0$ by (b) and $P(A',A) > 0$ by (a). So $A' \sim_P A$ by Theorem 6(b). ¤

Given a statement A of L_ω, we shall hereafter use 'D_A' for any statement A' for which clauses (a)–(d) of Theorem 33 hold.

Theorem 34. *For every statement B of L_ω such that $B \sim_P A$, $P(D_A,B) = 1$.*

Proof: Let B be an arbitrary statement such that $B \sim_P A$, in which case $B \vee D_A \sim_P A$ by Theorem 10, Theorem 34(d), and the assumption on B. Now

$$\text{(i) } P(A,D_A) \leq P(A,A \vee (D_A \vee B))$$

by Theorem 34(b),

$$\text{(ii) } P(A,A \vee (D_A \vee B)) \leq P(A,D_A \vee B)$$

by Lemma 4(tt), this since $P(D_A \vee B,A) = 1$ by Theorem 34(a) and Lemma 4(aa), and

$$\text{(iii) } P(A,D_A \vee B) \leq P(A,D_A)$$

by Lemma 4(tt) and Theorem 34(a). So $P(A,D_A) = P(A,D_A \vee B)$. But

$$P(A,D_A \vee B) = P(A,D_A) \times P(D_A,D_A \vee B)$$

by Lemma 4(ss) and Theorem 34(a). Hence $P(D_A,D_A \vee B) = 1$, since $P(A,D_A) > 0$ by Theorem 34(c). But

$$P(D_A,D_A \vee B) + P(B,D_A \vee B) - P(D_A \wedge B,D_A \vee B) = 1$$

by Lemma 4(y) and **RP2**. So $P(D_A \wedge B,D_A \vee B)$ equals $P(B,D_A \vee B)$, which is larger than 0 since $B \sim_P D_A$. But by **RP4**, IE, and Lemma 4(tt),

$$P(D_A \wedge B,D_A \vee B) = P(D_A,B) \times P(B,D_A \vee B).$$

So $P(D_A,B) = 1$. ¤

Theorem 35. (a) $M_{P,D_A}(B) = P(B,D_A)$;
$M_{P,D_A}(\Lambda B \wedge C) = \lim_{n\to\infty} M_{P,D_A}((B_1 \wedge \ldots \wedge B_n) \wedge C)$, i.e. M_{P,D_A} *meets* **AP**Λ.

Proof: (a) $M_{P,D_A}(B) = \sup_{C\in L} \{Q_P(B \wedge C,D_A) : C \sim_P D_A\}$. Let then C be P-commensurable with D_A. Then $D_A \gtrsim_P B \wedge C$ by Theorems 9(a) and 5. If $D_A >_P B \wedge C$, then

$$P(B \wedge C,(B \wedge C) \vee D_A) = 0,$$

and hence

$$\text{(i) } P(D_A,(B \wedge C) \vee D_A) = 1$$

by Lemma 4(hh) and

$$\text{(ii) } P(B \wedge C,(B \wedge C) \vee D_A) = P(B \wedge C,D_A)$$

by Theorem 6(a). If $D_A \sim_P B \wedge C$, then $P(D_A,B \wedge C) = 1$ by Theorem 34 and hence

$$\text{(i) } P(D_A,(B \wedge C) \vee D_A) = 1$$

by Lemma 4(kk) and

$$\text{(ii) } P(B \wedge C,(B \wedge C) \vee D_A) = P(B \wedge C,D_A)$$

by Lemma 4(ss). But $P(B \wedge C,D_A) \leq P(B,D_A)$ by Lemma 4(d) and hence

$$P(B \wedge C,D_A) \leq P(B \wedge D_A,D_A)$$

by **RP4** and **RP2**. So $M_{P,D_A}(B) = P(B,D_A)$.

(b) By Theorems 32(a) and 35(a). ¤

So, with a Type I-ordered class of measure functions for L declared *Type V-ordered* if it meets conditions **O5**, **O6**, and

O4' *For every member M of $\mathcal{M}$ there exists a statement A of L such that $M = M^A$ and $M(A) = 1$,*

(so that a class of measure functions is Type V-ordered if it meets **O4'**, **O5**, and **O6**) we obtain this fifth Representation Theorem:

Theorem 36. *Every relative probability function for L_ω can be represented by means of a Type V-ordered class of measure functions for L_ω.*

But by Theorem 35(b) the measure functions representing P meet constraint **AP**Λ of Chapter 2, Section 1, and so, given **O6**, are absolute probability functions for L_ω. Consequently, what a relative probability function P for L_ω is represented by is a well-ordered class of absolute probability functions for L_ω, every one of which is the restriction of P to some statement D of L_ω.

And the converse of Theorem 36:

Theorem 37. *Every Type V-ordered class of absolute probability functions for L_ω represents a relative probability function for L_ω.*

Proof: Let $\mathcal{M}$ be a Type V-ordered class of absolute probability functions for L_{ω} and $P_{\mathcal{M}}$ be generated by $\mathcal{M}$. Since $\mathcal{M}$ satisfies the conditions of a Type I-ordered class, $P_{\mathcal{M}}$ meets **RP1–RP7** by Theorem 23. As for **RP**Λ,

$$
\begin{aligned}
&P_{\mathcal{M}}(\Lambda A, B) \\
&\quad = M^B(\Lambda A \wedge B) / M^B(B) && \textbf{(GEN)} \\
&\quad = \lim_{n\to\infty} M^B((A_1 \wedge \ldots \wedge A_n) \wedge B) / M^B(B) && \textbf{(AP}\Lambda\textbf{)} \\
&\quad = \lim_{n\to\infty} [M^B((A_1 \wedge \ldots \wedge A_n) \wedge B) / M^B(B)] \\
&\quad = \lim_{n\to\infty} P_{\mathcal{M}}(A_1 \wedge \ldots \wedge A_n, B).
\end{aligned}
$$

So $P_{\mathcal{M}}$ meets **RP**Λ. And by Theorem 24 $\mathcal{M}$ represents $P_{\mathcal{M}}$. ¤

Chapter 5

The Recursive Definability of Probability Functions

A probability function is at times fully determined by its restriction to a certain subset of its arguments. This chapter investigates when this is the case. Of particular interest are those cases in which the restriction in question can be independently characterised, i.e. as other than just the restriction to a certain class of arguments. We shall investigate primarily probability functions defined on sets of statements. The corresponding results for probability functions defined on statements are then easily obtained by virtue of the one-to-one correspondences noted in Chapter 1, Section 3.

In Section 1 we establish that absolute probability functions for L_P, L_Q, and L_ω can be recursively defined from their restrictions to finite sets of atomic statements. Section 2 introduces the auxiliary function F_P with the help of which one can characterise absolute probability functions on finite sets of atomic statements by language-independent constraints, hence other than as restrictions of absolute probability functions for L_P, L_Q, or L_ω. This then gives the result that every absolute probability function for sets of atomic statements of L_P (of L_Q, of L_ω) has a unique extension to an absolute probability function for L_P (for L_Q, for L_ω), an extension which is recursively definable.

We then show in Section 3 that there are no significant recursive definability results for relative probability functions with one exception, proved in Section 4, namely that any relative probability function for L_P meeting constraint **DL1** extends to just one relative probability function for L_ω meeting constraint **RPA2**, a result first alluded to in Chapter 2.

Section 1. The Recursive Definability of Absolute Probability Functions

Truth-value assignments, and hence two-valued absolute probability functions, are fully determined by their values for atomic statements; more precisely, if P is a two-valued absolute probability function, then $P(X)$ is a numerical function of the values that P assumes for the singleton sets of the atomic statements occurring in X. This feature extends to the larger family of functions, the two-valued ones among them, meeting

If X and Y are finite sets of atomic statements, then $P(X \cup Y) \times P(X \cap Y) = P(X) \times P(Y)$,

i.e. the functions under which the atomic statements of L_P are stochastically independent[1] of one another. For all of these functions $P(X)$ is a numerical function of the values of P for the singleton sets of the atomic statements. But this is not the case with absolute probability functions in general, as we will show later.

Extending to Infinite Sets of Statements

Looking for definability results that hold for all absolute probability functions, we begin with the observation that any unary function P defined on the sets of statements, both finite and infinite, of L_P and meeting constraints **AS1**, **AS2**, **AS¬**, **AS∧** and **ASω**, can be defined from its restriction P' to the finite sets of statements of L_P thus:

$$P(X) = \inf_{X'} \{P'(X') : X' \subseteq_f X\},$$

which, given that, for any X and Y, $P(X \cup Y) \leq P(X)$ (cf. Lemma 10(b)), proves to be the same as

$$P(X) = \begin{cases} P'(X) & \textit{if X is finite} \\ \inf_{X'} \{P'(X') : X' \subseteq_f X\} & \textit{if X is infinite.} \end{cases}$$

Moreover, any absolute probability function P' defined on the finite sets of statements of L_P can be extended by means of this definition to a function P, defined on all sets of statements; or, in other words, every absolute probability function P' on the finite sets of statements is the restriction of a function P defined on all sets of statements. For P, so defined, obviously meets **AS1** and **AS2**. P also meets **AS¬**, since

$$\begin{aligned} &\inf_{X'} \{P'(X') : X' \subseteq_f X\} \\ &\quad = \inf_{X'} \{P'(X' \cup \{A\}) + P'(X \cup \{\neg A\}) : X' \subseteq_f X\} && (\mathbf{AS\neg}) \\ &\quad = \inf_{X'} \{P'(X' \cup \{A\}) : X' \subseteq_f X\} + \inf_{X'} \{P'(X \cup \{\neg A\}) : X' \subseteq_f X\} \\ &\quad = \inf_{X'} \{P'(X') : X' \subseteq_f X \cup \{A\}\} + \inf_{X'} \{P'(X') : X' \subseteq_f X \cup \{\neg A\}\} && \text{(Lemma 10(b))} \end{aligned}$$

And P meets **AS∧**, since

$$\begin{aligned} &\inf_{X'} \{P'(X') : X' \subseteq_f X \cup \{A \wedge B\}\} \\ &\quad = \inf_{X'} \{P'(X' \cup \{A \wedge B\}) : X' \subseteq_f X\} && \text{(Lemma 10(b))} \\ &\quad = \inf_{X'} \{P'(X' \cup \{A,B\}) : X' \subseteq_f X\} && (\mathbf{AS\wedge}) \\ &\quad = \inf_{X'} \{P'(X') : X' \subseteq_f X \cup \{A,B\}\} && \text{(Lemma 10(b))} \end{aligned}$$

The same result holds for absolute probability functions defined on sets of statements of the languages L_Q and L_ω. Take the constraint **AS∀**. It is met by P, if P is defined as above from P', a function on the finite sets of statements of L_Q; for, writing '$\mathrm{I}((\forall x)A)$' instead of '$\mathrm{I}((\forall x)A, \mathcal{D}_0 \cup \mathcal{D}_\omega)$' for the set of substitution instances $A(T/x)$ of $(\forall x)A$ with terms T drawn from $\mathcal{D}_0 \cup \mathcal{D}_\omega$,

$$\begin{aligned} &\inf_{X'} \{P'(X') : X' \subseteq X \cup \{(\forall x)A\}\} \\ &\quad = \inf_{X'} \{P'(X' \cup \{(\forall x)A\}) : X' \subseteq X\} && \text{(Lemma 10(b))} \\ &\quad = \inf_{X'} \{\inf_{Y'} \{P'(X' \cup Y') : Y' \subseteq_f \mathrm{I}((\forall x)A)\} : X' \subseteq X\} && (\mathbf{AS\forall}) \\ &\quad = \inf_{Y'} \{\inf_{X'} \{P'(X' \cup Y') : X' \subseteq X\} : Y' \subseteq_f \mathrm{I}((\forall x)A)\} \\ &\quad = \inf_{Y'} \{\inf_{X'} \{P'(X') : X' \subseteq X \cup Y'\} : Y' \subseteq_f \mathrm{I}((\forall x)A)\} && \text{(Lemma 10(b))} \end{aligned}$$

And a similar demonstration shows that if P is defined from an absolute probability function P' on the finite sets of statements of L_ω, then P meets **AS**ω. So,

Theorem 1. *Any absolute probability function on the finite sets of statements of L_P, L_Q, or L_ω extends uniquely to an absolute probability function defined on all sets of statements of those languages.*

Extending from Restrictions to Atomic Statements

Next, we prove that any absolute probability function on the finite sets of statements can be recovered from its restriction to sets of atomic statements. So, let P' be an absolute probability function on the finite sets of statements of L_P, i.e. a unary function from finite sets of statements to the reals that meets constraints **AS1**, **AS2**, **AS**$\neg$, and **AS**$\wedge$. Starting with the restriction of P' to the finite sets of atomic statements we will recursively define a function P on the finite sets of statements of L_P. The definition will be recursive on the complexity c of finite sets X of statements. We first explain c for statements of L_P:

$$\mathrm{c}(A) = \begin{cases} 0 & \textit{if } A \textit{ is atomic} \\ \mathrm{c}(B) + 1 & \textit{if } A \textit{ is a negation } \neg B \\ \mathrm{c}(B) + \mathrm{c}(C) + 1 & \textit{if } A \textit{ is a conjunction } B \wedge C. \end{cases}$$

As for $X = \{A_1, A_2, ..., A_n\}$ a finite set of statements of L_P we define:

$$\mathrm{c}(X) = \begin{cases} 0 & \textit{if } X \textit{ is } \varnothing \\ \mathrm{c}(A_1) + ... + \mathrm{c}(A_n) & \textit{otherwise.} \end{cases}$$

We also assume that the denumerably many statements of L_P are arranged in a fixed so-called *alphabetical* order. In the following definition A is the alphabetically latest compound member of X when $\mathrm{c} > 0$.

$$P(X) = \begin{cases} P'(X) & \textit{if } \mathrm{c}(X) = 0 \\ P(X-\{A\}) - P((X-\{A\}) \cup \{B\}) & \textit{if } \mathrm{c}(X) > 0 \textit{ and } A \textit{ is of the sort } \neg B \\ P((X-\{A\}) \cup \{B,C\}) & \textit{if } \mathrm{c}(X) > 0 \textit{ and } A \textit{ is of the sort } B \wedge C \end{cases}$$

But P' itself meets the totality of constraints that enter into the definition of P. For, trivially, when c equals 0, P' agrees with its own restriction to the sets of atomic statements; and the recursive clauses amount to a subset of the constraints that are instances of **AS**$\neg$ and **AS**$\wedge$, constraints which P' of course meets by assumption. So P, being the only function satisfying the clauses of the definition, is the same function as P'. It is also clear that, which particular ordering of the statements of L_P is declared to be the alphabetical ordering of these statements, is immaterial, since no reference is made to any further features of the ordering besides its being an ordering.

In Chapter 1, Section 3 it was shown that the absolute probability functions for L_P and those for finite sets of statements of L_P are one-to-one correlated. As a consequence, the

A	$P(A)$	A	$P(A)$
$E_1 \wedge \neg E_1$	0	$E_1 \wedge \neg E_2$	$1/2 - r$
$E_1 \vee \neg E_1$	1	$\neg(E_1 \wedge \neg E_2)$	$1/2 + r$
E_1	1/2	$\neg E_1 \wedge E_2$	$1/2 - r$
$\neg E_1$	1/2	$\neg(\neg E_1 \wedge E_2)$	$1/2 + r$
E_2	1/2	$\neg E_1 \wedge \neg E_2$	r
$\neg E_2$	1/2	$\neg(\neg E_1 \wedge \neg E_2)$	$1 - r$
$E_1 \wedge E_2$	r	$E_1 \equiv E_2$	$2r$
$\neg(E_1 \wedge E_2)$	$1 - r$	$\neg(E_1 \equiv E_2)$	$1 - 2r$

Table 1

restrictions of the absolute probability functions for L_P to atomic statements and conjunctions of atomic statements are one-to-one correlated with the restrictions to finite sets of atomic statements of the absolute probability functions on finite sets of statements of L_P. We therefore have:

Theorem 2. (a) *Any absolute probability function for the finite sets of statements of* L_P *can be recursively defined from its restriction to sets of atomic statements;*
(b) *Any absolute probability function for the statements of* L_P *can be recursively defined from its restriction to atomic statements and their conjunctions.*

This result cannot be further improved upon. The absolute probability of the conjunction of two or more atomic statements is not always a numerical function of, and hence determined by, the absolute probabilities of those statements. For proof, suppose E_1 and E_2 to be two distinct atomic statements; and suppose each of the sixteen statements in Table 1 to stand for all the statements that are logically equivalent to it. The table delivers an absolute probability function[2] for each real number r such that $0 < r < 1/2$. So, r_1 and r_2 being two distinct choices of that r, let P_1 be the function P such that

$$P(E_1 \wedge E_2) = r_1$$

and P_2 be the one such that

$$P(E_1 \wedge E_2) = r_2.$$

The reader will note that, although the two functions P_1 and P_2 agree on each of the two atomic statements E_1 and E_2, they nonetheless disagree on their conjunction $E_1 \wedge E_2$. So, the absolute probability of the conjunction of two or more *atomic* statements is not always a numerical function of their absolute probabilities. So, *absolute probability functions are recursively definable from a certain point on, but from that point on only.*

Extending from Quantifier-Free Statements to Statements of L_Q

The complexity of the statements of L_Q will be measured by a pair of functions, the second of which is essentially the complexity function introduced above for the statements of L_P:

$$\mathrm{c}(A) = \begin{cases} 0 & \textit{if } A \textit{ is atomic} \\ 1 & \textit{if } A \textit{ is a quantification } (\forall x)B \\ \mathrm{c}(B)+1 & \textit{if } A \textit{ is a negation } \neg B \\ \mathrm{c}(B)+\mathrm{c}(C)+1 & \textit{if } A \textit{ is a conjunction } B \wedge C \end{cases}$$

The quantifier height qh of a statement of L_Q is defined thus:

$$\mathrm{qh}(A) = \begin{cases} 0 & \textit{if } A \textit{ is a quantifier-free statement} \\ \mathrm{qh}(B) & \textit{if } A \textit{ is a negation } \neg B \\ \max\{\mathrm{qh}(B),\mathrm{qh}(C)\} & \textit{if } A \textit{ is a conjunction } B \wedge C \\ \mathrm{qh}(B(T/x))+1 & \textit{if } A \textit{ is a quantification } (\forall x)B \end{cases}$$

The quantifier height $\mathrm{qh}(X)$ of a finite set X of statements is 0 if $X = \varnothing$, otherwise $\mathrm{qh}(X) = \max\{\mathrm{qh}(A) : A \in X\}$, while the connective complexity $\mathrm{c}(X)$ is understood to be the sum of the complexities of those members of X whose quantifier height is the same as the maximum height in X, i.e. $\mathrm{c}(X) = \Sigma\{\mathrm{c}(A) : A \in S \text{ and } \mathrm{qh}(A) = \mathrm{qh}(X)\}$.

Finally, we define the rank $\mathrm{r}(X)$ of a finite set of statements of L_Q as the pair $\langle \mathrm{qh}(X),\mathrm{c}(X)\rangle$, and the rank order among finite sets of statements by stipulating:

$$\mathrm{r}(X) < \mathrm{r}(X')$$

if

$$\text{(i)} \quad \mathrm{qh}(X) < \mathrm{qh}(X')$$

or

$$\text{(ii)} \quad \mathrm{qh}(X) = \mathrm{qh}(X') \textit{ but } \mathrm{c}(X) < \mathrm{c}(X').$$

Now let P' be an absolute probability function for the finite sets of statements of L_Q. We recursively define a function P from the restriction of P' to the sets of quantifier-free statements of L_Q, A being the alphabetically latest compound member of X of quantifier height $\mathrm{qh}(X)$.

$$P(X) = \begin{cases} P'(X) & \textit{if } \mathrm{qh}(X) = 0 \\ P(X-\{A\}) - P((X-\{A\}) \cup \{B\}) & \textit{if } \mathrm{qh}(X) > 0 \textit{ and } A \textit{ is } \neg B \\ P((X-\{A\}) \cup \{B,C\}) & \textit{if } \mathrm{qh}(X) > 0 \textit{ and } A \textit{ is } B \wedge C \\ \inf_{Y'}\{P((X-\{A\}) \cup Y') : Y' \subseteq_{\mathrm{f}} \mathrm{I}((\forall x)B)\} & \textit{if } \mathrm{qh}(X) > 0 \textit{ and } A \textit{ is } (\forall x)B \end{cases}$$

The function P' itself meets the totality of constraints that enter into the definition of P. For, trivially, when qh(X) equals 0, P' agrees with its own restriction to the sets of quantifier free statements; and the recursive clauses amount to a subset of the constraints that are instances of **AS¬**, **AS∧**, and **AS∀**, constraints which P' of course meets by assumption. So P, being the only function satisfying the clauses of the definition, is the same function as P'. Hence the first part of the following theorem. The second part is justified by the correspondence between the absolute probability functions for the finite sets of statements of L_Q and the absolute probability functions for the statements of L_Q.

Theorem 3. (a) *Any absolute probability function for the finite sets of statements of L_Q can be recursively defined from its restriction to sets of quantifier free statements;*
(b) *Any absolute probability function for the statements of L_Q can be recursively defined from its restriction to quantifier free statements.*

Since the quantifier free statements of L_Q are the statements of L_P, Theorem 3 can be combined with Theorem 2 to yield the result that *any absolute probability function for the finite sets of statements of L_Q can be recursively defined from its restriction to the sets of atomic statements.*

Extending from Finitary Statements to Statements of L_ω

In a similar way we can recursively define any absolute probability function for the finite sets of statements of L_ω from its restriction to the finite sets of finitary statements, i.e. from the finite sets of statements of L_P. In the following definition P' is an absolute probability function for the finite sets of statements of L_ω. The definition is again recursive, but this time the recursion is transfinite. The recursion parameter is an ordered pair consisting of the *infinite conjunction height ih* and the finite connective complexity *c*. We first define c for statements of L_ω:

$$\mathrm{c}(A) = \begin{cases} 0 & \textit{if } A \textit{ is atomic} \\ 1 & \textit{if } A \textit{ is an infinite conjunction } \mathbf{\Lambda B} \\ \mathrm{c}(B) + 1 & \textit{if } A \textit{ is a negation } \neg B \\ \mathrm{c}(B) + \mathrm{c}(C) + 1 & \textit{if } A \textit{ is a conjunction } B \wedge C \end{cases}$$

For statements of L_P, c(A) coincides with the complexity of A as defined before. The *infinite conjunction height* of a statement of L_ω is a, possibly transfinite, ordinal number. The supremum in clause four of the definition is the maximum of the heights of statements in X if there is such a maximum; otherwise it is the first limit ordinal greater than the infinite conjunction height of any member of X.

$$\mathrm{ih}(A) = \begin{cases} 0 & \textit{for } A \textit{ a statement of } L_P \\ \mathrm{ih}(B) & \textit{for } A \textit{ a negation } \neg B \\ \max\{\mathrm{ih}(B), \mathrm{ih}(C)\} & \textit{for } A \textit{ a conjunction } B \wedge C \\ \sup\{\mathrm{ih}(B_i) : i = 1, 2, \ldots\} & \textit{for } A \textit{ an infinitary conjunction } \mathbf{\Lambda B} \end{cases}$$

The infinite conjunction height ih(X) of a finite set of statements is 0 if $X = \varnothing$, otherwise ih(X) $= \max\{\text{ih}(A) : A \in X\}$, while the connective complexity c(X) is understood to be the sum of the complexities of those members of X whose infinite conjunction height is the same as the maximum height in X, i.e. $\text{c}(X) = \Sigma\{\text{c}(A) : A \in X \text{ and ih}(A) = \text{ih}(X)\}$.

Finally, we define the rank r(X) of a finite set X of statements of L_ω as the pair $\langle \text{ih}(X), \text{c}(X) \rangle$, and we introduce the rank order among finite sets of statements by stipulating:

$$\text{r}(X) < \text{r}(X')$$

if

$$\text{(i)} \quad \text{ih}(X) < \text{ih}(X')$$

or

$$\text{(ii)} \quad \text{ih}(X) = \text{ih}(X') \textit{ but } \text{c}(X) < \text{c}(X').$$

For the definition of a function P from the restriction to L_P of an absolute probability function P' for L_ω we invoke again the alphabetical order of statements, this time the statements of L_ω. As there are non-denumerably many statements of L_ω, an implicit appeal is made here to the axiom of choice. Also, A is again the alphabetically latest compound member of X with infinite conjunction height ih(X).

$$P(X) = \begin{cases} P'(X) \quad \textit{if } \text{ih}(X) = 0 \\ P(X - \{A\}) - P((X - \{A\}) \cup \{B\}) \quad \textit{if } \text{ih}(X) > 0 \textit{ and } A \textit{ is } \neg B \\ P((X - \{A\}) \cup \{B, C\}) \quad \textit{if } \text{qh}(X) > 0 \textit{ and } A \textit{ is } B \wedge C \\ \inf_{Y'} \{P((X - \{A\}) \cup Y') : Y' \subseteq_{\text{f}} \text{S}(\boldsymbol{B})\} \quad \textit{if } \text{qh}(X) > 0 \textit{ and } A \textit{ is } \Lambda \boldsymbol{B} \end{cases}$$

The function P' itself meets the totality of constraints that enter into the definition of P. For, trivially, when the infinite conjunction height equals 0 P' agrees with its own restriction to the sets of statements of L_P; and the recursive clauses amount to a subset of the constraints that are instances of **AS¬**, **AS∧**, and **ASΛ**, constraints which P' of course meets by assumption. So, P being the only function satisfying the clauses of the definition, is the same function as P'.

Theorem 4. (a) *Any absolute probability function for the finite sets of statements of L_ω can be recursively defined from its restriction to sets of finitary statements, i.e. of statements of L_P;* (b) *Any absolute probability function for the statements of L_ω can be recursively defined from its restriction to L_P.*

Since the finitary statements of L_ω are the statements of L_P, Theorem 4 can be combined with Theorem 2 to yield the result that *any absolute probability function for the finite sets of statements of L_ω can be recursively defined from its restriction to the sets of atomic statements.*

Section 2. The Auxiliary Function F_P

Theorem 2 can be strengthened considerably since the restrictions of absolute probability functions to finite sets of atomic statements can be independently characterised, i.e. other than as restrictions of absolute probability functions for finite sets of statements for L_P. While it is natural to wonder whether the constraints that do not mention connectives, i.e. constraints **AS1** and **AS2**, are sufficient for the purpose, it is also immediately clear that they are not: from **AS1** and **AS2** alone one cannot even deduce that $P(X) \leq 1$. In order to arrive at a satisfactory characterisation we need to enlist an auxiliary function F_P that takes two finite sets of statements as arguments. The definition is a recursive one on the cardinality of the second argument. Note that no assumption is made in the definition about the language from which the statements are drawn.

Definition of F_P. *Let P be any function from the finite sets of statements of a language L to the reals; let X and Y be finite sets of statements; and, in the event that Y is non-empty, let A be the alphabetically latest member of Y. Then*

$$F_P(X,Y) = \begin{cases} P(X) & \textit{if } Y = \varnothing \\ F_P(X,Y-\{A\}) - F_P(X \cup \{A\},Y-\{A\}) & \textit{otherwise} \end{cases}$$

Lemma 8(b) in Appendix I shows that for any member A of Y, whether the alphabetically latest one or not,

$$F_P(X,Y) = F_P(X,Y-\{A\}) - F_P(X \cup \{A\},Y-\{A\}),$$

and, more generally, that

$$F_P(X,Y \cup \{A\}) = F_P(X,Y) - F_P(X \cup \{A\},Y).$$

Also note that when P meets **AS1, AS2, AS¬**, and **AS∧**, then, meaning by $Y^{\neg}$ the set whose members are the negations of the members of Y, we get

$$F_P(X,Y) = P(X \cup Y^{\neg})$$

by Lemma 10(i), and meaning by $\vee Y$ that disjunction of the members of Y in which the disjuncts appear in alphabetical order, or $A \wedge \neg A$, in case Y is $\varnothing$, we have

$$F_P(X,Y) = P(X) - P(X \cup \{\vee Y\}),$$

this by Lemma 10(r). We will also make use of the following identities concerning F_P which presuppose that P meets **AS¬, AS∧, AS∀**, or **ASΛ**.

Constraints on F_P

Theorem 5. *Let P be a function on the finite sets of statements of L_P, L_Q, or L_ω which meets constraint* **AS¬**. *Then:*

(a) $F_P(X \cup \{\neg A\},Y) = F_P(X,Y \cup \{A\})$;

(b) $F_P(X,Y \cup \{\neg A\}) = F_P(X \cup \{A\},Y)$.

Proof: (a) By induction on the cardinality of Y.

Basis: $Y = \varnothing$. Then

$P(X \cup \{\neg A\}) = P(X) - P(X \cup \{A\})$ by **AS¬** and hence $F_P(X \cup \{\neg A\},Y) = F_P(X,Y \cup \{A\})$ by the basic clause in the definition of F_P.

Inductive Step: card(Y) > 0. Let B be an arbitrary member of Y and let Y' be $Y-\{B\}$. Then:

$$\begin{aligned} &F_P(X \cup \{\neg A\},Y' \cup \{B\}) \\ &\quad = F_P(X \cup \{\neg A\},Y') - F_P(X \cup \{\neg A\} \cup \{B\},Y') && \text{(Lemma 8(b))} \\ &\quad = F_P(X,Y' \cup \{A\}) - F_P(X \cup \{B\},Y' \cup \{A\}) && \text{(Ind. Hyp.)} \\ &\quad = F_P(X,Y' \cup \{B\} \cup \{A\}) && \text{(Lemma 8(b))} \end{aligned}$$

Hence $F_P(X \cup \{\neg A\},Y) = F_P(X,Y \cup \{A\})$.

$$\begin{aligned} &\text{(b) } F_P(X,Y \cup \{\neg A\}) \\ &\quad = F_P(X,Y) - F_P(X \cup \{\neg A\},Y) && \text{(Lemma 8(b))} \\ &\quad = F_P(X,Y) - F_P(X,Y \cup \{A\}) && \text{((a))} \\ &\quad = F_P(X \cup \{A\},Y) && \text{(Lemma 8(b))} \end{aligned}$$ ¤

Theorem 6. *Let P be a function on the finite sets of statements of* L_P, L_Q, *or* L_ω *which meets constraint* AS∧. *Then:*

(a) $F_P(X \cup \{A \wedge B\},Y) = F_P(X \cup \{A,B\},Y)$;

(b) $F_P(X,Y \cup \{A \wedge B\}) = F_P(X,Y \cup \{A\}) + F_P(X \cup \{A\},Y \cup \{B\}$.

Proof: (a) By induction on the cardinality of Y.

Basis: $Y = \varnothing$. Then $P(X \cup \{A \wedge B\}) = P(X \cup \{A,B\})$ by **AS∧**, hence $F_P(X \cup \{A \wedge B\},Y) = F_P(X \cup \{A,B\},Y)$ by the basic clause in the definition of F_P.

Inductive Step: card(Y) > 0. Let C be an arbitrary member of Y and let Y' be $Y-\{C\}$. Then:

$$\begin{aligned} &F_P(X \cup \{A \wedge B\},Y' \cup \{C\}) \\ &\quad = F_P(X \cup \{A \wedge B\},Y') - F_P(X \cup \{A \wedge B\} \cup \{C\},Y') && \text{(Lemma 8(b))} \\ &\quad = F_P(X \cup \{A,B\},Y') - F_P(X \cup \{A,B\} \cup \{C\},Y') && \text{(Ind. Hyp.)} \\ &\quad = F_P(X \cup \{A,B\},Y' \cup \{C\}) && \text{(Lemma 8(b))} \end{aligned}$$

Hence $F_P(X \cup \{A \wedge B\},Y) = F_P(X \cup \{A,B\},Y)$.

$$\begin{aligned} &\text{(b) } F_P(X,Y \cup \{A \wedge B\}) \\ &\quad = F_P(X,Y) - F_P(X \cup \{A \wedge B\},Y) && \text{(Lemma 8(b))} \\ &\quad = F_P(X,Y) - F_P(X \cup \{A,B\},Y) && \text{((a))} \\ &\quad = F_P(X,Y) - F_P(X \cup \{A\},Y) + F_P(X \cup \{A\},Y \cup \{B\}) && \text{(Lemma 8(b))} \\ &\quad = F_P(X,Y \cup \{A\}) + F_P(X \cup \{A\},Y \cup \{B\} && \text{(Lemma 8(b))} \end{aligned}$$ ¤

Theorem 7. *Let P be a function on the finite sets of statements of* L_Q *which meets constraint* AS∀. *Then*

(a) $F_P(X \cup \{(\forall x)A\},Y) = \underset{n\to\infty}{\text{limit}}\, F_P(X \cup \{A(t_1/x),\ldots,A(t_n/x)\},Y)$;

(b) $F_P(X,Y \cup \{(\forall x)A\}\}) =$

$\underset{n\to\infty}{\text{limit}}\, [F_P(X,Y \cup \{A(t_1/x)\}) + F_P(X \cup \{A(t_1/x)\},Y \cup \{A(t_2/x)\}) + \ldots$

$$+ F_P(X \cup \{A(t_1/x),...,A(t_{n-1}/x)\}, Y \cup \{A(t_n/x)\})].$$

Proof of Theorem 7, which is like that of Theorem 8, is omitted.

Theorem 8. *Let P be a function on the finite sets of statements of L_ω which meets constraint* **ASΛ**, *and let A be the sequence $\langle A_1, A_2, ..., A_n, ... \rangle$. Then*
(a) $F_P(X \cup \{\Lambda A\}, Y) = \lim_{n\to\infty} F_P(X \cup \{A_1,...,A_n\}, Y)$;
(b) $F_P(X, Y \cup \{\Lambda A\}\}) =$
$\lim_{n\to\infty} [F_P(X, Y \cup \{A_1\}) + F_P(X \cup \{A_1\}, Y \cup \{A_2\}) + ...$
$$+ F_P(X \cup \{A_1,...,A_{n-1}\}, Y \cup \{A_n\})].$$

Proof: (a) By induction on the cardinality of Y.
Basis: $Y = \varnothing$. Then $P(X \cup \{\Lambda A\}) = \lim_{n\to\infty} P(X \cup \{A_1,...,A_n\})$ by **ASΛ**. Hence
$$F_P(X \cup \{\Lambda A\}, Y) = \lim_{n\to\infty} F_P(X \cup \{A_1,...,A_n\}, Y)$$
by the basic clause in the definition of F_P.
Inductive Step: card(Y) > 0. Let C be an arbitrary member of Y and let Y' be $Y-\{C\}$. Then:
$F_P(X \cup \{\Lambda A\}, Y' \cup \{C\})$
$= F_P(X \cup \{\Lambda A\}, Y') - F_P(X \cup \{C\} \cup \{\Lambda A\}, Y')$ (Lemma 8(b))
$= \lim_{n\to\infty} F_P(X \cup \{A_1,...,A_n\}, Y') - \lim_{n\to\infty} F_P(X \cup \{C\} \cup \{A_1,...,A_n\}, Y')$
(Ind. Hyp.)
$= \lim_{n\to\infty} F_P(X \cup \{A_1,...,A_n\}, Y' \cup \{C\})$ (Lemma 8(b))
Hence $F_P(X \cup \{\Lambda A\}, Y) = \lim_{n\to\infty} F_P(X \cup \{A_1,...,A_n\}, Y)$.
(b) $F_P(X, Y \cup \{\Lambda A\})$
$= F_P(X, Y) - F_P(X \cup \{\Lambda A\}, Y)$ (Lemma 8(b))
$= F_P(X, Y) - \lim_{n\to\infty} F_P(X \cup \{A_1,...,A_n\}, Y)$ ((a))
$= \lim_{n\to\infty} [F_P(X, Y) - F_P(X \cup \{A_1,...,A_n\}, Y)]$
$= \lim_{n\to\infty} [F_P(X, Y \cup \{A_1\}) + F_P(X \cup \{A_1\}, Y \cup \{A_2\}) + ...$
$+ F_P(X \cup \{A_1,...,A_{n-1}\}, Y \cup \{A_n\})]$ (Lemma 8(c)) ¤

Characterising Absolute Probability Functions for Finite Sets of Atomic Statements

With F_P defined, we introduce the new constraint

AS1' $0 \leq F_P(X,Y)$,

and we show that the set of constraints **AS1**, **AS2**, **AS¬**, and **AS∧** is equivalent to the set consisting of **AS1'**, **AS2**, **AS¬**, and **AS∧**. That any function for the finite sets of statements of L_P that meets **AS1'** meets **AS1** is obvious since $P(X) = F_P(X,Y)$ by the definition of F_P. So, assuming that P meets **AS1**, **AS2**, **AS¬**, and **AS∧**, we need to show that P meets **AS1'**. But this follows immediately from **AS1** and the identity

$$F_P(X,Y) = P(X \cup Y^{\neg})$$

mentioned above. Constraints **AS1'** and **AS2** have the desired feature of characterising those functions for the finite sets of atomic statements that are restrictions of absolute probability functions for the finite sets of statements of L_P. We show this by establishing that the functions for finite sets of atomic statements that meet **AS1'** and **AS2** can be recursively extended to absolute probability functions for the finite sets of statements of L_P.

Theorem 9. *Let P' be a function for the finite sets of atomic statements of L_P which meets constraints* **AS1'** *and* **AS2**; *let P be recursively defined by*

$$P(X) = \begin{cases} P'(X) & \textit{if } \mathrm{c}(X) = 0 \\ P(X-\{A\}) - P((X-\{A\}) \cup \{B\}) & \textit{if } \mathrm{c}(X) > 0 \textit{ and } A \textit{ is of the sort } \neg B \\ P((X-\{A\}) \cup \{B,C\}) & \textit{if } \mathrm{c}(X) > 0 \textit{ and } A \textit{ is of the sort } B \wedge C. \end{cases}$$

Then P is a function for the finite sets of statements of L_P which meets **AS1'**, **AS2**, **AS¬**, *and* **AS∧**.

Proof:

AS2: $P(\varnothing) = P'(\varnothing) = 1$ by the definition of P and the fact that P' meets **AS2**.

AS¬: (i) $\neg A \notin X$. Then proof is by induction on the complexity c of X:

Basis: $\mathrm{c}(X) = 0$, i.e. the statements in X are atomic statements and $\neg A$ is perforce the alphabetically latest compound member of $X \cup \{\neg A\}$. Then

$$P(X \cup \{\neg A\}) = P(X) - P(X \cup \{A\})$$

by clause 2 of the definition.

Inductive Step: $\mathrm{c}(X) > 0$. Let B be the alphabetically latest compound member of $X \cup \{\neg A\}$.

Case 1: B is $\neg A$. Then $P(X \cup \{\neg A\}) = P(X) - P(X \cup \{A\})$ by clause 2 of the definition.

Case 2: B is a statement other than $\neg A$.

Case 2.1: B is some other negation $\neg C$ of L_P. Let X' be $X-\{\neg C\}$. Then

$$P(X' \cup \{\neg C\} \cup \{\neg A\}) = P(X' \cup \{\neg C\}) - P(X' \cup \{\neg C\} \cup \{A\})$$

by the definition of P and the inductive hypothesis.

Case 2.2: B is some conjunction $C \wedge D$ of L_P. Let X' be $X-\{C \wedge D\}$. Then

$$P(X' \cup \{C \wedge D\} \cup \{\neg A\}) = P(X' \cup \{C \wedge D\}) - P(X' \cup \{C \wedge D\} \cup \{A\})$$

by the definition of P and the inductive hypothesis.

(ii) $\neg A \in X$. Then $P(X \cup \{\neg A\}) = P(X)$ and by (i):

$$\begin{aligned} &P(X \cup \{A\}) \\ &\quad = P((X-\{\neg A\}) \cup \{A,\neg A\}) \\ &\quad = P((X-\{\neg A\}) \cup \{A\}) - P((X-\{\neg A\}) \cup \{A,A\}) \\ &\quad = 0 \end{aligned}$$

Hence $P(X \cup \{\neg A\}) = P(X) - P(X \cup \{A\})$.

AS∧: Similarly.

AS1': If $X \cap Y \neq \varnothing$, $F_P(X,Y) = 0$ by Lemma 8(a), and hence $0 \leq F_P(X,Y)$. So, assume that $X \cap Y = \varnothing$. Proof is then by induction on the complexity c of $X \cup Y$.

Basis: $\mathrm{c}(X \cup Y) = 0$, i.e. the statements in X and in Y are atomic statements. Then $F_P(X,Y)$

$= F_{P'}(X,Y)$ and hence $0 \leq F_P(X,Y)$ by the assumption that P' meets **AS1'**.
Inductive Step: $c(X \cup Y) > 0$. Let A be the alphabetically latest statement in $X \cup Y$.
Case 1: A is $\neg B$.
Case 1.1: $A \in X$. Let X' be $X-\{\neg B\}$. Then $F_P(X' \cup \{\neg B\},Y) = F_P(X,Y \cup \{B\})$ by Theorem 5(a). Hence $0 \leq F_P(X,Y)$ by the inductive hypothesis.
Case 1.2: $A \in Y$. Let Y' be $Y-\{\neg B\}$. Then $F_P(X,Y' \cup \{\neg B\}) = F_P(X \cup \{B\},Y)$ by Theorem 5(b). Hence $0 \leq F_P(X,Y)$ by the inductive hypothesis.
Case 2: A is $B \wedge C$.
Case 2.1: $A \in X$. Let X' be $X-\{B \wedge C\}$. Then $F_P(X' \cup \{B \wedge C\},Y) = F_P(X \cup \{B,C\},Y)$ by Theorem 6(a). Hence $0 \leq F_P(X,Y)$ by the inductive hypothesis.
Case 2.2: $A \in Y$. Let Y' be $Y-\{B \wedge C\}$. Then

$$F_P(X,Y' \cup \{B \wedge C\}) = F_P(X,Y' \cup \{A\}) + F_P(X \cup \{A\},Y' \cup \{B\})$$

by Theorem 6(b). Hence $0 \leq F_P(X,Y)$ by the inductive hypothesis. ¤

AS1' and **AS2** together characterise those functions on the finite sets of atomic statements of L_P which are the restrictions of absolute probability functions on the finite sets of statements of L_P. But atomic statements of L_P are just statements which are treated as not truth-functionally composite; hence they can be any statements whatsoever, whose possible internal structure is ignored. We conclude that **AS1'** and **AS2** are the appropriate constraints to characterise absolute probability functions on the finite sets of statements of an arbitrary language of uninterpreted statements. **AS1'** and **AS2** are the most general and language-independent constraints for absolute probability functions, constraints which regulate only the relationships between probabilities assigned to the various finite sets of statements without regard to any effect that the meanings of statements might have on those probabilities. The constraint **AS**ω brings the probabilities of infinite sets of statements in relation to those of finite sets and obviously also has this very general character.

The study of these general constraints for absolute probability functions will be taken up in Chapter 7, when we will motivate constraints for probability functions by linking probabilities to degrees of logical consequence.

Extending from Quantifier-Free Statements to Statements of L_Q

In a similar vein, Theorem 3 can be strengthened by taking as the starting point of the recursive definition not the restriction of a function already known to be an absolute probability function for L_Q but an arbitrary absolute probability function for the finite sets of quantifier-free statements of L_Q, i.e. for the finite sets of statements of L_P.

Theorem 10. *Let P_0 be a function for the finite sets of quantifier-free statements of L_Q which meets constraints* **AS1'**, **AS2**, **AS**$\neg$, and **AS**$\wedge$; *and let P be recursively defined by*

$$P(X) = \begin{cases} P_0(X) & \textit{if } \mathrm{qh}(X) = 0 \\ P(X-\{A\}) - P((X-\{A\}) \cup \{B\}) & \textit{if } \mathrm{qh}(X) > 0 \textit{ and } A \textit{ is } \neg B \\ P((X-\{A\}) \cup \{B,C\}) & \textit{if } \mathrm{qh}(X) > 0 \textit{ and } A \textit{ is } B \wedge C \\ \underset{n\to\infty}{\mathrm{limit}}\, P((X-\{A\}) \cup \{B(\mathrm{t}_1/x),\ldots,B(\mathrm{t}_n/x)\}) & \textit{if } \mathrm{qh}(X) > 0 \textit{ and } A \textit{ is } (\forall x)B \end{cases}$$

Then P is a function for the finite sets of statements of L_Q *which meets* **AS1'**, **AS2**, **AS¬**, **AS∧**, *and* **AS∀**.

We omit the proof, which is similar to that of Theorem 11 below. Because of the one-to-one correspondence between absolute probability functions for the finite sets of statements of L_Q and those for the sets of statements of L_Q and because of a similar correspondence in the case of L_P we also have it that any absolute probability function for the quantifier-free statements of L_Q can be recursively extended to a full absolute probability function for L_Q.

Extending from Finitary Statements to Statements of L_ω

An analogous result can be obtained for L_ω:

Theorem 11. *Let* P_0 *be a function for the finite sets of finitary statements of* L_ω *which meets constraints* **AS1'**, **AS2**, **AS¬**, *and* **AS∧**; *and let P be recursively defined by*

$$P(X) = \begin{cases} P_0(X) & \textit{if } \mathrm{ih}(X) = 0 \\ P(X-\{A\}) - P((X-\{A\}) \cup \{B\}) & \textit{if } \mathrm{ih}(X) > 0 \textit{ and } A \textit{ is } \neg B \\ P((X-\{A\}) \cup \{B,C\}) & \textit{if } \mathrm{ih}(X) > 0 \textit{ and } A \textit{ is } B \wedge C \\ \underset{n\to\infty}{\mathrm{limit}}\, P((X-\{A\}) \cup \{B_1,\ldots,B_n\}) & \textit{if } \mathrm{ih}(X) > 0 \textit{ and } A \textit{ is } \Lambda\mathbf{B} \end{cases}$$

Then P is a function for the finite sets of statements of L_ω *which meets* **AS1'**, **AS2**, **AS¬**, **AS∧**, *and* **ASΛ**.

Proof:

AS2: $P(\varnothing) = P_0(\varnothing) = 1$ by the definition of P and the fact that P_0 meets **AS2**.

AS¬: (i) $\neg A \notin X$. Then proof is by induction on the rank r of $X \cup \{\neg A\}$.

Basis: $\mathrm{ih}(X \cup \{\neg A\}) = 0$, i.e. $\neg A$ and the statements in X belong to L_P. Then $P(X \cup \{\neg A\}) = P_0(X \cup \{\neg A\})$, $P(X) = P_0(X)$, and $P(X \cup \{A\}) = P_0(X \cup \{A\})$. Hence $P(X \cup \{\neg A\}) = P(X) - P(X \cup \{A\})$ by assumption.

Inductive Step: ih > 0. Let B be the alphabetically latest compound member of $X \cup \{\neg A\}$ with $\mathrm{ih}(B) = \mathrm{ih}(X \cup \{\neg A\})$.

Case 1: B is $\neg A$.

Then $P(X \cup \{\neg A\}) = P(X) - P(X \cup \{A\})$ by clause 2 of the definition.

Case 2: B is a statement other than $\neg A$. Then $B \in X$, $\mathrm{ih}(B) = \mathrm{ih}(X \cup \{\neg A\}) = \mathrm{ih}(X)$, and B is the alphabetically latest compound member of X with $\mathrm{ih}(B) = \mathrm{ih}(X)$.

Case 2.1: B is some other negation $\neg C$ of L_ω. Let X' be $X-\{\neg C\}$. Then:

$$P(X' \cup \{\neg C\} \cup \{\neg A\}) = P(X' \cup \{\neg C\}) - P(X' \cup \{\neg C\} \cup \{A\})$$

by the definition of P and the inductive hypothesis.

Case 2.2: B is some conjunction $C \wedge D$ of L_ω. Let X' be $X-\{C \wedge D\}$. Then

$$P(X' \cup \{C \wedge D\} \cup \{\neg A\}) = P(X' \cup \{C \wedge D\}) - P(X' \cup \{C \wedge D\} \cup \{A\})$$

by the definition of P and the inductive hypothesis.

Case 2.3: B is some infinitary conjunction $\Lambda\boldsymbol{C}$ of L_ω, with $\mathrm{S}(\boldsymbol{C})$ the set of elements of the sequence $\boldsymbol{C}$. Let X' be $X-\{\Lambda\boldsymbol{C}\}$. Then

$$P(X' \cup \{\Lambda\boldsymbol{C}\}) = P(X' \cup \{\Lambda\boldsymbol{C}\} \cup \{A\}) + P(X' \cup \{\Lambda\boldsymbol{C}\} \cup \{\neg A\})$$

by the definition of P and the inductive hypothesis.

(ii) $\neg A \in X$. Then $P(X \cup \{\neg A\}) = P(X)$ and by (i):

$P(X \cup \{A\})$
$= P((X-\{\neg A\}) \cup \{A,\neg A\})$
$= P((X-\{\neg A\}) \cup \{A\}) - P((X-\{\neg A\}) \cup \{A,A\})$
$= 0$

Hence $P(X \cup \{\neg A\}) = P(X) - P(X \cup \{A\})$.

AS$\wedge$: Similarly.

ASΛ: (i) $\Lambda\boldsymbol{A} \notin X$. Then proof is by induction on the rank r of $X \cup \{\Lambda\boldsymbol{A}\}$.

Basis: $\mathrm{r}(X \cup \{\Lambda\boldsymbol{A}\}) = \langle 1,0\rangle$, $\mathrm{ih}(X) = 0$ and $\mathrm{ih}(X \cup \{\Lambda\boldsymbol{A}\}) = 1$, i.e. all the statements in X as well as the elements of the sequence $\boldsymbol{A}$ belong to L_P. Then $\Lambda\boldsymbol{A}$ is the alphabetically latest member of $X \cup \{\Lambda\boldsymbol{A}\}$ with the same infinite conjunction height as X. So, by the definition of P, $P(X \cup \{\Lambda\boldsymbol{A}\}) = \lim_{n\to\infty} P(X \cup \{A_1,\ldots,A_n\})$.

Inductive Step: $\mathrm{r}(X \cup \{\Lambda\boldsymbol{A}\}) > \langle 1,0\rangle$. Let B be the alphabetically latest compound member of $X \cup \{\Lambda\boldsymbol{A}\}$ with $\mathrm{ih}(B) = \mathrm{ih}(X \cup \{\Lambda\boldsymbol{A}\})$.

Case 1: B is $\Lambda\boldsymbol{A}$. Then $P(X \cup \{\Lambda\boldsymbol{A}\}) = \lim_{n\to\infty} P(X \cup \{A_1,\ldots,A_n\})$ by clause 4 of the definition of P.

Case 2: B is not $\Lambda\boldsymbol{A}$. Then $B \in X$, $\mathrm{ih}(B) = \mathrm{ih}(X \cup \{\Lambda\boldsymbol{A}\}) = \mathrm{ih}(X)$, and B is the alphabetically latest compound member of X with $\mathrm{ih}(B) = \mathrm{ih}(X)$.

Case 2.1: B is some negation $\neg C$ of L_ω. Let X' be $X-\{\neg C\}$.

$$P(X' \cup \{\neg C\} \cup \{\Lambda\boldsymbol{A}\}) = \lim_{n\to\infty} P(X' \cup \{\neg C\} \cup \{A_1,\ldots,A_n\})$$

by the definition of P and the inductive hypothesis.

Case 2.2: B is some conjunction $C \wedge D$ of L_ω. Let X' be $X-\{C \wedge D\}$. Then

$$P(X' \cup \{C \wedge D\} \cup \{\Lambda\boldsymbol{A}\}) = \lim_{n\to\infty} P(X' \cup \{C \wedge D\} \cup \{A_1,\ldots,A_n\})$$

by the definition of P and the inductive hypothesis.

Case 2.3: B is some infinitary conjunction $\Lambda\boldsymbol{C}$ of L_ω, other than $\Lambda\boldsymbol{A}$, and let X' be $X-\{\Lambda\boldsymbol{C}\}$. Then:

$$P(X' \cup \{\Lambda\boldsymbol{C}\} \cup \{\Lambda\boldsymbol{A}\}) = \lim_{n\to\infty} P(X' \cup \{\Lambda\boldsymbol{C}\} \cup \{A_1,\ldots,A_n\})$$

by the definition of P and the inductive hypothesis.

(ii) $\Lambda\boldsymbol{A} \in X$. Then $P(X \cup \{\Lambda\boldsymbol{A}\}) = P(X)$ and by (i):

$$\begin{aligned}
&\underset{n\to\infty}{\text{limit}}\ P(X \cup \{A_1,\ldots,A_n\})\\
&\quad= \underset{n\to\infty}{\text{limit}}\ P((X-\{\Lambda\boldsymbol{A}\}) \cup \{\Lambda\boldsymbol{A}\} \cup \{A_1,\ldots,A_n\})\\
&\quad= \underset{n\to\infty}{\text{limit}}\ \underset{m\to\infty}{\text{limit}}\ P((X-\{\Lambda\boldsymbol{A}\}) \cup \{A_1,\ldots,A_m\} \cup \{A_1,\ldots,A_n\})\\
&\quad= \underset{n\to\infty}{\text{limit}}\ P((X-\{\Lambda\boldsymbol{A}\}) \cup \{A_1,\ldots,A_n\})\\
&\quad= P((X-\{\Lambda\boldsymbol{A}\}) \cup \{\Lambda\boldsymbol{A}\})\\
&\quad= P(X \cup \{\Lambda\boldsymbol{A}\})
\end{aligned}$$

AS1': If $X \cap Y \neq \varnothing$, $F_P(X,Y) = 0$ by Lemma 8(a), and hence $0 \leq F_P(X,Y)$. So, assume that $X \cap Y = \varnothing$. Proof is then by induction on the rank r of $X \cup Y$.

Basis: $\text{ih}(X \cup Y) = 0$, i.e. the statements in X and in Y are statements of L_P. Then $F_P(X,Y) = F_{P_0}(X,Y)$ and hence $0 \leq F_P(X,Y)$ by the assumption that P_0 meets **AS1'**.

Inductive Step: $\text{ih}(X \cup Y) > 0$. Let A be the alphabetically latest statement in $X \cup Y$ with $\text{ih}(A) = \text{ih}(X \cup Y)$.

Case 1: A is $\neg B$.

Case 1.1: $A \in X$. Let X' be $X-\{\neg B\}$. Then $F_P(X' \cup \{\neg B\},Y) = F_P(X,Y \cup \{B\})$ by Theorem 5(a). Hence $0 \leq F_P(X,Y)$ by the inductive hypothesis.

Case 1.2: $A \in Y$. Let Y' be $Y-\{\neg B\}$. Then $F_P(X,Y' \cup \{\neg B\}) = F_P(X \cup \{B\},Y)$ by Theorem 5(b). Hence $0 \leq F_P(X,Y)$ by the inductive hypothesis.

Case 2: A is $B \wedge C$.

Case 2.1: $A \in X$. Let X' be $X-\{B \wedge C\}$. Then $F_P(X' \cup \{B \wedge C\},Y) = F_P(X \cup \{B,C\},Y)$ by Theorem 6(a). Hence $0 \leq F_P(X,Y)$ by the inductive hypothesis.

Case 2.2: $A \in Y$. Let Y' be $Y-\{B \wedge C\}$. Then

$$F_P(X,Y' \cup \{B \wedge C\}) = F_P(X,Y' \cup \{A\}) + F_P(X \cup \{A\},Y' \cup \{B\})$$

by Theorem 6(b). Hence $0 \leq F_P(X,Y)$ by the inductive hypothesis.

Case 3: A is $\Lambda\boldsymbol{B}$, where $\boldsymbol{B}$ is the sequence $\langle B_1, B_2, \ldots, B_n, \ldots\rangle$.

Case 3.1: $A \in X$. Let X' be $X-\{\Lambda\boldsymbol{B}\}$. Then

$$F_P(X' \cup \{\Lambda\boldsymbol{B}\},Y) = \underset{n\to\infty}{\text{limit}}\ F_P(X \cup \{A_1,\ldots,A_n\}),Y)$$

by Theorem 8(a). Hence $0 \leq F_P(X,Y)$ by the inductive hypothesis.

Case 3.2: $A \in Y$. Let Y' be $Y-\{\Lambda\boldsymbol{B}\}$. Then

$$\begin{aligned}
&F_P(X,Y' \cup \{\Lambda A\}) =\\
&\underset{n\to\infty}{\text{limit}}\ [F_P(X,Y' \cup \{B_1\}) + F_P(X \cup \{B_1\},Y \cup \{B_2\}) + \ldots\\
&\qquad\qquad + F_P(X \cup \{B_1,\ldots,B_{n-1}\},Y \cup \{B_n\})]
\end{aligned}$$

by Theorem 8(b). Hence $0 \leq F_P(X,Y)$ by the inductive hypothesis. ¤

So, for every absolute probability function for the finite sets of statements of L_P there exists one extension, and one only, to an absolute probability function for the finite sets of statements of L_ω. By Theorem 1 such a function can be further extended to an absolute probability function for all sets of statements of L_ω by means of this definition:

$$P_\omega(X) = \inf_{X'} \{P(X') : X' \subseteq_f X\}.$$

Probability Functions on Statements and on Finite Sets of Statements

Given the results in Section 3 of Chapter 1 concerning the one-to-one correlation between absolute probability functions for L and absolute probability functions for the finite sets of statements of L, L being L_P, L_Q, or L_ω, we also have the following:

Theorem 12. (a) *If P is any absolute probability function for L_Q, then P can be recursively defined from its restriction to quantifier-free statements;*
(b) *If P is any absolute probability function for L_ω, then P can be recursively defined from its restriction to finitary statements, i.e. to statements of L_P;*
(c) *If P is any absolute probability function for L_P, then P can be recursively extended to an absolute probability function P_ω for L_ω. And P_ω is then the only absolute probability function for L_ω that has P as its restriction to the statements of L_P.*

Section 3. The Recursive Definability of Relative Probability Functions

Like absolute probability functions, relative probability functions are definable from certain restrictions, but to a much lesser extent. A fairly uninteresting definability result for relative probability functions P for L_P can be obtained from the fact that the C-restriction P_C of P is an absolute probability function, provided that C is P-normal (Theorem 2 in Chapter 1, Section 2). We recall that if C is P-abnormal, then $P(A,C) = 1$ for every statement A. And if C is any P-normal statement, then P_C can be recursively defined from its restriction to atomic statements and conjunctions of atomic statements. As a consequence P itself can be recursively defined from its restriction, *as regards the first argument only*, to atomic statements and conjunctions of atomic statements *and* the specification of the class of P-normal statements.

Analogously, a relative probability function for L_ω can be recursively defined from its restriction, *as regards the first argument*, to statements of L_P; and a relative probability function for L_Q can be recursively defined from its restriction, *as regards the first argument*, to quantifier-free statements, i.e. statements of L_P. In both cases P-normal statements must be specifiable as such.

The results obtained for absolute probability functions in the previous section are immediately transferable to Kolmogorov relative probability functions. In Chapter 3, Section 2, we saw that these relative probability functions correspond one-to-one to absolute probability functions. Given any Kolmogorov function P, its T-restriction P_{T} is an absolute probability function and P is definable from P_{T} thus:

A	$P^1(A)$	$P^2(A)$	$P^3{}_r(A)$	A	$P^1(A)$	$P^2(A)$	$P^3{}_r(A)$
$\bot$	0	0	0	$E_1 \wedge \neg E_2$	1	0	0
$\top$	1	1	1	$\neg(E_1 \wedge \neg E_2)$	0	1	1
E_1	1	0	r	$\neg E_1 \wedge E_2$	0	1	0
$\neg E_1$	0	1	$1-r$	$\neg(\neg E_1 \wedge E_2)$	1	0	1
E_2	0	1	r	$\neg E_1 \wedge \neg E_2$	0	0	$1-r$
$\neg E_2$	1	0	$1-r$	$\neg(\neg E_1 \wedge \neg E_2)$	1	1	r
$E_1 \wedge E_2$	0	0	r	$E_1 \equiv E_2$	0	0	1
$\neg(E_1 \wedge E_2)$	1	1	$1-r$	$\neg(E_1 \equiv E_2)$	1	1	0

Table 2

$$P(A,B) = \begin{cases} P_\top(A \wedge B) \,/\, P_\top(B) & \textit{if } P_\top(B) = 0 \\ 1 & \textit{otherwise.} \end{cases}$$

Hence, drawing on the discussion of Section 1, we can arrive at definability and extendability results for Kolmogorov functions.

Theorem 13. *Any Kolmogorov relative probability function P for L_P can be recursively defined from its restriction to atomic statements, conjunctions of atomic statements, and* $\top$.

Proof: By Theorem 2(b). ¤

Theorem 14. *Any Kolmogorov relative probability function P for L_ω can be recursively defined from its restriction to L_P.*

Proof: By Theorem 4(b). ¤

And Theorem 11 yields:

Theorem 15. *If P is any Kolmogorov relative probability function for L_P, then P can be recursively extended to a Kolmogorov relative probability function P_ω for L_ω.*

But relative probability functions generally are not representable by a single absolute function, as has been extensively argued in Chapter 4. Hence extendability results along the lines of the last three theorems are not to be had in the general case. Moreover, the second argument of a relative probability function behaves quite differently from the first, and a relative probability function for L_P can, in general, not be recovered from its restriction (as regards both arguments) to atomic statements and conjunctions of atomic statements. For proof of this we will introduce two relative probability functions for L_P whose restrictions to atomic statements

		B															
	$P(A,B)$	$\bot$	$\top$	E_1	$\neg E_1$	E_2	$\neg E_2$	$E_1 \wedge E_2$	$\neg(E_1 \wedge E_2)$	$E_1 \wedge \neg E_2$	$E_1 \supset E_2$	$\neg E_1 \wedge E_2$	$E_2 \supset E_1$	$\neg E_1 \wedge \neg E_2$	$E_1 \vee E_2$	$E_1 \equiv E_2$	$\neg(E_1 \equiv E_2)$
	$\bot$	1	0	0	0	0	0	0	0	0	0	0	0	0	0	0	0
	$\top$	1	1	1	1	1	1	1	1	1	1	1	1	1	1	1	1
	E_1	1	1	1	0	0	1	1	1	1	0	0	1	0	1	r	1
	$\neg E_1$	1	0	0	1	1	0	0	0	0	1	1	0	1	0	$1-r$	0
	E_2	1	0	0	1	1	0	1	0	0	1	1	0	0	0	r	0
	$\neg E_2$	1	1	1	0	0	1	0	1	1	0	0	1	1	1	$1-r$	1
	$E_1 \wedge E_2$	1	0	0	0	0	0	1	0	0	0	0	0	0	0	r	0
A	$\neg(E_1 \wedge E_2)$	1	1	1	1	1	1	0	1	1	1	1	1	1	1	$1-r$	1
	$E_1 \wedge \neg E_2$	1	1	1	0	0	1	0	1	1	0	0	1	0	1	0	1
	$E_1 \supset E_2$	1	0	0	1	1	0	1	0	0	1	1	0	1	0	1	0
	$\neg E_1 \wedge E_2$	1	0	0	1	1	0	0	0	0	1	1	0	0	0	0	0
	$E_2 \supset E_1$	1	1	1	0	0	1	1	1	1	0	0	1	1	1	1	1
	$\neg E_1 \wedge \neg E_2$	1	0	0	0	0	0	0	0	0	0	0	0	0	0	$1-r$	0
	$E_1 \vee E_2$	1	1	1	1	1	1	1	1	1	1	1	1	0	1	r	1
	$E_1 \equiv E_2$	1	0	0	0	0	0	1	0	0	0	0	0	1	0	1	0
	$\neg(E_1 \equiv E_2)$	1	1	1	1	1	1	0	1	1	1	1	1	0	1	0	1

Table 3

and conjunctions of atomic statements are identical. We first specify in Table 2 three absolute probability functions P^1, P^2, and $P^3{}_r$; E_1 and E_2 are atomic statements. And, this done, let P_r be defined as follows:

$$\mathrm{P}_r(A,B) = \begin{cases} P^1(A \wedge B) \,/\, P^1(B) & \textit{if } P^1(B) \neq 0 \\ P^2(A \wedge B) \,/\, P^2(B) & \textit{if } P^1(B) = 0 \textit{ but } P^2(B) \neq 0 \\ P^3{}_r(A \wedge B) \,/\, P^3{}_r(B) & \textit{if } P^1(B) = 0,\, P^2(B) = 0, \textit{ but } P^3{}_r(B) \neq 0 \\ 1 \textit{ otherwise.} \end{cases}$$

The resulting functions $\mathrm{P}_r(A,B)$ are relative probability functions for L_P by Theorem 23 of Chapter 4.[3] Their values are listed in Table 3. As far as statements whose atomic components are E_1 and E_2 are concerned, these functions differ from one another only when the second argument is $E_1 \equiv E_2$ or is logically equivalent to $E_1 \equiv E_2$. For example, since $P^1(E_1 \equiv E_2) = P^2(E_1 \equiv E_2) = 0$, and $P^3{}_r(E_1 \wedge (E_1 \equiv E_2)) = P^3{}_r(E_1 \wedge E_2) = r$, and $P^3{}_r(E_1 \equiv E_2) = r + (1 - r) = 1$, $\mathrm{P}_r(E_1,E_1 \equiv E_2) = r$. For distinct choices of r, say r_1 and r_2, P_{r_1} and P_{r_2} are therefore different functions. On the other hand, their restrictions to atomic statements and conjunctions of atomic statements are identical. Relative probability functions for L_P can therefore not be recovered from their restrictions to atomic statements and conjunctions of atomic statements.

The reader will note that, although the functions P_{r_1} and P_{r_2} disagree on the pair

$$\langle E_1 \wedge E_2, E_1 \equiv E_2 \rangle,$$

they nonetheless agree on the forty-five pairs $\langle A,B \rangle$ where A is $E_1 \wedge E_2$, E_1, or E_2, and B is a compound of E_1 and E_2 other than $E_1 \equiv E_2$. So, the probability $P(A \wedge B,C)$ of the conjunction of two *atomic* statements A and B relative to some other compound C of A and B is not always a numerical function of the probabilities of $A \wedge B$, A, and B relative to compounds of A and B that are in any sense less complex than C itself. So, $P(A \wedge B,C)$ is not recursively definable from those other probabilities on. The result also holds true when just one atomic statement or the conjunction of more than two serves as the first argument of P. So, relative probability functions are not recursively definable to any further extent.

When A is not logically false, the absolute probability of A is the sum, hence a numerical function, of the absolute probabilities of the state-descriptions A_1, A_2, ..., and A_n to whose disjunction A is logically equivalent. Now let P be an arbitrary relative probability function, B be a P-normal statement, and B_1, B_2, and B_m be the state-descriptions to whose disjunction B is logically equivalent. Then $P(A,B)$ is likewise the sum, hence, a numerical function, of $P(A_1,B)$, $P(A_2,B)$, ..., and $P(A_n,B)$ (see Chapter 1, Section 1). But consider the two state-descriptions, $E_1 \wedge E_2$ and $\neg E_1 \wedge \neg E_2$, to whose disjunction $E_1 \equiv E_2$ is logically equivalent. $\mathrm{P}_r(E_1 \wedge E_2,E_1 \wedge E_2)$ and $\mathrm{P}_r(E_1 \wedge E_2,\neg E_1 \wedge \neg E_2)$ equal 1 and 0, respectively, whereas $\mathrm{P}_r(E_1 \wedge E_2,E_1 \equiv E_2)$ equals r, a real number we presumed to be larger than 0 but smaller than 1. So, $P(A,B)$ is not always a numerical function of $P(A,B_1)$, $P(A,B_2)$, ..., and $P(A,B_m)$.

Section 4. Relative Probability Functions for L_P Meeting DL1 can be Extended to Relative Probability Functions for L_ω Meeting RPΛ2

In Section 1 we saw that every absolute probability function for L_P is the restriction to L_P of exactly one absolute probability function for L_ω. In the relative case the restriction to L_P of any probability function for L_ω is of course a probability function for L_P and, we shall see, every probability function for L_P that meets constraint **DL1** is the restriction to L_P of some probability function for L_ω. But distinct relative probability functions for L_ω can have the same restriction to L_P. We shall identify a subclass of the relative probability functions for L_ω which stand in one-to-one correspondence to their restrictions to L_P and which can be recursively defined from them.

A constraint from Chapter 2 will play a role in this, to wit:

$$\textbf{RP}\Lambda\textbf{2} \qquad P(A,\Lambda B) = \begin{cases} \lim_{n\to\infty} P(A,B_1 \wedge \ldots \wedge B_n) \;\; \textit{if} \\ \qquad \lim_{m\to\infty} \lim_{n\to\infty} P(B_1 \wedge \ldots \wedge B_n, B_1 \wedge \ldots \wedge B_m) = 1 \\ 1 \;\; \textit{otherwise.} \end{cases}$$

It was shown in Chapter 2, Section 2, that if P is a relative probability function for L_ω, then its restriction to L_P is sure to meet constraint **DL1**.[4] Our task here is to prove the converse of that result. In fact, we will show not only that every relative probability function for L_P which meets **DL1** is the restriction to L_P of a relative probability function for L_ω, but also that it is the restriction of a relative probability function of L_ω that meets **RPΛ2**.

With an eye to that, suppose P_0 a relative probability function for L_P that meets **DL1**. Then by Theorem 31 of Chapter 4 P_0 is represented by means of a certain well-ordered class $\mathcal{M}_{P_0}$ of absolute probability functions for L_P. We recursively extend every absolute probability function P' in that class to an absolute probability function P'_ω for L_ω in accordance with Theorem 12(c) in Section 2.

Theorem 16. *Let P_0 be a relative probability function for L_P that meets* **DL1**, *let $\mathcal{M}_{P_0}$ be the class of absolute probability functions for L_P that represents P_0, and let $\mathcal{M}_\omega$ be the class $\{P'_\omega : P' \in \mathcal{M}_{P_0}\}$ of extensions to L_ω. Then $\mathcal{M}_\omega$ is a Type V-ordered class of absolute probability functions for L_ω.*

Proof: (i) By Theorem 36 in Chapter 4 the members of $\mathcal{M}_\omega$ are absolute probability functions for L_ω. (ii) $\mathcal{M}_{P_0}$ is well-ordered and, the ordering of $\mathcal{M}_\omega$ being the same by definition as that of $\mathcal{M}_{P_0}$, $\mathcal{M}_\omega$ is well-ordered, i.e. meets **O5**. (iii) Since $\mathcal{M}_{P_0}$ is Type IV-ordered by Theorem 30 in Chapter 4, it meets **O4**. So, for any absolute probability function P' for L_P in $\mathcal{M}_{P_0}$, $\sup\{P'(C) : P' = P'^{C}\} = 1$ by **O4, AP2,** and IE. Hence there exists a sequence $C_1, \ldots, C_n, \ldots$ of statements of L_P such that (i) $P' = P'^{C_n}$ for every n from 1 on and (ii) $\lim_{n\to\infty} P'(C_n) = 1$. Hence by Lemma 1(t) $\lim_{n\to\infty} P'(C_1 \vee \ldots \vee C_n) = 1$, and hence $P'(\mathrm{V}C) = 1$ by Theorem 2 of

Chapter 2, where $C = \langle C_1, ..., C_n, ...\rangle$. $P' = P'^{VC}$ and so $\mathcal{M}_\omega$ meets **O4'**. ¤

Now let $P_{\mathcal{M}_\omega}$ be defined by $\mathcal{M}_\omega$ in accordance with **GEN** of Chapter 2, Section 2:

$$P_{\mathcal{M}_\omega}(A,B) = \begin{cases} 1 \text{ if } M(B) = 0 \text{ for every member } M \text{ of } \mathcal{M}_\omega \\ M^B(A \wedge B) \,/\, M^B(B) \text{ otherwise.} \end{cases}$$

Theorem 17. *$P_{\mathcal{M}_\omega}$ constitutes a relative probability function for L_ω meeting constraint* **RPΛ2**.

Proof: Given Theorem 37 in Chapter 4, it only needs to be established that $P_{\mathcal{M}_\omega}$ meets **RPΛ2**. So suppose that

$$\lim_{m\to\infty} \lim_{n\to\infty} P_{\mathcal{M}_\omega}(B_1 \wedge ... \wedge B_n, B_1 \wedge ... \wedge B_m) = 1.$$

There then exists a number m such that

$$\lim_{n\to\infty} P_{\mathcal{M}_\omega}(B_1 \wedge ... \wedge B_n, B_1 \wedge ... \wedge B_m) > 0,$$

i.e. such that

$$\lim_{n\to\infty} [P'_\omega{}^{B_1 \wedge ... \wedge B_m}(B_1 \wedge ... \wedge B_n) \,/\, P'_\omega{}^{B_1 \wedge ... \wedge B_m}(B_1 \wedge ... \wedge B_m)] > 0,$$

and hence

$$P'_\omega{}^{B_1 \wedge ... \wedge B_m}(B_1 \wedge ... \wedge B_n) > 0$$

for all $n \geq m$. Hence, by the well-orderedness of $\mathcal{M}_\omega$ and Lemma 1(a),

$$P'_\omega{}^{B_1 \wedge ... \wedge B_m} = P'_\omega{}^{B_1 \wedge ... \wedge B_n} \textit{ for every } n \geq m.$$

But $P'_\omega{}^{B_1 \wedge ... \wedge B_m}(\Lambda \boldsymbol{B})$

$= \lim_{n\to\infty} P'_\omega{}^{B_1 \wedge ... \wedge B_m}(B_1 \wedge ... \wedge B_n)$,

> 0.

Hence $P'_\omega{}^{\Lambda B} = P'_\omega{}^{B_1 \wedge ... \wedge B_m}$. So,

$P_{\mathcal{M}_\omega}(A,\Lambda B)$

$= P'_\omega{}^{B_1 \wedge ... \wedge B_m}(A \wedge \Lambda \boldsymbol{B}) \,/\, P'_\omega{}^{B_1 \wedge ... \wedge B_m}(\Lambda \boldsymbol{B})$

$= \lim_{n\to\infty} P'_\omega{}^{B_1 \wedge ... \wedge B_m}(A \wedge (B_1 \wedge ... \wedge B_n)) \,/\, \lim_{n\to\infty} P'_\omega{}^{B_1 \wedge ... \wedge B_m}(\Lambda \boldsymbol{B})$

$= \lim_{n\to\infty} [P'_\omega{}^{B_1 \wedge ... \wedge B_m}(A \wedge (B_1 \wedge ... \wedge B_n)) \,/\, P'_\omega{}^{B_1 \wedge ... \wedge B_m}(\Lambda \boldsymbol{B})]$,

$= \lim_{n\to\infty} P_{\mathcal{M}_\omega}(A,B_1 \wedge ... \wedge B_n)$.

If, on the other hand,

$$\lim_{m\to\infty} \lim_{n\to\infty} P_{\mathcal{M}_\omega}(B_1 \wedge ... \wedge B_n, B_1 \wedge .... B_m) \neq 1,$$

then $P'_\omega(B_1 \wedge ... \wedge B_m) = 0$ for every P'_ω in $\mathcal{M}_\omega$ and every number m. Hence

$$P'_\omega(\Lambda \boldsymbol{B}) = \lim_{m\to\infty} P'_\omega(B_1 \wedge ... \wedge B_m) = 0 \text{ for every } P'_\omega \text{ in } \mathcal{M}_\omega.$$

Hence $P_{\mathcal{M}_\omega}(A,\Lambda \boldsymbol{B}) = 1$. So, $P_{\mathcal{M}_\omega}$ meets **RPΛ2**. ¤

But P_0 is the restriction to L_P of $P_{\mathcal{M}_\omega}$. So:

Theorem 18. *Any relative probability function for L_P that meets constraint* **DL1** *is the restriction to L_P of a relative probability function for L_ω that meets constraint* **RPΛ2**.

Chapter 6

Families of Probability Functions Characterised by Equivalence Relations

In this chapter we identify a number of families of relative probability functions for L, L being L_P, L_Q, or L_ω, and investigate their classification. The unifying feature of this brief study is that the constraints by means of which a family of functions is characterised are throughout formulated in terms of equivalence relations on statements, i.e. relations that are reflexive, symmetrical, and transitive. Each of the constraints in fact stipulates that two such equivalence relations coincide. We begin with a list of equivalence relations which will play a role in this chapter. All but the first relation are induced by P, P any relative probability function for L. A and B are of course arbitrary statements of L.

Equivalence Relations Among Statements of L

Equivalence relation $\mathbf{R_1}$:

A and B are logically equivalent

or, what comes to the same,

$A \equiv B$ *is logically true.*

Equivalence relation $\mathbf{R_2}$:

$P(A,C) = P(B,C)$ *for every statement* C *of* L,

or, as explained in Section 2 of Chapter 1,

A and B are (left–) indiscernible under P,

of which an equivalent formulation is (see Lemma 4(xx)):

$P(A \equiv B,C) = 1$ *for every statement* C *of* L.

Equivalence relation $\mathbf{R_3}$:

$P(C,A) = P(C,B)$ *for every statement* C *of* L

or, for short,

A and B are right-indiscernible under P,

of which by Lemma 4(o) an equivalent characterisation is

$$P(A,B) = P(B,A) = 1.$$

Equivalence relation $\mathbf{R_4}$:

$$P(A,A \vee B) = P(B,A \vee B),$$

which is an equivalence relation by Lemma 4(uu).

Equivalence relation $\mathbf{R_5}$:

W	$P_1(W)$	$P_2(W)$
$E_1 \wedge E_2 \wedge E_3$	2/5	0
$E_1 \wedge E_2 \wedge \neg E_3$	1/5	0
$E_1 \wedge \neg E_2 \wedge E_3$	1/5	0
$E_1 \wedge \neg E_2 \wedge \neg E_3$	1/5	0
$\neg E_1 \wedge E_2 \wedge E_3$	0	2/3
$\neg E_1 \wedge E_2 \wedge \neg E_3$	0	1/3
$\neg E_1 \wedge \neg E_2 \wedge E_3$	0	0
$\neg E_1 \wedge \neg E_2 \wedge \neg E_3$	0	0

Table 1

$$P(A,A \vee B) > 0 \textit{ and } P(B,A \vee B) > 0$$

or, using the notation introduced in Chapter 4 for this, i.e. Császár's, relation,

$$A \sim_P B,$$

which has already been shown to be an equivalence relation in Theorem 8 of Chapter 4.

Equivalence relation $\mathbf{R_6}$:

$$P(\bot,A) = P(\bot,B).$$

By Lemma 4(k) this amounts to

A and B are both P-normal or else both P-abnormal.

Equivalence relation $\mathbf{R_7}$:

$$P(A \equiv B,\top) = 1.$$

This relation is reflexive, since $A \equiv A$ is a logical truth, and consequently $P(A \equiv A,\top) = 1$ by Theorem 3(a) in Chapter 1, Section 3, and it is transitive by Lemma 4(yy). Moreover, Lemma 4(ac) establishes that it is equivalent to

Either A and B are right-indiscernible under P or their negations are.

Equivalence relation $\mathbf{R_8}$:

$$P(A,\top) = P(B,\top)$$

Equivalence relation $\mathbf{R_9}$:

$$P(A,\top) = 0 \textit{ if and only if } P(B,\top) = 0.$$

Comparative Strength of Equivalence Relations

Our next task is to establish a number of facts about the relative strength of these relations. In the course of this we shall repeatedly make use of a particular relative probability function under which the nine relations $\mathbf{R_1}$–$\mathbf{R_9}$ have different extensions. Suppose E_1, E_2, and

E_3 to be three distinct atomic statements. Table 1 lists the values of two absolute probability functions P_1 and P_2 for the 8 state-descriptions in E_1, E_2, and E_3.[1] As we have seen in Chapter 1, all values of P_1 and P_2 are thereby determined, the absolute probability of a statement A being the sum of the absolute probabilities of the state-descriptions whose disjunction is logically equivalent to A.

And, this done, let P be defined as follows:

$$P(A,B) = \begin{cases} P_1(A \wedge B) / P_1(B) & \text{if } P_1(B) \neq 0 \\ P_2(A \wedge B) / P_2(B) & \text{if } P_1(B) = 0 \text{ but } P_2(B) \neq 0 \\ 1 & \text{otherwise.} \end{cases}$$

The resulting function is a relative probability function for L_P by Theorem 23 of Chapter 4.

As to the relationship of inclusion among the relations $\mathbf{R_1}$–$\mathbf{R_9}$, note first that if A is logically equivalent to B, then A is left-indiscernible from B, this by Theorem 3(c) in Chapter 1, Section 2. So logical equivalence is at least as strong as left-indiscernibility. That logical equivalence is actually stronger than left-indiscernibility is documented by the pair of statements $E_1 \vee E_2$ and T. Indeed, since $P_1(E_1 \vee E_2) = P_2(E_1 \vee E_2) = 1$, $P(E_1 \vee E_2, C) = 1$ for every statement C. But $P(T,C) = 1$ for every statement C by Lemma 4(m). So $E_1 \vee E_2$ is left-indiscernible under P from T, but $E_1 \vee E_2$ is not logically equivalent to T. So, $\mathbf{R_1}$ (logical equivalence) is stronger than $\mathbf{R_2}$ (left-indiscernibility).

$\mathbf{R_2}$ (left-indiscernibility) is stronger than $\mathbf{R_3}$ (right-indiscernibility), this by Lemma 4(r) and the observation that the pair of statements E_1 and T are right-indiscernible under P but not left-indiscernible under P. For, since $P_1(E_1) = 1$, it follows from the definition of P and Lemma 2(d) that $P(C,E_1) = P(C,T) = P_1(C)$ for every statement C, while $P(E_1,\neg E_1) = 0 \neq P(T,\neg E_1)$.

By Lemma 4(kk) and IE $P(A,A \vee B) = P(B,A \vee B) = 1$ if $P(A,B) = P(B,A) = 1$. So, $\mathbf{R_3}$ (right-indiscernibility) is at least as strong as $\mathbf{R_4}$. Since

$$P(E_1 \wedge \neg E_2 \wedge E_3, (E_1 \wedge \neg E_2 \wedge E_3) \vee (E_1 \wedge \neg E_2 \wedge \neg E_3)) = 1/2$$

and

$$P(E_1 \wedge \neg E_2 \wedge \neg E_3, (E_1 \wedge \neg E_2 \wedge E_3) \vee (E_1 \wedge \neg E_2 \wedge \neg E_3)) = 1/2,$$

but

$$P(E_1 \wedge \neg E_2 \wedge E_3, E_1 \wedge \neg E_2 \wedge \neg E_3) = 0,$$

$\mathbf{R_3}$ (right-indiscernibility) is stronger than $\mathbf{R_4}$.

If $P(A,A \vee B) = P(B,A \vee B)$, then, by Lemma 4(hh), both $P(A,A \vee B) > 0$ and $P(B,A \vee B) > 0$, i.e. $A \sim_P B$. So, $\mathbf{R_5}$ obtains when $\mathbf{R_4}$ does. Moreover, $\mathbf{R_4}$ is actually stronger than $\mathbf{R_5}$ (Császár's relation), since

$$P(\neg E_1 \wedge E_2, (\neg E_1 \wedge E_2) \vee (\neg E_1 \wedge E_2 \wedge E_3)) = 1,$$

but

$$P(\neg E_1 \wedge E_2 \wedge E_3,(\neg E_1 \wedge E_2) \vee (\neg E_1 \wedge E_2 \wedge E_3)) = 2/3.$$

By Lemma 4(zz) $P(A \equiv B,T) = 1$ if A is right-indiscernible from B. On the other hand, the two statements $\neg E_1 \wedge E_2$ and $\neg E_1 \wedge \neg E_2$ are not both P-normal, nor both P-abnormal, since $P(\bot,\neg E_1 \wedge E_2) = 0$ but $P(\bot,\neg E_1 \wedge \neg E_2) = 1$. But

$$P((\neg E_1 \wedge E_2) \equiv (\neg E_1 \wedge \neg E_2),T) = 1,$$

since $(\neg E_1 \wedge E_2) \equiv (\neg E_1 \wedge \neg E_2)$ is logically equivalent to E_1 and $P(E_1,T) = P_1(E_1) = 1$. So $\mathbf{R_3}$ (right-indiscernibility) is stronger than $\mathbf{R_7}$.

If $A \sim_P B$, then, by Lemma 4(nn) and the transitivity of $\sim_P$, A and B are either both P-normal or both P-abnormal. Hence, if $\mathbf{R_5}$ holds, so does $\mathbf{R_6}$. On the other hand, E_1 and $\neg E_1 \wedge E_2$ are both P-normal, since $P(\bot,E_1) = P(\bot,\neg E_1 \wedge E_2) = 0$; but, since

$$P(\neg E_1 \wedge E_2,E_1 \vee (\neg E_1 \wedge E_2)) = 0,$$

it is not the case that $E_1 \sim_P \neg E_1 \wedge E_2$. So, $\mathbf{R_5}$ (Császár's relation) is stronger than $\mathbf{R_6}$.

For the pair of statements $E_1 \wedge E_2$ and T, we have $E_1 \wedge E_2 \sim_P T$, since

$$P(E_1 \wedge E_2,(E_1 \wedge E_2) \vee T) = P_1(E_1 \wedge E_2) = 3/5 > 0,$$

but not

$$P((E_1 \wedge E_2) \equiv T,T) = 1,$$

while for the pair of statements $\neg E_1 \wedge E_2$ and $\neg E_1 \wedge \neg E_2$ we have

$$P((\neg E_1 \wedge E_2) \equiv (\neg E_1 \wedge \neg E_2),T) = 1,$$

since $(\neg E_1 \wedge E_2) \equiv (\neg E_1 \wedge \neg E_2)$ is logically equivalent to E_1 and $P(E_1,T) = P_1(E_1) = 1$, but not

$$\neg E_1 \wedge E_2 \sim_P \neg E_1 \wedge \neg E_2,$$

since

$$P(\neg E_1 \wedge \neg E_2,(\neg E_1 \wedge \neg E_2) \vee (\neg E_1 \wedge E_2)) = P(\neg E_1 \wedge \neg E_2,\neg E_1) = P_2(\neg E_1 \wedge \neg E_2) = 0.$$

Hence neither of the two relations $\mathbf{R_5}$ (Császár's relation) and $\mathbf{R_7}$ implies the other.

By Lemma 4(ww) the relation $P(A \equiv B,T) = 1$ implies $P(A,T) = P(B,T)$. But for the pair of statements $E_2 \wedge \neg E_3$ and $\neg E_2 \wedge E_3$ we have $P(E_2 \wedge \neg E_3,T) = P(\neg E_2 \wedge E_3,T) = 1/5$, while

$$P((E_2 \wedge \neg E_3) \equiv (\neg E_2 \wedge E_3),T) = P(E_2 \equiv E_3,T) = 3/5 \neq 1.$$

So $\mathbf{R_7}$ is a stronger relation than $\mathbf{R_8}$.

If $P(A,T) = P(B,T)$, then of course $P(A,T) = 0$ *if and only if* $P(B,T) = 0$. But while both $P(T,T) \neq 0$ and $P(E_1 \wedge E_2,T) \neq 0$, $P(T,T) \neq P(E_1 \wedge E_2,T)$, which shows the relation $\mathbf{R_8}$ is stronger than $\mathbf{R_9}$.

Finally, note also that while $P((\neg E_1 \wedge E_2) \equiv (\neg E_1 \wedge \neg E_2),T) = 1$, as we have seen, one of the two statements $\neg E_1 \wedge E_2$ and $\neg E_1 \wedge \neg E_2$ is P-normal, the other P-abnormal. Hence $\mathbf{R_7}$

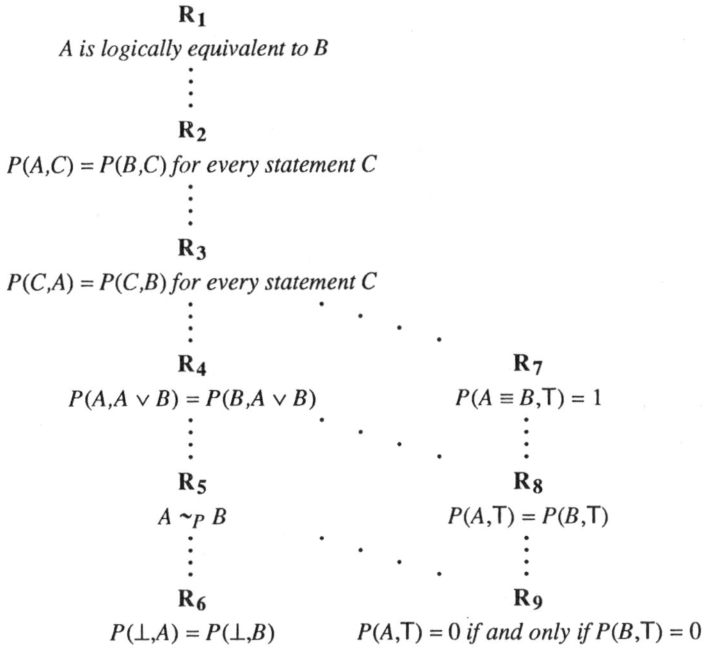

Diagram 1

does not imply $\mathbf{R_6}$. And while $E_1 \wedge E_2 \sim_P \top$, $\mathrm{P}(E_1 \wedge E_2,\top) \neq \mathrm{P}(\top,\top)$ so that $\mathbf{R_5}$ (Császár's relation) does not imply $\mathbf{R_8}$.

Consequently, the Popper function P illustrates that in general the equivalence relations $\mathbf{R_1}$–$\mathbf{R_9}$ do not coincide and that they are partially ordered by strength, as shown in Diagram 1.

Families of Relative Probability Functions

While these equivalence relations are generally distinct from each other, they are not distinct under every relative probability function. The fact that two or more of these relations coincide can serve to characterise various classes of relative probability functions. Some of these classes have already been encountered, especially in Chapter 3, others have not yet been explicitly considered.

The condition

If $P(A,C) = P(B,C)$ for every statement C, then A is logically equivalent to B,

according to which the first two relations coincide under P, characterises the class of Carnap 1952 functions, i.e. the class of those relative probability functions of Popper's that meet constraint **RC52**

If A is P-abnormal, then A is logically false.

It follows from the results of Chapter 4 that only statements that are logically false belong to the $\geq_P$ - last group of statements. An equivalent formulation of **RC52** is

If P(A,B) = 1 for every statement B, then A is logically true,

this by Lemma 4(mm).

Theorem 1. *Let P be a relative probability function for L. Then P meets constraint* **RC52** *if and only if it meets*

If P(A,C) = P(B,C) for every statement C, then A is logically equivalent to B.

Proof: Suppose *first* that P meets **RC52**, and let A and B be left-indiscernible under P. Then $P(A \equiv B,C) = 1$ for every statement C and hence $A \equiv B$ is logically true, i.e. A is logically equivalent to B. Suppose *next* that A is logically equivalent to B, if left-indiscernible from B and let $P(A,B) = 1$ for every statement B. By Lemma 4(m) A is then left-indiscernible from T and therefore logically equivalent to T. But in that case A is itself logically true. ¤

Kolmogorov functions, i.e. the functions meeting constraint **RK**,

*If P(A,*T*) = 0, then A is P-abnormal,*

generate just two $\sim_P$ - equivalence classes of statements: the statements A such that $A \sim_P$ T and the statements A such that $A \sim_P \perp$, i.e. the class of P-normal statements and the class of P-abnormal statements. Kolmogorov functions are also characterised by the condition

If A and B are right-indiscernible, then they are left-indiscernible,

so that under any Kolmogorov function left- and right-indiscernibility coincide. The proof of this result can be simplified by noting that **RK** can also be formulated as

*If P(A,*T*) = 1, then P(A,B) = 1 for every statement B,*

this in virtue of **RP3** and Lemmas 4(t) and 4(mm).

Theorem 2. *Let P be a relative probability function for L. Then P meets constraint* **RK** *if and only if P meets*

If P(C,A) = P(C,B) for every statement C, then P(A,C) = P(B,C) for every statement C.

Proof: Suppose *first* that P meets **RK**, and let A and B be right-indiscernible under P. Then, as we have seen, $P(A \equiv B,\text{T}) = 1$ and hence $P(A \equiv B,C) = 1$ for every statement C, i.e. A and B are left-indiscernible under P. Suppose *next* that A and B are left-indiscernible if right-indiscernible under P and assume that $P(A,\text{T}) = 1$. In that case, though, $P(A,\text{T}) = P(\text{T},A) = 1$ by Lemma 4(m) and A is right-indiscernible from T. By the supposition A is therefore left-indiscernible from T so that, by Lemma 4(m) again, $P(A,B) = 1$ for every statement B. ¤

The proof shows that when P meets **RK**, the relation $\mathbf{R_7}$ also coincides with $\mathbf{R_2}$ and $\mathbf{R_3}$. And since $P(\perp,A) = P(\perp,B)$ comes to the same as *A and B are both P-normal or both P-abnormal,* relation $\mathbf{R_6}$ coincides with $\mathbf{R_5}$, and also with $\mathbf{R_9}$. Moreover, $\mathbf{R_4}$ coincides with $\mathbf{R_8}$ so that the original nine distinct equivalence relations collapse into just four such relations.

The next class of relative probability functions is one that has not been characterised before; it consists of those functions P for which, for every statement B, the B-restriction $P_B(A)$ is a 2- or many-valued truth-value function. These functions can be characterised in terms of equivalence relations by the constraint

If $P(A,A \vee B) = P(B,A \vee B)$, *then* A *and* B *are right-indiscernible under* P.

Theorem 3. *Let* P *be a relative probability function for* L. *Then* P *meets the constraint*

If $P(A,A \vee B) = P(B,A \vee B)$, *then* $P(C,A) = P(C,B)$ *for every statement* C,

if and only if $P(A,B)$ *as a function of* A *is a 2- or many-valued truth-value function.*

Proof: Suppose first that any statements A and B are right-indiscernible under P, if $P(A,A \vee B) = P(B,A \vee B)$. Given that by **RP3** $P(\neg A,C)$ is a function of $P(A,C)$, all that needs to be shown is that if $P(B,C) = P(B',C)$, then $P(A \wedge B,C) = P(A \wedge B',C)$. So, assume that $P(B,C) = P(B',C)$. If $P(B,C) = P(B',C) = 0$, then $P(A \wedge B,C) = P(A \wedge B',C) = 0$ by Lemma 4(e). So, let $P(B,C) > 0$. Then $P(B \wedge C,C) > 0$ by Lemma 4(u) and hence $P((B \wedge C) \vee (B' \wedge C),C) > 0$ by Lemma 4(aa).

$$P(B \wedge C,C) = P(B \wedge C,(B \wedge C) \vee (B' \wedge C)) \times P((B \wedge C) \vee (B' \wedge C),C)$$

by Lemma 4(jj) and IE. Similarly,

$$P(B' \wedge C,C) = P(B' \wedge C,(B \wedge C) \vee (B' \wedge C)) \times P((B \wedge C) \vee (B' \wedge C),C).$$

Hence

$$P(B \wedge C,(B \wedge C) \vee (B' \wedge C)) = P(B' \wedge C,(B \wedge C) \vee (B' \wedge C));$$

and consequently, by the supposition, $P(D,B \vee C) = P(D,B' \vee C)$ for every statement D, and, in particular, $P(A,B \vee C) = P(A,B' \vee C)$. So, $P(A \wedge B,C) = P(A \wedge B',C)$ by **RP4**. Suppose next that $P(A,D)$ as a function of A *is* a truth-value function for every statement D and assume that $P(A,A \vee B) = P(B,A \vee B)$. Then $P(C \wedge A,A \vee B) = P(C \wedge B,A \vee B)$, and hence $P(C,A) = P(C,B)$ by **RP4** and IE. ¤

The constraint

If $A \sim_P B$, *then* $P(A,A \vee B) = P(B,A \vee B)$

singles out the class of 2-valued Popper functions.

Theorem 4. *Let* P *be a relative probability function for* L. *Then* P *meets the constraint*

If $A \sim_P B$, *then* $P(A,A \vee B) = P(B,A \vee B)$

if and only if P *takes only* 0 *and* 1 *as its values.*

Proof: Suppose first that, for any statements A and B, if $A \sim_P B$, then $P(A,A \vee B) = P(B,A \vee B)$ and assume that $P(A,B) = r$, where $0 < r < 1$. Then $A \wedge B \sim_P B$; for

$$P(A \wedge B,(A \wedge B) \vee B) = P(A,B) > 0$$

by IE and

$$P(B,(A \wedge B) \vee B) = P(B,B) = 1 > 0$$

by IE and **RP2**. But $P(A \wedge B,B) \neq P(B,B)$ by Lemma 4(u), contradicting the supposition. So, r must equal 0 or 1. Suppose next that P is a 2-valued function and assume that $A \sim_P B$. Then $P(A,A \vee B) > 0$ and hence $P(A,A \vee B) = 1$; and $P(B,A \vee B) = 1$ similarly. So, $P(A,A \vee B) = P(B,A \vee B)$. ¤

It is clear that in this case the two relations $\mathbf{R_8}$ and $\mathbf{R_9}$ also coincide. So, $\mathbf{R_3}$, $\mathbf{R_4}$, and $\mathbf{R_5}$ are collapsed, and so are $\mathbf{R_7}$, $\mathbf{R_8}$, and $\mathbf{R_9}$.

The constraint

$$\textit{If } P(\bot,A) = P(\bot,B), \textit{ then } A \sim_P B$$

does not single out a further class of relative probability functions for L_P; rather, it constitutes yet another way of characterising the functions we have associated with Kolmogorov. For, according to this constraint, all P-normal statements belong to the same $\sim_P$ - equivalence class. The list of equivalent constraints can be continued with

$$\textit{If } P(A \equiv B,\top) = 1, \textit{ then } A \textit{ is right-indiscernible from } B,$$

$$\textit{If } P(A,\top) = P(B,\top), \textit{ then } P(A,A \vee B) = P(B,A \vee B),$$

and

$$\textit{If } P(A,\top) = 0 \textit{ if and only if } P(B,\top) = 0, \textit{ then } A \sim_P B.$$

Two more classes of relative probability functions are obtained by identifying equivalence relations in the right column of Diagram 1, namely:

$$\textit{If } P(A,\top) = P(B,\top), \textit{ then } P(A \equiv B,\top) = 1,$$

and

$$\textit{If } P(A,\top) = 0 \textit{ if and only if } P(B,\top) = 0, \textit{ then } P(A,\top) = P(B,\top).$$

The first of these conditions, under which $\mathbf{R_7}$ and $\mathbf{R_8}$ coincide, characterises relative probability functions whose $\top$-restriction is a truth-value function (2-valued or many-valued), while the second, which identifies $\mathbf{R_9}$ with $\mathbf{R_8}$, singles out the relative probability functions with a 2-valued $\top$-restriction, so that $\mathbf{R_9}$ coincides with $\mathbf{R_7}$ as well.

Further families of relative probability functions can be obtained by combining constraints. For Carnap 1950 functions we combine the constraints for Carnap 1952 and for Kolmogorov functions. Hence they are characterised by

$$\textit{If } P(C,A) = P(C,B) \textit{ for every statement } C, \textit{ then } A \textit{ is logically equivalent to } B,$$

according to which $\mathbf{R_3}$ and $\mathbf{R_1}$ coincide. When P is a Carnap 1950 function, $\mathbf{R_1}$, $\mathbf{R_2}$, $\mathbf{R_3}$ and $\mathbf{R_7}$ coincide, as do $\mathbf{R_4}$ and $\mathbf{R_8}$, and $\mathbf{R_5}$, $\mathbf{R_6}$ and $\mathbf{R_9}$.

Combining the constraint for Kolmogorov functions

$$\textit{If } P(C,A) = P(C,B) \textit{ for every statement } C, \textit{ then } P(A,C) = P(B,C) \textit{ for every statement } C$$

either with the constraint for 2-valued functions

$$\textit{If } A \sim_P B, \textit{ then } P(A,A \vee B) = P(B,A \vee B)$$

or with the constraint for functions with a 2-valued T-restriction

$$\textit{If } P(A,\mathsf{T}) = 0 \textit{ if and only if } P(B,\mathsf{T}) = 0, \textit{ then } P(A,\mathsf{T}) = P(B,\mathsf{T})$$

results in the class of 2-valued Kolmogorov functions. But these, as we saw in Chapter 1, are 2-valued truth-value functions, with $P(A,B)$ the truth-value of $B \supset A$. The relations $\mathbf{R_2}$–$\mathbf{R_9}$ all coincide.

It is noteworthy that there are no 2-valued Carnap 1950 functions, functions for which all nine equivalence relations would coincide. Indeed the constraint

$$\textit{If } P(\perp,A) = P(\perp,B), \textit{ then } A \textit{ is logically equivalent to } B$$

is not satisfiable, since by Lemmas 4(t) and 4(ll) at least one of E_1 and $\neg E_1$ is P-normal, but neither is logically equivalent to T.

Part II

Probability Logic

Introduction to Part II

Part Two investigates the connection between probability functions and semantic notions. The idea of probabilistic semantics as an alternative to classical semantics was first introduced by Field in (1977) and was pursued by several writers, notably Leblanc in (1983a). It is the idea of explaining semantic notions, such as logical truth and logical consequence, in terms of probability functions. The idea is prompted by Popper's succeeding to formulate constraints for probability functions that make no use themselves of any semantic notions, in contrast to the statement counterparts of Kolmogorov's constraints which crucially involve the mutual substitutability of logically equivalent statements. We have been favouring Popper's approach and we have already noted some of the links between probability and semantic notions. So, Theorem 1 in Chapter 1 shows that if A is a logical truth then $P(A) = 1$ for every absolute probability function P, and Theorem 3 that if A is logically true $P(A,C) = 1$ for every statement C and every relative probability function P, and if A logically implies B, then $P(A,C) \leq P(B,C)$ for every statement C and every relative probability function P. It is not difficult to extend these results to

A is logically true if and only if $P(A) = 1$ for every absolute probability function P,

A is logically true if and only if $P(A,C) = 1$ for every statement C and every relative probability function P,

A logically implies B if and only if $P(A,C) \leq P(B,C)$ for every statement C and every relative probability function P.

In all these results logical truth and logical implication are defined in the usual truth-functional way. Field's idea, prompted by these results, is to read the right hand side in each case as an alternative definition of the respective semantic notions. So,

$$P(A) = 1 \textit{ for every } P$$

is stipulated to mean that

A is logically true in the probabilistic sense,

and

$$P(A,C) \leq P(B,C) \textit{ for every } C \textit{ and } P$$

is stipulated to mean that

A logically implies B in the probabilistic sense.

The above results then amount to stating the coincidence of the truth-functional and the probabilistic semantic notions, and they can be used to obtain soundness and completeness results relative to the probabilistically defined semantic notions.

What interests us at this point is not so much the details of the proofs of these results but the significance and explanatory relevance of probabilistic semantics. The usual truth-functional semantics is transparent in this respect: The definition of the semantic notions, such as logical implication, is based on the thought that these notions are invariants across all *possible* ways of assigning truth-values to the statements of the language, and therefore are invariants across all (possible) interpretations. And the possible ways of assigning truth-values, i.e. the possible interpretations, are dictated by the truth-functional meaning of the sentential connectives, i.e. by the manner in which these connectives contribute to the truth-conditions of statements. The range of possible truth-value assignments is the range of interpretations of the language.

Probability semantics also defines semantic notions as invariants across interpretations, the interpretations now being probability functions. But the reason why the semantic concepts can be defined in this way and the nature of the interpretations are left rather obscure. An initially plausible suggestion, but one not actually championed by Field, is that probability functions constitute generalisations of truth-value assignments. An absolute probability function, to concentrate on the simpler case, is then a function from statements to 'truth-values', i.e. the real numbers in the interval [0,1]. Probability, so understood, is a generalisation of truth and falsity. $P(A) = 1$ means that A is true under P, $P(A) = 0$ that A is false; and A is logically true if A is true under every function P.

But probability logic should then be a variety of many-valued logic and the probability of a compound statement should be a function of the probabilities of its atomic components, which, leaving aside a few exceptions, is of course not the case. Now we do not want to claim that a function on statements that generalises truth and falsity has to be recursively definable from its values for atomic statements. But there is a need for an account of the role that the meanings of sentential connectives play in constraining the range of probability functions. One plausible view relates the probability $P(A)$ to the degree of belief in A. An absolute probability function must then be *coherent* in the sense of De Finetti (1937), i.e. it must allow a rational agent to believe simultaneously the various statements A, B, ... of the language to the respective extents $P(A)$, $P(B)$, ... Obviously, coherence in this sense is closely related to the consistency of a class of beliefs, and consistency is of course a semantic notion. It seems therefore that even when the constraints imposed on probability functions make no use of semantic concepts, such concepts motivate and justify these constraints. If so, the meanings of the connectives affect the constraints on probability functions via semantic notions, and to define the latter in terms of probability functions amounts to conceptual circularity. In this situation the availability of a convincing alternative account of the significance of probability functions and the contribution of the sentential connectives to the constraints imposed on probability functions is sufficient reason for abandoning probabilistic semantics in Field's sense.

Argued in the chapters to follow is that absolute probability functions generalise the semantic notions of logical truth and logical falsity, while relative probability functions generalise that of logical consequence. These claims amount to an endorsement of Carnap's views, as put forward in Carnap (1950). As regards absolute probability functions, this means that

$$P(A) = 1$$

amounts to

A is logically true (or *A is necessary*),

and that

$$P(A) = 0$$

amounts to

A is logically false (or *A is impossible*),

while a probability value between 0 and 1 bespeaks a statement that is neither necessary nor impossible. This understanding of an absolute probability function, however, would suit only a Carnap one: Popper's absolute probability functions in general do not assign the value 1 just to logical truths. So, following Leblanc's practice in (1983b), we rather talk of an absolute probability function as assigning degrees of necessity *relative to a set of assumptions or presuppositions.*

As for the contribution of the connectives '¬', '∧', '∨', etc. to the constraints placed on absolute probability functions, it is in our opinion best understood if one studies not those functions themselves but certain binary functions G on sets of statements that are interdefinable with them. Called here *Gentzen functions*, they generalise Gentzen's notion of multiple conclusion consequence, $G(X,Y) = 1$ now meaning that the members of Y (disjunctively understood) are a logical consequence of those of Y (conjunctively understood). The characterisation of our Gentzen functions then employs constraints for the connectives that are generalisations of the introduction rules (on the Left and on the Right) for connectives employed by Gentzen.

In the case of relative probability functions, too, a better understanding is achieved by considering binary functions on sets of statements, in fact those introduced in Chapter 1, Section 3. We argue that $P(\{A\},\varnothing)$ is the degree to which A is implied by a certain set $S_P(\varnothing)$ of statements, the background assumptions associated with P. And $P(\{A\},X)$ generally is the degree to which A is implied by $X \cup S_P(X)$, where $S_P(X)$ is the result of modifying the background assumptions so that they become logically compatible with X.

The constraints for connectives that are used to characterise these relative probability functions on sets of statements can be seen as generalisations of Gentzen's introduction and elimination rules for the calculus NJ of (single-conclusion) consequence, in other words, they are generalisations of Natural Deduction rules.

Chapter 7

Absolute Probability Functions Construed as Representing Degrees of Logical Truth

In this chapter we investigate the connection between probability and semantic notions in the case of absolute probability functions. We pursue the thought that absolute probability functions generalise logical necessity so that for a central category of functions the value of a function P for a statement A can be interpreted as the degree of necessity assigned to A. And in the general case $P(A)$ can be understood as the degree of necessity of A relative to certain assumptions specific to P.[1]

Our approach is somewhat indirect. We take as our starting point the relation of logical consequence in the multiple-conclusion sense, i.e. a relation between two sets of statements.[2] We then investigate how this relation can be generalised to become a matter of degrees of consequence. We characterise a group of binary functions – to be called *Gentzen functions* – which take sets of statements as their arguments so that $G(X,Y)$ is the degree to which the statements in X (conjunctively understood) entail the statements in Y (disjunctively understood). Interestingly, the simplest way of characterising Gentzen functions employs constraints that are formulated in terms of their complements

$$F(X,Y) = 1 - G(X,Y),$$

to be called *consistency functions*. We then proceed to obtain from the binary consistency functions unary functions by setting the second argument at $\varnothing$. These unary functions assign degrees of logical necessity to their arguments and they turn out to be absolute probability functions.

In the first two sections we look at consistency functions for arbitrary languages. The characterising constraints are then general constraints which make no reference to features of the syntax of the language. As a by-product we obtain a language-independent characterisation of absolute probability functions; the constraints in question can be formulated without using conjunction, negation, or any other connective.

Section 3 deals with those consistency functions for L_P which correspond to Carnap absolute probability functions and yields an autonomous characterisation of the latter. Section 4 introduces the notion of an assumption set already alluded to. We show that $P(A)$ is the degree to which the assumption set of P implies A. In the final two sections we deal with consistency functions for L_ω and for L_Q, respectively.

Section 1. Degrees of Logical Consequence in the Multiple-Conclusion Sense

To begin with finite languages, i.e. languages with just a finite number of statements, let L be such a language and assume it to come with a semantics which allows us to characterise, among the statements of L, those that are *logically necessary*. This is usually accomplished by stipulating a range of *interpretations* or *models* and specifying how statements are to be evaluated relative to them. As a result, some of the statements of L evaluate as *true* (*in the model*) or, as the matter is often put, are *designated*.[3] A simple example of a model is of course a truth-value assignment in propositional logic. Given that each atomic statement has one of the two truth-values 1 (*true*) and 0 (*false*) under such an assignment, the truth-value of every complex statement is truth-functionally determined, and designated under such an assignment are the statements with truth-value 1.

A statement of L counts as necessary if it is designated in every model. So what matters about a model as far as semantics is concerned is not how designated statements are determined but only which subset of the set of statements of L is designated in that model. For our purposes there is no need to distinguish between models in which the same statements are designated. In fact, we will focus our attention on the subsets of L, ignoring the question for exactly which model or models the statements in such a subset are the designated ones. What is important, though, is whether for a given set W of statements of L there exists a model which has just the statements in W as its designated statements. When there is no such model, then W is an *impossible* subset of L. (Note that for W to be *possible* it is not sufficient that for some model every statement in W be designated; it is also required that for the model in question *only* the statements in W be designated.)

Gentzen Functions and Consistency Functions

Shifting our attention now, as anticipated, to logical relations, we introduce a type of function $F(X,Y)$ – X and Y sets of statements of L – which assigns to the pair X and Y of sets of statements the degree to which it is possible that every member of X is designated while none of Y is, or, if you like, the degree of consistency of the assumption that every member of X is designated while none of Y is. $F(X,Y) = 1$ signifies that it is necessary that every member of X be designated but none of Y be. And $F(X,Y) = 0$ signifies that it is impossible for every member of X to be designated while none of Y is.

The second binary function on sets of statements of L, defined by

$$G(X,Y) = 1 - F(X,Y),$$

then signifies the degree of logical implication of Y (disjunctively understood) by X (conjunctively understood). In particular, $G(X,Y) = 1$ has the significance that when all the statements in X are designated, at least one of the statements in Y must be designated. This is the notion of multiple-conclusion consequence which is formalised in Gentzen's *Sequenzen-*

Kalkül.[4] $G(X,Y)$ then gives the degree to which X implies Y (in the multiple-conclusion sense).

An obvious way of obtaining degrees of necessity of statements is to introduce a weight function α which associates 0 with every *impossible* subset of L and a positive real number with every *possible* subset of L in such a way that these weights add up to 1. The function α thus meets the following two constraints:

W1 $0 \leq \alpha(W)$ *for every subset W of L*

W2 $\Sigma_{W \subseteq L}\{\alpha(W)\} = 1$

The degree of necessity of a statement is then the sum of weights of those subsets of L of which the statement is a member. $F(X,Y)$, on the other hand, is calculated by adding up the weights of those subsets W of L that contain every statement in X and none of the statements in Y:

W3 $F(X,Y) = \Sigma_{W \subseteq L}\{\alpha(W) : X \subseteq W \text{ and } Y \subseteq (L-W)\}$,

so that, as expected, X logically implies Y in the multiple-conclusion sense if and only if

$$\Sigma_{W \subseteq L}\{\alpha(W) : X \subseteq W \text{ and } Y \subseteq (L-W)\} = 0.$$

The functions F that can be obtained from weight functions α can also be implicitly characterised, as we are going to show next. The constraints are these:

F1 $0 \leq F(X,Y)$

F2 $F(\emptyset,\emptyset) = 1$

F3 $F(X,Y \cup \{A\}) + F(\{A\} \cup X,Y) = F(X,Y)$

Binary functions F on sets of statements of L which meet these constraints will be called *consistency functions*, while the functions G will be referred to as *Gentzen functions.*

Theorem 1. *Let L be a finite language, let α be a function on the subsets W of L meeting* **W1** *and* **W2**, *and let F be defined by* **W3**. *Then F meets constraints* **F1–F3**.

Proof:

F1: By **W1**.

F2: $\Sigma_{W \subseteq L}\{\alpha(W) : \emptyset \subseteq W \text{ and } \emptyset \subseteq (L-W)\} = 1$ by **W2**. So $F(\emptyset,\emptyset) = 1$ by **W3**.

F3: For any W, $X \subseteq W$ and $Y \subseteq (L-W)$ if and only if either $X \subseteq W$ and $Y \cup \{A\} \subseteq (L-W)$ or $\{A\} \cup X \subseteq W$ and $Y \subseteq (L-W)$. ¤

Theorem 2. *If F meets constraints* **F1–F3,** *there then exists a weight function α such that $F(X,Y) = \Sigma_{W \subseteq L}\{\alpha(W) : X \subseteq W \text{ and } Y \subseteq (L-W)\}$.*

Proof: Suppose F meets constraints **F1–F3**, and let $\alpha_F(W) = F(W,L-W)$. Then $\alpha_F(W) \geq 0$ by **F1**. And

$\Sigma_{W \subseteq L}\{\alpha_F(W)\}$

$= \Sigma_{W \subseteq L}\{F(W,L-W)\}$,

$= \Sigma_{W \subseteq L}\{F(W,L-W) : \varnothing \subseteq W \text{ and } \varnothing \subseteq L-W\}$

$= F(\varnothing,\varnothing)$ (Lemma 12(c))

$= 1$ **(F2)**

So α_F is a weight function. Furthermore,

$F(X,Y)$

$= \Sigma_{W \subseteq L}\{F(W,L-W) : X \subseteq W \text{ and } Y \subseteq (L-W)\}$ (Lemma 12(c))

$= \Sigma_{W \subseteq L}\{\alpha_F(W) : X \subseteq W \text{ and } Y \subseteq (L-W)\}$ (Def. α_F) ¤

Theorems 1 and 2 thus guarantee that the characterisation **W1–W3** of consistency functions in terms of weight functions and the one in terms of constraints **F1–F3** are equivalent. Because of the identity

$$G(X,Y) = 1 - F(X,Y)$$

F1–F3 also serve to characterise Gentzen functions and hence degrees of logical consequence.

Multiple-Conclusion Consequence

The implicit characterisation of consistency functions corresponds to the implicit characterisation of multiple-conclusion consequence relations $\vdash$ according to Gentzen's approach:[5]

M1 *If* $X \cap Y \neq \varnothing$, *then* $X \vdash Y$ *(Overlap)*

M2 *If* $X \vdash Y$, *then* $Z \cup X \vdash Y$ *and* $X \vdash Y \cup Z$ *(Thinning)*

M3 *If* $X \vdash Y \cup \{A\}$ *and* $\{A\} \cup X \vdash Y$, *then* $X \vdash Y$ *(Cut)*

Now let $\vdash_F$ be the binary relation on subsets of L which holds in the limiting case in which the degree of logical consequence is 1, i.e. when $G(X,Y) = 1$ or, equivalently, when $F(X,Y) = 0$:

$$X \vdash_F Y \textit{ if and only if } F(X,Y) = 0.$$

Then $\vdash_F$ is a multiple-conclusion consequence relation.

Theorem 3. *If F meets constraints* **F1–F3**, *then the relation $\vdash_F$ satisfies* **M1–M3**.

Proof:

M1: Suppose $X \cap Y \neq \varnothing$. Then $F(X,Y) = 0$ by Lemma 12(a), i.e. $X \vdash_F Y$.

M2: Suppose $X \vdash_F Y$, i.e. $F(X,Y) = 0$. Then both $F(\{A\} \cup X,Y) \leq 0$ and $F(X,Y \cup \{A\}) \leq 0$ by Lemma 12(d). Hence $F(\{A\} \cup X,Y) = 0$ and $F(X,Y \cup \{A\}) = 0$ by **F1**.

M3: Suppose $X \vdash_F Y \cup \{A\}$ and $\{A\} \cup X \vdash_F Y$, i.e. $F(X,Y \cup \{A\}) = 0$ and $F(\{A\} \cup X,Y) = 0$. Then $F(X,Y) = 0$ by **F3**. Hence $X \vdash_F Y$. ¤

To sum up, any function F that meets **F1–F3** for a finite language L is tantamount to a semantics for L, i.e. it yields, in the shape of

$$G(X,Y) = 1 - F(X,Y),$$

degrees of logical consequence in the multiple conclusion sense and thereby also everything that a semantics contributes, namely the family of *impossible* subsets W of L (the subsets W with $F(W,L-W) = 0$) and hence the extension of the consequence relation $\vdash_F$. Since F contains all the information that a semantics furnishes we say that F *provides a semantics*. So, any one binary function on the statements of L that meets constraints **F1** to **F3** provides a semantics for L.

Language-Independent Constraints for Absolute Probability Functions

Having concentrated on the case $F(X,Y) = 0$, let us now turn to necessary and impossible statements (according to F) and to degrees of necessity. Given the present terminology, a necessary statement is one that is a member of every *possible* subset of L. The degree of necessity of such a statement should be 1. Degrees of necessity less than 1, but larger than 0, should be attributed to statements belonging to some, but not all, *possible* subsets of L; and statements belonging to no *possible* subset of L should have a zero degree of necessity. So, degrees of necessity are given by the function P defined thus:

$$P(A) = G(\varnothing,\{A\}) = F(\{A\},\varnothing).$$

But we claim that degrees of necessity are absolute probabilities. So we obtain the following general characterisation of a probability function for a finite language L:

P is an absolute probability function for L if and only if there exists a consistency function F for L such that for every statement A of L

$$P(A) = F(\{A\},\varnothing).$$

From such a function P the consistency function F cannot be retrieved,[6] and hence P is not sufficient for the specification of the weight function α. However, if we turn to the function that attributes degrees of necessity to *sets* of statements, say,

$$P(X) = F(X,\varnothing),$$

then we find this function to contain all the information that F contains. For let F_P be the auxiliary function introduced in Chapter 5, Section 2:

$$F_P(X,Y) = \begin{cases} P(X) & \textit{if } Y = \varnothing \\ F_P(X,Y-\{A\}) - F_P(X \cup \{A\},Y-\{A\}) & \textit{otherwise,} \end{cases}$$

where A is the alphabetically latest member of Y.

Now F itself meets all the constraints that enter into the definition of F_P. For $F(X,\varnothing) = P(X)$ by the definition of P, while the instances of the second, recursive, clause of the definition of F_P are instances of **F3**. Hence F_P is the same function as F. And P can be characterised by the constraints

AS1' $\quad 0 \leq F_P(X,Y)$

AS2 $\quad P(\varnothing) = 1,$

AS1' being equivalent to **F1**, and **AS2** to **F2**, with **F3** implicit in the definition of F_P from P.

We take P to be a probability function and, as anticipated in Chapter 5, Section 2, regard **AS1'** and **AS2** as the most general constraints for absolute probability functions for finite sets of statements, constraints which make no reference at all to the syntax of L. So, we declare that

P is an absolute probability function for the finite sets of statements of L if and only if P meets **AS1'** *and* **AS2**.

Section 2. Consistency Functions for Infinite Languages

As we turn to languages with infinitely many statements, we no longer think of a consistency function as defined by means of a weight function α. For with L infinite, there will be non-denumerably many subsets W of L, and hence questions about non-denumerably infinite sums would have to be addressed in connection with **W3**. In any case, the purpose of discussing the weight function was to motivate the implicit characterisation of consistency functions and this characterisation itself does not presuppose that the language L consist of only finitely many statements. So, the constraints **F1–F3** remain in force when L is infinite.

We must add to these constraints, however, in order to fix the relationship between the values of F for finite and for infinite sets of statements. Hence the following further constraints:

FωR $F(X,Y) = \inf_{Y'} \{F(X,Y') : Y' \subseteq_f Y\}$

FωL $F(X,Y) = \inf_{X'} \{F(X',Y) : X' \subseteq_f X\}$

FωR and **FωL** ensure that consistency functions and hence degrees of consequence are in effect determined by their finite parts. We do not believe that there is an incontrovertible argument for the adoption of **FωR** and **FωL**. Rather, an understanding of degrees of consequence is conceivable which allows the degree of implication of a statement A by an infinite set X of statements to be higher than the upper limit of the degrees of implication of A by finite subsets of X. Our opting for **FωR** and **FωL** is in line with the adoption in Chapter 1, Section 3, of constraint **ASω**.

FωR and **FωL** hold for finite sets X and Y by virtue of Lemma 12(e). As for infinite X and Y, **FωR** and **FωL** together have the character of a definition by which any consistency function defined on the finite subsets of an infinite language L can be extended to a full function on the sets of statements of L.

Theorem 4. *Let F' be a binary function on the finite sets of statements of L that meets constraints* **F1–F3** *and let F be the extension of F' to all sets of statements of L by means of the definition*

$$F(X,Y) = \begin{cases} F'(X,Y) & \textit{if } X \textit{ and } Y \textit{ are finite} \\ \inf_{X'} \{\inf_{Y'} \{F'(X',Y') : Y' \subseteq_f Y\} : X' \subseteq_f X\} & \textit{otherwise} \end{cases}$$

Then F meets **F1**–**F3**.

Proof:

F1 and **F2**: Obvious.

F3: $\inf_{X'} \{\inf_{Y'} \{F'(X',Y') : Y' \subseteq_f Y \cup \{A\}\} : X' \subseteq_f X\}$

$$+ \inf_{X'} \{\inf_{Y'} \{F'(X',Y') : Y' \subseteq_f Y\} : X' \subseteq_f \{A\} \cup X\}$$

$$= \inf_{X'} \{\inf_{Y'} \{F'(X',Y' \cup \{A\}) : Y' \subseteq_f Y\} : X' \subseteq_f X\}$$

$$+ \inf_{X'} \{\inf_{Y'} \{F'(\{A\} \cup X',Y') : Y' \subseteq_f Y\} : X' \subseteq_f X\} \qquad \text{(Lemma 12(e))}$$

$$= \inf_{X'} \{\inf_{Y'} \{F'(X',Y' \cup \{A\}) + \{F'(\{A\} \cup X',Y') : Y' \subseteq_f Y\} : X' \subseteq_f X\}$$

$$= \inf_{X'} \{\inf_{Y'} \{F'(X',Y') : Y' \subseteq_f Y\} : X' \subseteq_f X\} \qquad \textbf{(F3 for } F'\text{)}$$ ¤

With consistency functions now taking infinite sets of statements as arguments, we can no longer expect that in the limiting case $F(X,Y) = 0$, i.e. $G(X,Y) = 1$, X logically implies Y in the multiple-conclusion sense. For $F(W,L-W)$ can be other than 0 in at most denumerably many cases, i.e. for at most denumerably many W, if $F(\varnothing,\varnothing)$ is to be 1 rather than infinite. And so $W \vdash L-W$ would have to hold in all but denumerably many cases. This would surely be an unsatisfactory restriction. Consequently, the relationship between logical consequence and degrees of consequence must be more complicated than in the finite case; logical consequence cannot simply be a limiting case of degree of consequence. This means that in general logical consequence cannot be characterised by a single consistency function, but rather, that a class $\boldsymbol{F}$ of such functions is required so that $X \vdash Y$ when $F(X,Y) = 0$ for every function F in $\boldsymbol{F}$. For any single consistency function F in the class $\boldsymbol{F}$ the relation $F(X,Y) = 0$ amounts to a proper extension of the consequence relation determined by $\boldsymbol{F}$.

For any given semantics, i.e. any family of subsets of L regarded as *possible* under that semantics, there exists a class of functions which characterises the consequence relation $\vdash$ determined by the semantics. As we just saw, in general no single consistency function F attributes to pairs $\langle X,Y \rangle$ the value 0 just when $X \vdash Y$. But, given a function F with $F(X,Y) = 0$ when $X \vdash Y$, there exists, for any $\langle Z,Z' \rangle$ with $F(Z,Z') = 0$ and $Z \nvdash Z'$, a consistency function F' such that $F'(X,Y) = 0$ whenever $X \vdash Y$, but $F'(Z,Z') > 0$. For, since $Z \nvdash Z'$, there must be a *possible* subset W of L with $Z \subseteq W$ and $Z' \subseteq L-W$. The function F' defined thus

$$F'(X,Y) = \begin{cases} r \times F(X,Y) & \textit{if not both } X \subseteq W \textit{ and } Y \subseteq L-W \\ r \times F(X,Y) + (1-r) & \textit{if } X \subseteq W \textit{ and } Y \subseteq L-W, \end{cases}$$

where $0 < r < 1$, is easily seen to be a consistency function. $F'(Z,Z')$ equals $1 - r$ and is therefore larger than 0. Moreover, when $X \vdash Y$, $F'(X,Y) = 0$. For in that case it cannot be that $X \subseteq W$ and $Y \subseteq L-W$, since W is *possible*. Hence $F'(X,Y) = F(X,Y)$; but $F(X,Y) = 0$, when

$X \vdash Y$. Consequently, given a multiple-conclusion consequence relation $\vdash$, there exists a class $\boldsymbol{F}$ of consistency functions so that $X \vdash Y$ if and only if $F(X,Y) = 0$ for every F in $\boldsymbol{F}$.

Conversely, any class $\boldsymbol{F}$ of consistency functions on L characterises a multiple-conclusion relation $\vdash$ on L, if $\vdash$ is defined as

$$X \vdash Y \textit{ if and only if } F(X,Y) = 0 \textit{ for every consistency function } F \textit{ in } \boldsymbol{F}.$$

For it is quite obvious that $\vdash$, so defined, meets the constraints **M1–M3**.

If the number of consistency functions in $\boldsymbol{F} = \{F_1, F_2, ..., F_n, ...\}$ is countable, then a single function is capable of specifying the consequence relation characterised by $\boldsymbol{F}$. For, since the constraints **F1–F3**, **FωR**, and **FωL** are linear, any linear combination of the functions in $\boldsymbol{F}$, say,

$$F(X,Y) = a_1 \times F_1(X,Y) + a_2 \times F_2(X,Y) + ... + a_n \times F_n(X,Y) + ...$$

will be a consistency function, as long as the real numbers $a_1, a_2, ..., a_n, ...$ add up to 1. And if those real numbers are all positive, $F(X,Y) = 0$ if and only if $F_i(X,Y) = 0$ for each of $i = 1, 2, ..., n, ...$. If $\boldsymbol{F}$ is non-denumerable then there may be no single function for which $\langle X,Y \rangle$ has value 0 just when it has that value for every function in $\boldsymbol{F}$. In that case, every individual function F in $\boldsymbol{F}$ will assign the value 0 to some pairs $\langle X,Y \rangle$ of statements in cases in which $X \nvdash Y$; the relation $F(X,Y) = 0$ (or, equivalently, $G(X,Y) = 1$) is then a genuine extension of the relation $\vdash$.

The definition of absolute probability functions for L and absolute probability functions for the sets of statements of L from F in Section 1 remains unchanged. For the independent characterisation of absolute probability functions for the sets of statements of L we need of course the additional constraint

$$\textbf{AS}\omega \quad P(X) = \inf_{X' \subseteq L} \{P(X') : X' \subseteq_f X\},$$

of Chapter 1, Section 3. So, we say

P is an absolute probability function for the sets of statements of L if and only if P meets **AS1'**, **AS2**, *and* **ASω**.

Section 3. Carnap's Absolute Probability Functions Generalise Logical Truth

While the previous sections dealt with degrees of logical consequence and with consistency functions in a general manner, i.e. without paying attention to the structure of the language for which degrees of consequence are being defined, we deal in this section with the propositional language L_P and we characterise intrinsically degrees of consequence that correspond to the classical consequence relation for L_P, i.e. the relation of truth-functional implication. The consequence relation for truth-functional logic is compact. So, in the first instance, we are interested in consistency functions defined on finite sets of statements of L_P.

Gentzen-Like Constraints for Connectives

We obtain the appropriate constraints which have to be added to **F1–F3** by generalising Gentzen's introduction rules for connectives (on the Left and on the Right).[7] In keeping with the style we have adopted we do not give rules here, but rather corresponding constraints in conditional form. For the primitive connectives '¬' and '∧' these are:

M∧R *If* $X \vdash Y \cup \{A\}$ *and* $X \vdash Y \cup \{B\}$, *then* $X \vdash Y \cup \{A \wedge B\}$

M∧L *If* $\{A,B\} \cup X \vdash Y$, *then* $\{A \wedge B\} \cup X \vdash Y$

M¬R *If* $\{A\} \cup X \vdash Y$, *then* $X \vdash Y \cup \{\neg A\}$

M¬L *If* $X \vdash Y \cup \{A\}$, *then* $\{\neg A\} \cup X \vdash Y$

Less simple but better suited to our purposes is this version of **M∧R**, equivalent in the presence of **M2** and **M3**:

M∧R* *If* $\{A\} \cup X \vdash Y \cup \{B\}$ *and* $X \vdash Y \cup \{A,B\}$ *and* $\{B\} \cup X \vdash Y \cup \{A\}$, *then* $X \vdash Y \cup \{A \wedge B\}$

This constraint shares with **M¬R**, **M¬L**, and **M∧L** the feature that the immediate sub-statements of the compound statement in the consequent occur all of them in every instance of $\vdash$ which occurs in the antecedent. Constraints for '∨' and '⊃', defined connectives in our version of L_P, can easily be obtained from **M∧R**, **M∧L**, **M¬R**, and **M¬L**, and are listed here for completeness.

M∨R *If* $X \vdash Y \cup \{A,B\}$, *then* $X \vdash Y \cup \{A \vee B\}$

M∨L *If* $\{A\} \cup X \vdash Y$ *and* $\{B\} \cup X \vdash Y$, *then* $\{A \vee B\} \cup X \vdash Y$

M⊃R *If* $\{A\} \cup X \vdash Y \cup \{B\}$, *then* $X \vdash Y \cup \{A \supset B\}$

M⊃L *If* $X \vdash Y \cup \{A\}$ *and* $\{B\} \cup X \vdash Y$, *then* $\{A \supset B\} \cup X \vdash Y$

As obvious generalisations of the Gentzen constraints for connectives we obtain these constraints for consistency functions:

F∧R' $F(X,Y \cup \{A \wedge B\}) = F(\{A\} \cup X,Y \cup \{B\}) + F(X,Y \cup \{A,B\}) + F(\{B\} \cup X,Y \cup \{A\})$

F∧L $F(\{A \wedge B\} \cup X,Y) = F(\{A,B\} \cup X,Y)$

F¬R $F(X,Y \cup \{\neg A\}) = F(\{A\} \cup X,Y)$

F¬L $F(\{\neg A\} \cup X,Y) = F(X,Y \cup \{A\})$

One can also justify these constraints by employing the weight functions of Section 1.

Given **F3**, **F∧R'** reduces to the simpler constraint

F∧R $F(X,Y \cup \{A \wedge B\}) = F(X,Y \cup \{A\}) + F(\{A\} \cup X,Y \cup \{B\})$, which we will at times avail ourselves of. **F∧R** lacks the symmetry that **F∧R'** has. Of the two constraints for each connective only one needs to be postulated; the other one follows in each case with the help of **F3**. We will, however, assume all four constraints as axiomatic, since this will enable us to dispense with **F3** up to a point. Since '¬' and '∧' are our primitive connectives we do not need additional constraints for other connectives. However, it is instructive to see what the constraints for other connectives look like. They follow from the four constraints listed by invoking the definitions of those connectives:

F∨R $F(X,Y \cup \{A \vee B\}) = F(X,Y \cup \{A,B\})$

F∨L $F(\{A \vee B\} \cup X,Y) = F(\{A\} \cup X,Y) + F(\{B\} \cup X,Y \cup \{A\})$

F⊃R $F(X,Y \cup \{A \supset B\}) = F(\{A\} \cup X,Y \cup \{B\})$

F⊃L $F(\{A \supset B\} \cup X,Y) = F(X,Y \cup \{A\}) + F(\{A,B\} \cup X,Y)$

F≡R $F(X,Y \cup \{A \equiv B\}) = F(\{A\} \cup X,Y \cup \{B\}) + F(\{B\} \cup X,Y \cup \{A\})$

F≡L $F(\{A \equiv B\} \cup X,Y) = F(X,Y \cup \{A,B\}) + F(\{A,B\} \cup X,Y)$

We adopt **F1–F3**, **F∧R**, **F∧L**, **F¬R**, and **F¬L** as constraints to characterise consistency functions for L_P. And corresponding to each such consistency function we have a Gentzen function for L_P, defined by

$$G(X,Y) = 1 - F(X,Y).$$

Gentzen's calculus of sequents for propositional logic has the attractive characteristic that all instances of the consequence relation can be recursively obtained from the basic instances of the relation, i.e. those that involve only atomic statements. The instances of the consequence relation involving non-atomic statements are obtained by the introduction rules for connectives and the rules corresponding to **M2** and **M3**. Interestingly, Gentzen is then able to show that the Cut Rule is not required:[8] any instance of the consequence relation can be obtained from the basic instances by means of the introduction rules for connectives alone. This result is known as the *Cut-Elimination Theorem*.

Recursive Definition of Consistency Functions

Gentzen's Cut-Elimination Theorem has its parallel among the consistency functions for L_P. Any consistency function defined just on the finite sets of atomic statements of L_P can be recursively extended by means of the four constraints for the connectives '¬' and '∧' to a consistency function on all finite sets of statements of L_P. In other words, given a consistency function F_0 for the sets of atomic statements of L_P, there is just one extension of F_0 to all finite sets of statements of L_P that meets the constraints for connectives, and that extension is then guaranteed to meet **F1**, **F2**, and, in particular, **F3**, the counterpart of *Cut*.

The extension of F_0 in question can be obtained by a recursive definition. To this end the four constraints for connectives have to be turned into clauses of a definition, a definition that is analogous to that employed in Chapter 5, Section 1, to define a probability function on the finite sets of statements of L_P from its restriction to sets of atomic statements. We also employ the notion of an *alphabetic order* of the statements of L_P and that of the *complexity* c of a statement and of a finite set of statements of L_P, both introduced in Chapter 5, Section 1. Here then is the recursive definition of F, A there being the alphabetically latest compound member of $X \cup Y$.

$$F(X,Y) = \begin{cases} F_0(X,Y) & \text{if } c(X) + c(Y) = 0 \\ F(X-\{A\}, Y \cup \{B\}) & \text{if } c(X) + c(Y) > 0,\ A \text{ is } \neg B, \text{ and } A \in X \\ F(\{B\} \cup X, Y-\{A\}) & \text{if } c(X) + c(Y) > 0,\ A \text{ is } \neg B, \text{ and } A \notin X \\ F(\{B,C\} \cup (X-\{A\}), Y) & \text{if } c(X) + c(Y) > 0,\ A \text{ is } B \wedge C, \text{ and } A \in X \\ F(X,(Y-\{A\}) \cup \{B\}) + F(\{B\} \cup X,(Y-\{A\}) \cup \{C\}) & \text{if } c(X) + c(Y) > 0, \\ & A \text{ is } B \wedge C, \text{ and } A \notin X \end{cases}$$

We need to ascertain that the resulting function F meets all constraints **F1–F3**, **F∧R**, **F∧L**, **F¬R**, and **F¬R**. We also want to be assured that it makes no difference to the function obtained through the recursive definition, exactly which order of the statements of L_P is employed in the definition.

Theorem 5. *Let F_0 be a function on the finite sets of atomic statements of L_P which meets* **F1–F3**, *and let F be the recursively defined extension of F_0 to all finite sets of statements of L_P. Then F meets* **F∧R**, **F∧L**, **F¬R**, and **F¬L**.

Proof:

F¬L: By induction on $c = c(X) + c(Y) + c(\neg A)$.

Basis: $c = 1$. Then A as well as the statements in X and those in Y are atomic, and $\neg A$ is the alphabetically latest compound member of $X \cup Y \cup \{\neg A\}$. Then $F(\{\neg A\} \cup X,Y) = F(X,Y \cup \{A\})$ by clause 2 of the definition.

Inductive Step: $c > 1$. Let B be the alphabetically latest compound member of $X \cup Y \cup \{\neg A\}$.

Case 1: B is $\neg A$.

Case 1.1: $B \notin X$. Then $F(\{\neg A\} \cup X,Y) = F(X,Y \cup \{A\})$ by clause 2 of the definition.

Case 1.2: $B \in X$. Then

$$\begin{aligned} &F(\{\neg A\} \cup X,Y) \\ &\quad = F(X-\{\neg A\},Y \cup \{A\}) && \text{(Clause 2, Def. } F) \\ &\quad = F(X,Y \cup \{A\}) && \text{(Ind. Hyp.)} \end{aligned}$$

Case 2: B is some negation $\neg C$ other than $\neg A$.

Case 2.1: $B \in X$. Then

$$\begin{aligned} &F(\{\neg A\} \cup X,Y) \\ &\quad = F(\{\neg A\} \cup (X-\{B\}),Y \cup \{C\}) && \text{(Clause 2, Def. } F) \\ &\quad = F(X-\{B\},Y \cup \{C\} \cup \{A\}) && \text{(Ind. Hyp.)} \\ &\quad = F(X,Y \cup \{A\}) && \text{(Clause 2, Def. } F) \end{aligned}$$

Case 2.2: B is not a member of X, hence is a member of Y. Then

$F(\{\neg A\} \cup X, Y)$
$= F(\{C\} \cup \{\neg A\} \cup X, Y-\{B\})$ (Clause 3, Def. F)
$= F(\{C\} \cup X, (Y-\{B\}) \cup \{A\})$ (Ind. Hyp.)
$= F(X, Y \cup \{A\})$ (Clause 3, Def. F)

Case 3: B is some conjunction $C \wedge D$.

Case 3.1: B is in X. Then

$F(\{\neg A\} \cup X, Y)$
$= F(\{C,D\} \cup \{\neg A\} \cup (X-\{B\}), Y)$ (Clause 4, Def. F)
$= F(\{C,D\} \cup (X-\{B\}), Y \cup \{A\})$ (Ind. Hyp.)
$= F(X, Y \cup \{A\})$ (Clause 4, Def. F)

Case 3.2: B is not a member of X, hence is a member of Y. Then

$F(\{\neg A\} \cup X, Y)$
$= F(\{\neg A\} \cup X, (Y-\{B\}) \cup \{C\}) + F(\{C\} \cup \{\neg A\} \cup X, (Y-\{B\}) \cup \{D\})$
(Clause 5, Def. F)
$= F(X, (Y-\{B\}) \cup \{C\} \cup \{A\}) + F(\{C\} \cup X, (Y-\{B\}) \cup \{D\}) \cup \{A\})$
(Ind. Hyp.)
$= F(X, Y \cup \{A\})$ (Clause 5, Def. F)

F∧R, **F∧L**, and **F¬R**: Analogous proof. ¤

Theorem 6. *Let F_0 be a function on the finite sets of atomic statements of L_P which meets* **F1–F3**, *and let F be the recursively defined extension of F_0 to all finite sets of statements of L_P. Then F meets* **F1–F3**.

Proof:

F1: By induction on $c = c(X) + c(Y)$.

Basis: $c = 0$. Then the members of X and those of Y are atomic statements and $0 \le F(X,Y)$ by clause 1 of the definition of F and the assumption that F_0 meets **F1**.

Inductive Step: $n > 0$. Then $0 \le F(X,Y)$ by the inductive hypothesis and clauses 2, 3, 4, and 5 of the definition.

F2: Since F_0 meets **F2**, $F(\varnothing,\varnothing) = 1$ by clause 1 of the definition.

F3: By induction on $c = c(X) + c(Y) + c(A)$.

Basis: $c = 0$. Then A, the members of X, and those of Y are atomic statements and

$$F(X, Y \cup \{A\}) + F(\{A\} \cup X, Y) = F(X,Y)$$

by clause 1 of the definition and the assumption that F_0 meets **F3**.

Inductive Step: $c > 1$. Let B be the alphabetically latest compound member of $\{A\} \cup X \cup Y$.

Case 1: B is A.

Case 1.1: A is of the form $\neg C$. Then $F(X, Y \cup \{A\}) = F(\{C\} \cup X, Y)$ and $F(\{A\} \cup X, Y) =$

$F(X, Y \cup (C\})$ by Theorem 5. Hence $F(X, Y \cup \{A\}) + F(\{A\} \cup X, Y) = F(X, Y)$ by the inductive hypothesis.

Case 1.2: A is of the form $C \wedge D$. Then:

$$F(X, Y \cup \{A\}) = F(X, Y \cup \{C\}) + F(\{C\} \cup X, Y \cup \{D\})$$

and $F(\{A\} \cup X, Y) = F(\{C, D\} \cup X, Y)$ by Theorem 5. Hence

$$F(X, Y \cup \{A\}) + F(\{A\} \cup X, Y) = F(X, Y)$$

by the inductive hypothesis.

Case 2: $B \in X$.

Case 2.1: B is of the form $\neg C$. Then $F(X, Y \cup \{A\}) = F(X - \{B\}, Y \cup \{C\} \cup \{A\})$ and $F(\{A\} \cup X, Y) = F(\{A\} \cup (X - \{B\}), Y \cup \{C\})$ by Theorem 5. Hence $F(X, Y \cup \{A\}) + F(\{A\} \cup X, Y) = F(X, Y)$ by the inductive hypothesis and Theorem 5 again.

Case 2.2: B is of the form $C \wedge D$. Then $F(X, Y \cup \{A\}) = F(\{C, D\} \cup (X - \{B\}), Y \cup \{A\})$ and $F(\{A\} \cup X, Y) = F(\{A\} \cup \{C, D\} \cup (X - \{B\}), Y)$ by Theorem 5. Hence $F(X, Y \cup \{A\}) + F(\{A\} \cup X, Y) = F(X, Y)$ by the inductive hypothesis and Theorem 5 again.

Case 3: $B \in Y$. Then $F(X, Y \cup \{A\}) + F(\{A\} \cup X, Y) = F(X, Y)$ by similar reasoning. ¤

Theorems 5 and 6 make it clear that in the recursive definition of F the particular ordering of the statements of L_P appealed to, i.e. what we have called the alphabetic ordering, is immaterial. Any ordering of the statements could have been used, resulting in the same function F. A simple induction on the complexity $c = c(X) + c(Y)$ will convince the reader of this: If F' is defined recursively like F but with a different ordering O' of the statements of L_P substituting for the alphabetic ordering, F' will of course agree with F when X and Y are sets of atomic statements. When c is greater than 0, let A' be the compound member of $X \cup Y$ that is the latest according to O'. Then, depending on whether A' is a negation or a conjunction and whether it occurs in X or not, $F'(X, Y)$ will be expressed in accordance with one of the clauses 2, 3, 4, or 5 in terms of values of F' for arguments with lesser complexity than $c = c(X) + c(Y)$. But by the inductive hypothesis these values are identical with those of F for the same arguments.

So, for a given F_0, the recursive definition above yields, independently of the ordering of statements invoked, a single function F, a function that satisfies constraints **F1–F3** as well as the constraints for connectives **F∧R**, **F∧L**, **F¬R**, and **F¬L**. Hence, any binary function F on the finite sets of statements of L_P which satisfies **F1–F3**, **F∧R**, **F∧L**, **F¬R**, and **F¬L** is definable from its restriction F_0 to sets of atomic statements by a recursive definition of this sort. As already remarked, this result parallels Gentzen's Cut Elimination Theorem. The recursive definition of F from values of F_0 corresponds to the derivation of instances of multiple-conclusion consequence from basic sequents, derivations that employ only the introduction rules for connectives, but not *Cut*.

Consistency Functions, Logical Consequence, and Logical Independence

We can now turn to the relationship between logical consequence and the limiting case $F(X,Y) = 0$ of consistency functions for L_P. Since the constraints **F∧R'**, **F∧L**, **F¬R**, and **F¬L** are obviously generalisations of **M∧R***, **M∧L**, **M¬R**, and **M¬L**, which yield these Gentzen constraints when the value of the consistency function F is set to 0, it is easy to prove that if $X \vdash Y$, then $F(X,Y) = 0$.

Theorem 7. *Let F be a binary function on the finite sets of statements of* L_P *which meets*

(i) **F1–F3**, *and*

(ii) **F∧R**, **F∧L**, **F¬R**, *and* **F¬L**.

Then $F(X,Y) = 0$, *and hence* $G(X,Y) = 1$, *if* $X \vdash Y$.

Proof: By induction on the recursive specification of $\vdash$ by means of the Gentzen constraints **M1–M3**, **M∧R**, **M∧L**, **M¬R**, and **M¬L**.

Basis: $X \vdash Y$ holds by virtue of **M1**. Then $F(X,Y) = 0$ by Lemma 12(a).

Inductive Step: $X \vdash Y$ holds by virtue of one or another of the remaining constraints.

Case 1: The constraint is **M2**. If **M2** yields $Z \cup X \vdash Y$, then $F(X,Y) = 0$ by the inductive hypothesis; hence $F(Z \cup X,Y) = 0$ by Lemma 12(f). If **M2** yields $X \vdash Y \cup Z$, the argument is analogous.

Case 2: The constraint is **M3**. Then $F(X,Y \cup \{A\}) = 0$ and $F(\{A\} \cup X,Y) = 0$ by the inductive hypothesis and hence $F(X,Y) = 0$ by **F3**.

Case 3: The constraint is **M∧R**. Then $F(X,Y \cup \{A\}) = 0$ and $F(X,Y \cup \{B\}) = 0$ by the inductive hypothesis. It follows by Lemma 12(f) that $F(\{A\} \cup X,Y \cup \{B\}) = 0$. Hence $F(X,Y \cup \{A \wedge B\}) = 0$ by **F∧R**.

Cases 4, 5, and 6: When the constraint in question is **M∧L**, **M¬R**, or **M¬L**, $F(X,Y) = 0$ follows by similar reasoning. ¤

The converse result is not to be had. But so far the constraints on consistency functions do not mirror every feature of Gentzen's sequent calculus. For we have not taken account of the requirement that in the basic instances of $\vdash$, when X and Y are sets of atomic statements, $X \vdash Y$ only when X and Y overlap. The stipulation that $X \nvdash Y$ when $X \cap Y = \varnothing$ amounts to the assumption that the atomic statements of L_P are logically independent of one another. This notion of independence of a multiplicity of statements is a simple generalisation of the familiar idea that two statements A and B are logically independent when neither implies the other and, furthermore, they are neither contraries nor subcontraries. Hence, A and B are logically independent if none of $\{A\} \vdash \{B\}$, $\{B\} \vdash \{A\}$, $\{A,B\} \vdash \varnothing$, and $\varnothing \vdash \{A,B\}$ holds. Generalising this to arbitrary finite sets of statements, we have: the statements in Z are logically independent if and only if, for any subset Z' of Z, $Z' \nvdash (Z-Z')$.

The logical independence of the atomic statements of L_P translates into the following constraint for consistency functions:

FIND *If X and Y are finite sets of atomic statements and* $X \cap Y = \emptyset$, *then* $F(X,Y) \neq 0$.

Combined with **F1** this means that, for sets X and Y of atomic statements,

$$F(X,Y) = 0 \textit{ if and only if } X \cap Y \neq \emptyset.$$

Having added the constraint **FIND** it is reasonable to expect that the relation $\vdash_F$, defined as before by

$$X \vdash_F Y \textit{ if and only if } F(X,Y) = 0,$$

coincides with the relation $X \vdash Y$ characterised by Gentzen's system for multiple-conclusion consequence. After all, (1) F being required to meet **FIND** corresponds to the specification of basic sequents, and (2) the constraints **F∧R**, **F∧L**, **F¬R**, and **F¬L** were chosen as generalisations of Gentzen's introduction rules for connectives and, as we have stressed, the recursive definition of F from F_0 corresponds to the derivation of instances of multiple-conclusion consequence from basic sequents of Gentzen's sequent calculus. It is precisely this parallelism that the proof of the next theorem relies on.

Theorem 8. *Let F be a binary function on the finite sets of statements of* L_P *which meets*

(i) **F1–F3**,

(ii) **FIND**,

(ii) **F∧R**, **F∧L**, **F¬R**, *and* **F¬L**.

Then $X \vdash Y$ *if* $F(X,Y) = 0$.

Proof: By induction on $c = \text{c}(X) + \text{c}(Y)$.

Basis: $c = 0$. Then X and Y are sets of atomic statements. If $F(X,Y) = 0$, then $X \cap Y \neq \emptyset$ by **FIND**, and hence $X \vdash Y$ by **M1**.

Inductive Step: $c > 1$. Let A be the alphabetically latest compound member of $X \cup Y$ and assume $F(X,Y) = 0$.

Case 1: A is $\neg B$.

Case 1.1: $A \in X$. Then $F(X-\{\neg B\}, Y \cup \{B\}) = 0$ by **F¬L**. So, by the inductive hypothesis, $X-\{\neg B\} \vdash Y \cup \{B\}$, and so $X \vdash Y$ by **M¬L**.

Case 1.2: $A \notin X$, and hence $A \in Y$. Then, by **F¬R**, $F(\{B\} \cup X, Y-\{\neg B\}) = 0$. So, by the inductive hypothesis, $\{B\} \cup X \vdash Y-\{\neg B\}$, and so $X \vdash Y$ by **M¬R**.

Case 2: A is $B \wedge C$.

Case 2.1: $A \in X$. Then $F(\{B,C\} \cup (X-\{B \wedge C\}), Y) = 0$ by **F∧L**. So, by the inductive hypothesis, $\{B,C\} \cup (X-\{B \wedge C\}) \vdash Y$, and hence, by **M∧L**, $X \vdash Y$.

Case 2.2: $A \notin X$, hence $A \in Y$. Then, by **F∧R'**,

$$F(\{B\} \cup X, (Y-\{B \wedge C\}) \cup \{C\}) + F(X, (Y-\{B \wedge C\}) \cup \{B,C\}) + F(\{C\} \cup X, (Y-\{B \wedge C\}) \cup \{B\}) = 0.$$

So, by the inductive hypothesis,

$$\{B\} \cup X \vdash (Y-\{B \wedge C\}) \cup \{C\}, X \vdash (Y-\{B \wedge C\}) \cup \{B,C\},$$

and

$$\{C\} \cup X \vdash (Y-\{B \wedge C\}) \cup \{B\},$$

and so, by **M∧R** and **M3**, $X \vdash Y$. ¤

So, we have it that the consequence relation $\vdash_F$ associated with any single consistency function F for L_P which meets **FIND** coincides with the multiple-conclusion consequence relation $\vdash$ of truth-functional logic.

Theorem 9. *Let F be a binary function on the finite sets of statements of* L_P *which meets*

(i) **F1–F3**,

(ii) **FIND**,

(ii) **F∧R**, **F∧L**, **F¬R**, *and* **F¬L**.

Then $X \vdash Y$ *if and only if* $F(X,Y) = 0$.

Probability Functions and Logical Consequence

In Section 1 we remarked that associated with any consistency function F on a language L is a function P on the statements of L, defined by

$$P(A) = F(\{A\},\varnothing),$$

whose value for a given statement A can be interpreted as the degree of logical necessity of A. Also

$$P(X) = F(X,\varnothing)$$

defines a similar degrees-of-necessity function on sets of statements. We will now show that if F is one of the functions we have been discussing in this section, then the associated degrees-of necessity function P is an absolute probability function for L_P. Then we will go on to show that, conversely, for any absolute probability function P there is a consistency function F with which P is associated by the above identity.

Theorem 10. *Let F be a binary function on the finite sets of statements of* L_P *which meets*

(i) **F1–F3**,

(ii) **F∧R**, **F∧L**, **F¬R**, *and* **F¬L**,

and let $P(X)$ *be understood as* $F(X,\varnothing)$. *Then*

(a) *P meets the constraints* **AS1**, **AS2**, **AS¬**, and **AS∧**;

(b) *If F meets* **FIND** *as well, then P meets* **ASC**.

Proof:

(a) **AS1**: $0 \le F(X,\varnothing)$ by **F1**.

AS2: $F(\varnothing,\varnothing) = 1$ by **F2**.

AS¬: $F(X \cup \{\neg A\},\varnothing)$

$= F(X,\varnothing) - F(X,\{\neg A\})$ **(F3)**
$= F(X,\varnothing) - F(X \cup \{A\},\varnothing)$ **(F¬R)**

AS∧: $F(X \cup \{A \wedge B\},\varnothing) = F(X \cup \{A,B\},\varnothing)$ by **F∧L**.

(b) Assume $P(X) = 0$, i.e. $F(X,\varnothing) = 0$. Then $X \vdash \varnothing$ by Theorem 8 and hence X is logically inconsistent, since the multiple-conclusion consequence relation $\vdash$ is of course sound and complete relative to truth-functional implication. So P meets **ASC**. ¤

Conversely, for any absolute probability function for the finite sets of statements of L_P there exists a consistency function F, namely the auxiliary function F_P, introduced in Chapter 5, Section 2, by the definition

$$F_P(X,Y) = \begin{cases} P(X) & \text{if } Y = \varnothing \\ F_P(X,Y-\{A\}) - F_P(\{A\} \cup X, Y-\{A\}) & \textit{otherwise,} \end{cases}$$

where A is the alphabetically latest statement in Y.

F_P meets constraints **F1–F3** by Lemmas 10(j), 10(k) and 10(l), **F∧L** by Lemma 10(n), and **F¬R** by Lemma 10(m). And if P meets **ASC**, F_P meets **FIND**. For if X and Y are sets of atomic statements with $X \cap Y = \varnothing$, $X \cup Y^{\neg}$ is logically consistent, hence $P(X \cup Y^{\neg}) \neq 0$ by **ASC**, and hence $F_P(X,Y) \neq 0$ by Lemma 10(i).

So, since $P(X) = F_P(X,\varnothing)$, the consistency functions for the finite sets of statements of L_P and the absolute probability functions for the finite sets of statements of L_P are interdefinable, hence carry the same information. But to any absolute probability function P on the finite sets of statements of L_P there corresponds an absolute probability function P' for the statements of L_P, where $P'(A)$ equals $P(\{A\})$. And, as we have seen in Chapter 1, Section 4, P is a Carnap function if and only if P' is a Carnap function. So, to any consistency function F for L_P there corresponds an absolute probability function for L_P. And F meets **FIND** if and only if P' meets **AC** and P meets **ASC**.[9]

The functions that attribute degrees of logical necessity to the statements of L_P turn out to be Carnap probability functions. So, one interpretation of these probability functions is that they record degrees of necessity, with logical truths, and only logical truths, accorded the value 1 and logical falsehoods, but only logical falsehoods, accorded the value 0, while all other statements of L_P receive values between 0 and 1. This interpretation has often been advocated and Carnap himself entertained it.[10] A related interpretation of all of Popper's absolute probability functions for L_P will occupy us in the next section.

Given the correspondence between consistency functions and absolute probability functions and the results in Theorems 7 and 9 we arrive at this connection between logical consequence and absolute probability functions:

Theorem 11. (a) $X \vdash \{A\}$ *if and only if* $P(X \cup \{\neg A\}) = 0$ *for every absolute probability function for the finite sets of statements of* L_P;
(b) $X \vdash \{A\}$ *if and only if* $P'(A) = 1$ *for every absolute probability function* P' *for* L_P *such that* $P'(B) = 1$ *for every statement* $B \in X$.

Proof of (b): Suppose $P(X \cup \{\neg A\}) = 0$ and $P'(B) = 1$ for every $B \in X$. Then $P(X) = 1$ by Lemma 10(e) and $P(X \cup \{A\}) = 1$ by **AS¬**. Hence $P(\{A\}) = 1$ by Lemma 10(e). On the other hand, suppose $P(X \cup \{\neg A\}) \neq 0$ and $P'(B) = 1$ for every $B \in X$. Then $P(X) = 1$, but $P(X \cup \{A\}) \neq 1$ by **AS¬**. Hence $P(\{A\}) \neq 1$ by Lemma 10(e). ¤

From the point of view of probability semantics as commonly understood Theorem 11(b) is a soundness and completeness theorem about logical consequence in the probabilistic sense. The considerations of this and the next section lead us to a different understanding of the result.

Section 4. Assumption Sets of Absolute Probability Functions

Turning to the interpretation of consistency functions for L_P that do not meet the independence constraint **FIND**, we consider the statements A for which $F(\{A\},\varnothing) = 1$ or, what comes to the same, $F(\varnothing,\{A\}) = 0$, and so $G(\varnothing,\{A\}) = 1$. Let the set of these statements, for a given consistency function F, be designated by 'S' and let us call it the *assumption set* for F. The appellation goes back to Leblanc (1983); its aptness will emerge in what follows.

We observe first that, if P and P' are the absolute probability function on finite sets of statements and for statements of L_P, respectively, associated with F, then the set of statements A for which $P(\{A\}) = P'(A) = 1$ is identical with the assumption set for F. Therefore we also talk of the assumption sets of absolute probability functions.

Consistency Functions Defined on Infinite Sets of Statements of L_P

Since assumption sets are infinite sets, we will in this section consider consistency functions (and absolute probability functions) that are defined on all sets of statements of L_P. They meet the constraints

FωR $\quad F(X,Y) = \inf_{Y'} \{F(X,Y') : Y' \subseteq_f Y\}$

FωL $\quad F(X,Y) = \inf_{X'} \{F(X',Y) : X' \subseteq_f X\}$,

which were introduced in Section 2, in addition to **F1–F3**, **F∧R**, **F∧L**, **F¬R**, and **F¬L**. The extendability result obtained in Theorem 4 applies also to the constraints **F∧R**, **F∧L**, **F¬R**, and **F¬L**.

Theorem 12. *Let* F' *be a binary function on the finite sets of statements of* L_P *that meets constraints* **F1–F3**, **F∧R**, **F∧L**, **F¬R**, *and* **F¬L**, *and let* F *be the extension of* F' *to all sets of statements of* L_P *by means of the definition*

$$F(X,Y) = \begin{cases} F'(X,Y) & \text{if } X \text{ and } Y \text{ are finite} \\ \inf_{X'}\{\inf_{Y'}\{F'(X',Y') : Y' \subseteq_f Y\} : X' \subseteq_f X\} & \text{otherwise.} \end{cases}$$

Then F meets **F1–F3**, **F∧R**, **F∧L**, **F¬R**, **F¬L**, **FωR**, *and* **FωL**.

Proof: **F1–F3**: See Theorem 4.

F¬R: $F(X,Y \cup \{\neg A\})$

$= \inf_{X'}\{\inf_{Y'}\{F(X',Y') : Y' \subseteq_f Y \cup \{\neg A\}\} : X' \subseteq_f X\}$ (Def. F)

$= \inf_{X'}\{\inf_{Y'}\{F(X',Y' \cup \{\neg A\}) : Y' \subseteq_f Y\} : X' \subseteq_f X\}$ (Lemma 12(d))

$= \inf_{X'}\{\inf_{Y'}\{F(\{A\} \cup X',Y') : Y' \subseteq_f Y\} : X' \subseteq_f X\}$ (Theorem 8)

$= \inf_{X'}\{\inf_{Y'}\{F(X',Y') : Y' \subseteq_f Y\} : X' \subseteq_f \{A\} \cup X\}$ (Lemma 12(d))

$= F(\{A\} \cup X,Y)$ (Def. of F)

F¬L, **F∧R**, and **F∧L**: By analogous reasoning.

FωR and **FωL**: Obvious. ¤

Theorem 13. *Let F be a binary function on the sets of statements of L_P which meets*

(i) **F1–F3**,

(ii) **F∧R**, **F∧L**, **F¬R**, **F¬L**,

(iii) **FωR**, *and* **FωL**,

and let $S = \{A : F(\{A\},\varnothing) = 1\}$ *be the assumption set of F. Then*

(a) *If* $X \vdash Y$, *then* $F(X,Y) = 0$;

(b) $F(S \cup X,Y) = F(X,Y)$;

(c) *S is logically consistent;*

(d) *S is closed under logical consequence.*

Proof: (a) Suppose $X \vdash Y$. Then there exist finite subsets X' of X and Y' of Y such that $X' \vdash Y'$, since the consequence relation for L_P is compact. Hence $F(X',Y') = 0$ by Theorem 7, and so $F(X,Y) = 0$ by Lemma 12(f).

(b) By Lemma 13(b).

(c) $F(S,\varnothing) = 1$ by Lemma 12(i). Hence $S \nvdash \varnothing$ by (a). So, due to the soundness and completeness of $\vdash$ relative to truth-functional consequence, S is logically consistent.

(d) Suppose A is a logical consequence of S. Then $S \vdash \{A\}$ by the completeness of $\vdash$. Hence $F(S,\{A\}) = 0$ by (a) and hence $F(\varnothing,\{A\}) = 0$ by Lemma 13(b). But then $F(\{A\},\varnothing) = 1$ by **F2** and **F3** and hence $A \in S$. ¤

So, the assumption set of any consistency function for L_P, hence any absolute probability function for L_P, is logically consistent and closed under logical consequence. The converse result, that any set of statements of L_P which is logically consistent and closed under logical consequence is the assumption set of at least one absolute probability function, will be obtained later. Next, we show that the statements in the assumption set of a consistency function indeed play the role of assumptions.

Theorem 14. *Let F be a binary function on the sets of statements of L_P which meets*

(i) **F1–F3**,

(ii) **F∧R**, **F∧L**, **F¬R**, *and* **F¬L**,

(iii) **FωR** *and* **FωL**;

let S be the assumption set of F; and let X and Y be finite sets of statements of L_P. Then $F(X,Y) = 0$ if and only if $S \cup X \vdash Y$.

Proof: (a) Suppose $F(X,Y) = 0$. Then $F(\varnothing,\{\wedge X \supset \vee Y\}) = 0$ by Lemma 13(i), and so $F(\{\wedge X \supset \vee Y\},\varnothing) = 1$ by **F2** and **F3**. Hence the statement $\wedge X \supset \vee Y$ is a member of S. Therefore $S \cup X \vdash Y$ by the completeness of $\vdash$.
(b) Suppose $S \cup X \vdash Y$. Then $F(S \cup X,Y) = 0$ by Theorem 13(a). Hence $F(X,Y) = 0$ by Lemma 13(b). ¤

And as a consequence

$$G(X,Y) = 1 \textit{ if and only if } S \cup X \vdash Y.$$

So, the statements in S function as suppressed premises in the sense that for finite X and Y $G(X,Y)$ takes the limiting value 1 just when Y (disjunctively understood) follows from X together with S (both X and S conjunctively understood). And in terms of the associated absolute probability function P' for L_P, the function for which $P'(A) = F(\{A\},\varnothing)$, this means that $P'(A) = 1$ if and only if $S \vdash \{A\}$, i.e. if and only if A is a logical consequence of S. Consequently, an arbitrary Popper absolute probability function P' for L_P accords to a statement A not the degree of logical necessity of A, but the degree to which A follows from the assumption set S of P'. $P'(A) = 1$ signifies that A is a logical consequence of S, while $P'(A) = 0$ signifies that A is logically inconsistent with S. Hence we say that any absolute probability function P' for L_P generalises the property of being a logical consequence of a certain set of assumptions which is specific to P'.

Construction of a Consistency Function with a Given S as Assumption Set

If S is any logically consistent set of statements of L_P which is closed under logical consequence, then, as we will now show, S is the assumption set of at least one consistency function, hence at least one absolute probability function, for L_P. In the light of Theorem 14 such a consistency function must meet the condition

$$F(X,Y) = 0 \textit{ if and only if } S \cup X \vdash Y.$$

A consistency function F_0 for the finite sets of atomic statements of L_P which meets this condition can be obtained by the following stipulation:

Let X and Y be finite sets of atomic statements of L_P, and let A be an arbitrary atomic statement of L_P. Then

(i) $F_0(\varnothing,\varnothing) = 1$

(ii) *If* $S \cup X \vdash Y \cup \{A\}$, *then* $F_0(X,Y \cup \{A\}) = 0$ *and* $F_0(\{A\} \cup X,Y) = F_0(X,Y)$

(iii) If $S \cup X \cup \{A\} \vdash Y$, *then* $F_0(X,Y \cup \{A\}) = F_0(X,Y)$ *and* $F_0(\{A\} \cup X,Y) = 0$

(iv) *If neither* $S \cup X \vdash Y \cup \{A\}$ *nor* $S \cup X \cup \{A\} \vdash Y$, *then*

$$F_0(X,Y \cup \{A\}) = 1/2 \times F_0(X,Y) \text{ and } F_0(\{A\} \cup X,Y) = 1/2 \times F_0(X,Y)$$

F_0 is a consistency function for the atomic statements of L_P, which can be extended to a consistency function F for L_P by the definition in Section 3. That F satisfies the condition

$$F(X,Y) = 0 \text{ if and only if } S \cup X \vdash Y$$

can be established by proofs analogous to those of Theorem 7 and Theorem 8.

The Set of Logical Truths as Assumption Set

Let now S be the set of logical truths of L_P, i.e. $S = \{A : \varnothing \vdash \{A\}\}$. S is a subset of the assumption set of every consistency function F, since $F(\varnothing,\{A\}) = 0$ by Theorem 7, and hence $F(\{A\},\varnothing) = 1$ by **F2** and **F3** for every logically true A. And S is itself the assumption set of a consistency function F if and only if F meets the constraint **FIND**. For if F meets **FIND**, then $F(\varnothing,\{A\}) = 0$ if and only if $\varnothing \vdash \{A\}$. And if S is the assumption set of F and X and Y are finite sets of atomic statements such that $F(X,Y) = 0$, then $S \cup X \vdash Y$ by Theorem 14, hence $X \vdash Y$, since S consists of logical truths, and hence $X \cap Y \neq 0$, which means that F meets **FIND**.

Theorem 15. *The set* $S = \{A : \varnothing \vdash \{A\}\}$ *is the assumption set of a consistency function F if and only if F meets the constraint* **FIND**, *and the associated absolute probability function P for sets of statements meets* **ASC**.

Theorem 16. *Let X and Y be finite sets of statements of* L_P, *let F be a consistency function for* L_P *and S be the assumption set of F, and let F' be a consistency function for* L_P *that meets* **FIND**. *Then* $F'(S \cup X,Y) = 0$, *if* $F(X,Y) = 0$.

Proof: Suppose $F(X,Y) = 0$. Then $S \cup X \vdash Y$ by Theorem 14, and $F'(S \cup X,Y) = 0$ by Theorem 13(a). ¤

The statements in the assumption set S of a given consistency function F are also tacit assumptions in the sense that there exists a consistency function F' for L_P which meets **FIND** and is such that $F(X,Y)$ equals $F'(S \cup X,Y)$ apart from a proportionality factor. In the light of Theorem 13(b) this is obvious if F itself meets **FIND**. So, assume that F does not satisfy **FIND**, let r be any real number larger than 0 but smaller than 1, let F_1 be a consistency function for L_P which meets **FIND**, and define

$$F'(X,Y) = r \times F(X,Y) + (1 - r) \times [F_1(X,Y) - F_1(S \cup X,Y)] / [1 - F_1(S,\varnothing)].$$

Since F does not meet **FIND**, its assumption set S does not consist exclusively of logical truths, as is clear from Theorem 15. Hence, by Theorem 9 and **F3**, $F_1(\{A\},\varnothing) \neq 1$ for at least one statement A in S, and so $F_1(S,\varnothing) \neq 1$ by Lemma 12(i). Hence the definition of F' is sound. F' is a consistency function. For $F'(X,Y) \geq 0$ by Lemma 12(e), and $F'(\varnothing,\varnothing) = 1$, since F and F_1 meet **F2**. Since F' is a linear combination of F and F_1, F' meets **F3, F¬R, F¬L, F∧R,** and **F∧L**. Furthermore, F' meets **FIND**. For let X and Y be finite sets of atomic statements of L_P and assume that $F'(X,Y) = 0$. Then $F(X,Y) = 0$ and $F_1(X,Y) - F_1(S \cup X,Y) = 0$. Hence $F_1(S \cup X,Y) = 0$ by Theorem 16 and so $F_1(X,Y) = 0$. Hence $X \cap Y \neq \varnothing$, since F_1 meets **FIND**.

Now let X and Y be any sets of statements of L_P. Then

$$F'(S \cup X,Y) = r \times F(S \cup X,Y).$$

But

$$F(S \cup X,Y) = F(X,Y)$$

by Theorem 13(b). So,

$$F'(S,\varnothing) = r,$$

by **F2**, and hence

$$F(X,Y) = F'(S \cup X,Y) \,/\, F'(S,\varnothing).$$

So, apart from a proportionality factor, the degrees of consistency attributed by F are the degrees of consistency *relative to the assumption set of* F attributed by some consistency function which satisfies constraint **FIND**. Or, if we focus on the corresponding Gentzen function G, i.e. the function

$$G(X,Y) = 1 - F(X,Y),$$

we see that the degree of consequence assigned by G is, apart from a proportionality factor, the degree of implication, according to some Gentzen function G' whose complement F' meets **FIND**, of Y by the union of the statements in X and those in the assumption set S. The set S functions as a set of unstated premises or background assumptions.

Focusing on the associated absolute probability functions P for sets of statements of L_P we have the corresponding result that there exists an absolute probability function P' which meets **ASC** and is such that

$$P(X) = P'(X \cup S) \,/\, P'(S),$$

where S is the assumption set of P.

Section 5. Consistency Functions and Absolute Probability Functions for L_ω

The infinitary language L_ω was introduced in Chapter 2, Section 1, and the characteristic constraint for absolute probability functions for L_ω was given there as

AP∧ $\quad P(\wedge \boldsymbol{A} \wedge B) = \lim_{n\to\infty} P((A_1 \wedge \ldots \wedge A_n) \wedge B).$

Here $\boldsymbol{A}$ is, as before, an empty, finite, or ω-sequence of statements of L_ω and $A_1, A_2, \ldots, A_n, \ldots$ are the elements of the sequence $\boldsymbol{A}$. On the other hand, for absolute probability functions on the sets of statements of L_ω we introduced in the same section the constraint

ASΛ $P(X \cup \{\Lambda A\}) = \lim_{n\to\infty} P(X \cup \{A_1,...,A_n\})$.

We now need to formulate constraints for multiple-conclusion consequence which correspond to the right- and left-introduction rules for 'Λ':

MΛR *If* $X \vdash Y \cup \{A_i\}$ *for* $i = 1, 2, ...,$ *then* $X \vdash Y \cup \{\Lambda A\}$

MΛL *If* $S(A) \cup X \vdash Y$, *then* $\{\Lambda A\} \cup X \vdash Y$,

where S(A) is the set of elements of the sequence A. **MΛR** and **MΛL** simply generalise the constraints for '$\wedge$' to infinitary conjunctions.[11] The same kind of generalisation dictates the constraints for consistency functions. In fact, as we now know that the auxiliary functions F_P of Chapter 5 are consistency functions, we adopt the identities proved for them in Theorem 8 of Chapter 5.

FΛR $F(X,Y \cup \{\Lambda A\}) =$

$$\lim_{n\to\infty} [F(X,Y \cup \{A_1\}) + F(\{A_1\} \cup X,Y \cup \{A_2\} + ... + F(\{A_1,...,A_{n-1}\} \cup X,Y \cup \{A_n\}]$$

FΛL $F(\{\Lambda A\} \cup X,Y) = \lim_{n\to\infty} F(\{A_1,...,A_n\} \cup X,Y)$

So, the consistency functions under consideration in this section are the ones which meet **F1–F3, F$\wedge$R, F$\wedge$L, F$\neg$R, F$\neg$L, FΛR, FΛL, FωR,** and **FωL.** In one-to-one correspondence with the consistency functions for L_ω stand the absolute probability functions for the sets of statements of L_ω, unary functions which meet the constraints **AS1, AS2, AS$\neg$, AS$\wedge$, ASΛ,** and **ASω.** Given a consistency function F for L_ω, the associated absolute probability function P is again defined by

$$P(X) = F(X,\varnothing),$$

while for given P the auxiliary function F_P constitutes the associated consistency function for L_ω. That P meets **ASΛ**, if F meets **FΛL**, is trivial; and that F_P meets **FΛL** and **FΛR**, if P meets **ASΛ**, follows by Lemmas 10(o) and 12(j).

Some of the results obtained in Sections 3 and 4 for the consistency functions for L_P can be replicated for the consistency functions for L_ω. Corresponding to Theorems 5 and 6 we have the result that a consistency function for the finite sets of statements of L_ω can be recursively defined from its restriction F_0 to finite sets of atomic statements.

Theorem 17. *Let F_0 be a function on the finite sets of atomic statements of L_ω which meets* **F1–F3**. *Then F_0 can be recursively extended to a function F for all finite sets of statements of L_ω which meets* **F1–F3, F$\wedge$R, F$\wedge$L, F$\neg$R, F$\neg$L, FΛR,** *and* **FΛL.**

Proof: According to Theorems 9 and 11 of Chapter 5 any function P_0 for the finite sets of atomic statements of L_ω which meets **AS1'** and **AS2** can be recursively extended to a function P for the sets of statements of L_ω which meets **AS1', AS2, AS$\neg$, AS$\wedge$,** and **ASΛ.** But we have shown in Section 1 that a function P_0 which meets **AS1'** and **AS2** corresponds to a consistency function F_0 which meets constraints **F1–F3.** ¤

Theorem 7 has this counterpart:

Theorem 18. *Let F be a binary function on the sets of statements of* L_ω *which meets*

(i) **F1–F3,**

(ii) **F∧R, F∧L, F¬R, F¬L, FΛR, FΛL,**

(iii) **FωR,** *and* **FωL.**

Then $F(X,Y) = 0$, *and hence* $G(X,Y) = 1$, *if* $X \vdash Y$.

Proof: By transfinite induction on the recursive specification of ⊢ by means of the Gentzen constraints **M1–M3, M∧R, M∧L, M¬R, M¬L, MΛR,** and **MΛL.**

Basis: $X \vdash Y$ holds by virtue of **M1.** Then $F(X,Y) = 0$ by Lemma 12(a).

Inductive Step: $X \vdash Y$ holds by virtue of one or another of the remaining constraints.

Cases 1–6 concern the constraints **M2, M3, M∧R, M∧L, M¬R,** and **M¬L**; they are dealt with as in the proof of Theorem 7.

Case 7: The constraint is **MΛR.** Then $F(X,Y \cup \{A_i\}) = 0$ by the inductive hypothesis, this for $i = 1, 2, \ldots$. Hence, by Lemma 12(f), $F(X,Y \cup \{A_1\}) = 0$, $F(\{A_1\} \cup X,Y \cup \{A_2\}) = 0, \ldots,$ $F(\{A_1,\ldots,A_{i-1}\} \cup X,Y \cup \{A_i\}) = 0, \ldots$. Hence $F(X,Y \cup \{\Lambda A\}) = 0$ by **FΛR.**

Case 8: The constraint is **MΛL.** Then $F(S(A) \cup X,Y) = 0$ by the inductive hypothesis. Hence $\lim_{n\to\infty} F(\{A_1,\ldots,A_n\} \cup X,Y) = 0$ by **FωL** and Lemma 12(k) and so $F(\{\Lambda A\} \cup X,Y) = 0$ by **FΛL.** ¤

There are no consistency functions for L_ω such that $F(X,Y) = 0$ only if $X \vdash Y$. However, whenever $X \nvdash Y$, there exists a consistency function F for L_ω such that $F(X,Y) \neq 0$, as the following considerations show. Assume $X \nvdash Y$. By the soundness and completeness of ⊢ there exists then a truth-value assignment α such that $\alpha(A) = 1$ for every statement A in X and $\alpha(B) = 0$ for every statement B in Y. The two-valued function

$$F(U,V) = \begin{cases} 1 & \textit{if } \alpha(A) = 1 \textit{ for every } A \textit{ in } U \textit{ and } \alpha(B) = 0 \textit{ for every } B \textit{ in } V \\ 0 & \textit{otherwise} \end{cases}$$

is easily seen to be a consistency function for L_ω, and obviously $F(X,Y) = 1$.

The definition of and major results about assumption sets can easily be extended to absolute probability functions for L_ω and their associated consistency functions. If P is an absolute probability function for the sets of statements of L_ω, F_P the associated consistency function, and

$$S = \{A : P(\{A\}) = 1\}$$

its assumption set, then S is logically consistent and closed under logical consequence,

$$P(X \cup S) = P(X),$$

and

$$F_P(S \cup X,Y) = F_P(X,Y).$$

Proofs of these results are like those for Theorem 13. The procedure in Section 4 can also be used to show that any set of statements of L_ω which is logically consistent and closed under logical consequence is the assumption set of at least one absolute probability function for the sets of statements of L_ω.

The following theorem is the counterpart of Theorem 14. Note that Theorem 14, concerning L_P, applies to finite sets of statements, while Theorem 19, concerning L_ω, applies to countable sets of statements.

Theorem 19. *Let F be a binary function on the sets of statements of* L_ω *which meets*

(i) **F1–F3**,

(ii) **F∧R, F∧L, F¬R, F¬L, FΛR, FΛL**,

(iii) **FωR**, *and* **FωL**;

let S be the assumption set of F; and let X and Y be countable sets of statements. Then $F(X,Y) = 0$ *if and only if* $S \cup X \vdash Y$.

Proof: (a) Let **X** be a sequence of statements whose elements are the members of X, let similarly **Y** be a sequence of the members of Y, and suppose $F(X,Y) = 0$. Then $F(\{\Lambda\mathbf{X} \supset \mathrm{V}\mathbf{Y}\},\emptyset) = 1$ by Lemma 13(i), **F2**, and **F3**. Hence the statement $\Lambda\mathbf{X} \supset \mathrm{V}\mathbf{Y}$ is a member of S. Therefore $S \vdash \{\Lambda\mathbf{X} \supset \mathrm{V}\mathbf{Y}\}$, and hence $S \cup X \vdash Y$ by the Gentzen constraints for $\vdash$.

(b) Suppose $S \cup X \vdash Y$. Then $F(S \cup X,Y) = 0$ by Theorem 18. Hence $F(X,Y) = 0$. ¤

Consequently, if P is an absolute probability function for the sets of statements of L_ω,

$$P(X) = 1 \textit{ if and only if } S \vdash \{A\} \textit{ for every statement } A \textit{ in } X$$

and, for countable X,

$$P(X) = 0 \textit{ if and only if } X \text{ is logically inconsistent with } S.$$

I.e. P, restricted to countable sets of statements, attributes degrees of logical necessity relative to its assumption set S and the limiting values 1 and 0 signify logical necessity and logical inconsistency, respectively, relative to S.

Section 6. Consistency Functions and Absolute Probability Functions for L_Q

The language L_Q, which was introduced in Chapter 2, Section 3, has a denumerable number of terms $t_1, t_2, \ldots, t_n, \ldots$, of which countably many are individual terms and denumerably many, labelled *additional terms*, are not.[12] As before, L_Q^0 consists of those among the statements of L_Q which contain only individual terms.

Extending our results from previous sections, we will show in this section that the Gentzen functions for L_Q generalise the relation of multiple-conclusion consequence for L_Q. The results for L_Q differ from our earlier results for L_P and L_ω in that two consequence relations for L_Q need to be considered.[13] The two relations $\vdash_o$ and $\vdash_s$ were explained semantically in Chapter 2, Section 3 as single-conclusion consequence relations for L_Q. What is required in this section are their multiple-conclusion counterparts, whose definitions are obvious modifications of the earlier ones:

$X \vdash_o Y$
if and only if
for any domain D, at least one statement in Y is true
on every D-interpretation on which every statement in X is true.

and

$X \vdash_s Y$
if and only if
at least one statement in Y is true on every truth-value assignment to the atomic statements of L_Q
on which every statement in X is true.

For the syntactic characterisation of the two consequence relations we invoke appropriate constraints, which correspond to Gentzen's introduction rules for multiple-conclusion consequence. In the case of $\vdash_o$ they are

M∀R *If* $X \vdash Y \cup \{A(T/x)\}$, *then* $X \vdash Y \cup \{(\forall x)A\}$, *provided that T is foreign to X, to Y, and to* $(\forall x)A$

M∀L *If* $\{A(T/x)\} \cup X \vdash Y$, *then* $\{(\forall x)A\} \cup X \vdash Y$

$\vdash_o$ is the relation that meets **M∀R** and **M∀L** in addition to the Gentzen constraints mentioned earlier in this chapter. When the quantifier is interpreted substitutionally, a variant of **M∀R** is required which allows in effect the inference of a universally quantified statement from the set of all its instances, to wit:

M∀R$_s$ *If* $X \vdash Y \cup \{A(t_i/x)\}$ *for* $i = 1, 2, \ldots$, *then* $X \vdash Y \cup \{(\forall x)A\}$

And we use **M∀L** in the handier form

M∀L$_s$ *If* $I((\forall x)A) \cup X \vdash Y$, *then* $\{(\forall x)A\} \cup X \vdash Y$,

to which it is equivalent, given **M2**. ($I((\forall x)A)$ is, as before, the set of substitution instances of $(\forall x)A$.) $\vdash_s$ is then the relation which meets **M∀R** and **M∀L** in addition to the Gentzen constraints mentioned earlier in this chapter.

We note these results for the multiple conclusion consequence relations $\vdash_o$ and $\vdash_s$, analogous to the single-conclusion results in Chapter 2, Section 3.

(1) *If* $X \vdash_0 Y$, *then* $X \vdash_s Y$,

(2) *If the statements in X and in Y belong to* L_Q^0, *then* $X \vdash_0 Y$ *if and only if* $X \vdash_s Y$.

Our constraints for consistency functions, already formulated in Theorem 7, Chapter 5, for the auxiliary function F_P, are these:

F∀R $F(X,Y \cup \{(\forall x)A\}) =$

$$\lim_{n\to\infty} [F(X,Y \cup \{A(t_1/x)\}) + F(\{A(t_1/x) \cup X\},Y \cup \{A(t_2/x)\} + \ldots$$
$$+ F(\{A(t_1/x),\ldots,A(t_{n-1}/x)\} \cup X,Y \cup \{A(t_n/x)\})]$$

F∀L $F(\{(\forall x)A\} \cup X,Y) = \lim_{n\to\infty} F(\{A(t_1/x),\ldots,A(t_n/x)\} \cup X,Y)$

We also recall the constraints concerning '∀' for absolute probability functions for the sets of statements of L_Q and for the statements of L_Q:

AP∀ $P((\forall x)A \wedge B) = \lim_{n\to\infty} P((A(t_1/x) \wedge \ldots \wedge A(t_n/x)) \wedge B)$

AS∀ $P(X \cup \{(\forall x)A\}) = \lim_{n\to\infty} P(X \cup \{A(t_1/x),\ldots,A(t_n/x)\})$

So, the consistency functions for L_Q which we consider in this section are the ones which meet **F1–F3, F∧R, F∧L, F¬R, F¬L, F∀R, F∀L, FωR,** and **FωL**. They stand in one-to-one correspondence with the absolute probability functions for the sets of statements of L_Q, unary functions which meet the constraints **AS1', AS2, AS¬, AS∧, AS∀,** and **ASω**. For a given consistency function F for L_Q, the associated absolute probability function P is, as before, defined by

$$P(X) = F(X,\varnothing),$$

while for given P the auxiliary function F_P constitutes the associated consistency function for L_Q. That P meets **AS∀**, if F meets **F∀L** is trivial; and that F_P meets **F∀L** and **F∀R**, if P meets **AS∀**, follows by Lemma 10(o) and Lemma 12(j).

Theorem 20. *Let* F_0 *be a function on the finite sets of atomic statements of* L_Q *meeting* **F1–F3**. *Then* F_0 *can be recursively extended to a function* F *for all finite sets of statements of* L_Q *which meets* **F1–F3, F∧R, F∧L, F¬R, F¬L, F∀R,** *and* **F∀L**.

Proof: According to Theorems 9 and 10 of Chapter 5 any function P_0 for the finite sets of atomic statements of L_Q which meets **AS1'** and **AS2** can be recursively extended to a function P for the sets of statements of L_Q which meets **AS1', AS2, AS¬, AS∧,** and **AS∀**. But we have shown in Section 1 that a function P_0 which meets **AS1'** and **AS2** corresponds to a consistency function F_0 which meets constraints **F1–F3**. ¤

Theorem 21. *Let F be a binary function on the sets of statements of* L_Q *which meets*

(i) **F1–F3**,

(ii) **F∧R, F∧L, F¬R, F¬L, F∀R, F∀L**,

(iii) **FωR**, *and* **FωL**.

Then

(a) *If X and Y are any sets of statements of* L_Q *and* $X \vdash_S Y$*, then* $F(X,Y) = 0$*, and hence* $G(X,Y) = 1$*;*

(b) *If X and Y are sets of statements of* L_Q^0 *and* $X \vdash_0 Y$*, then* $F(X,Y) = 0$.

Proof: (a) By transfinite induction on the recursive specification of $\vdash_S$ by means of the Gentzen constraints **M1–M3, M∧R, M∧L, M¬R, M¬L,** $\mathbf{M\forall R_S}$, $\mathbf{M\forall L_S}$.

Basis: $X \vdash_S Y$ holds by virtue of **M1**. Then $F(X,Y) = 0$ by Lemma 12(a).

Inductive Step: $X \vdash_S Y$ holds by virtue of one or other of the remaining constraints.

Cases 1–6 concern the constraints **M2, M3, M∧R, M∧L, M¬R,** and **M¬L**; they are dealt with as in the proof of Theorem 7.

Case 7: The constraint is $\mathbf{M\forall R_S}$. Then $F(X,Y \cup \{A(t_i/x)\}) = 0$ by the inductive hypothesis, this for $i = 1, 2, \ldots$. Hence $F(X,Y \cup \{A(t_1/x)\}) = 0$, $F(\{A(t_1/x)\} \cup X,Y \cup \{A(t_2/x)\}) = 0$, ..., $F(\{A(t_1/x),\ldots,A(t_{i-1}/x)\} \cup X,Y \cup \{A(t_i/x)\}) = 0$, ... by Lemma 12(f). Hence $F(X,Y \cup \{(\forall x)A\}) = 0$ by **F∀R**.

Case 8: The constraint is $\mathbf{M\forall L_S}$. Then $F(\mathrm{I}((\forall x)A) \cup X,Y) = 0$ by the inductive hypothesis. Hence $\lim_{n\to\infty} F(\{A(t_1/x),\ldots,A(t_n/x)\} \cup X,Y) = 0$ by **FωL** and Lemma 12(l) and so $F(\{(\forall x)A\} \cup X,Y) = 0$ by **F∀L**.

(b) By (a) and remark (2) above. ¤

So, when F is a consistency function for L_Q, $F(X,Y) = 0$ when $X \vdash Y$ on either understanding of $\vdash$. The converse is not generally the case, and the consistency functions for which it is the case that $X \vdash Y$ when $F(X,Y)$ cannot be characterised autonomously.

Results about assumption sets paralleling those established in Section 4 can be obtained for the absolute probability functions for L_Q. But since no new considerations are involved, we shall not pursue the matter.

Chapter 8

Relative Probability Functions Construed as Representing Degrees of Logical Consequence

Having interpreted absolute probability functions as generalising the property of logical necessity, it is natural to construe relative probability functions as generalising the relation of logical implication. Since logical consequence is a relation between a set of statements and a single statement, it is convenient to concentrate on relative probability functions defined on sets of statements, rather than individual statements. It turns out that a relative probability function P for a language L encapsulates a consequence relation $\vdash_P$, defined

$$X \vdash_P A \text{ \textit{if and only if} } P(\{A\},X \cup Z) = 1 \text{ \textit{for every set} } Z \text{ \textit{of statements}},$$

and a function $S_P(X)$ whose value for a set X of statements as argument is a set of *assumptions for* X. In general $\vdash_P$ is not the relation $\vdash$ of logical consequence for L, but a weaker relation, namely

$$X \vdash_P A \text{ \textit{if and only if} } S_P(\{\bot\}) \cup X \vdash A,$$

where '$\bot$' denotes some logically inconsistent statement. The assumption sets $S_P(X)$ are ordered by set inclusion, $S_P(\varnothing)$ being the largest and $S_P(\{\bot\})$ the smallest assumption set. $S_P(X)$ is logically compatible with X, as long as X is P-normal. If X is P-abnormal, $S_P(X) = S_P(\{\bot\}) = \{A : \varnothing \vdash_P A\}$. And the interesting connection of relative probability with logical consequence and logical inconsistency is this:

$$P(\{A\},X) = 1 \text{ \textit{if and only if} } S_P(X) \cup X \vdash A.$$

and

$$P(\{A\},X) = 0 \text{ \textit{if and only if} } S_P(X) \cup X \text{ \textit{is logically incompatible with} } A.$$

We conclude that $P(\{A\},X)$ is to be interpreted as the degree to which the statements in X together with the assumptions for X have A as a consequence.

As in the preceding chapter, we begin by formulating constraints on functions which can claim to constitute generalisations of logical consequence, said constraints themselves obtained by suitably adapting constraints that correspond to rules of inference due to Gentzen. This time, though, the rules in question are those governing the relation of single-conclusion consequence, among them the familiar introduction and elimination rules of systems of Natural Deduction.

Taking as starting point language-independent constraints on single-conclusion consequence, we arrive at language-independent constraints for relative probability functions, i.e. at constraints for an arbitrary language L. In effect, we are treating L as initially uninterpreted, but then partially interpreted by any particular relative probability function which meets those constraints, with the result that $\vdash_P$ becomes the logical consequence relation for L.

Considered next are relative probability functions for L_P, in particular the Carnap 1952 ones. To the language-independent constraints we add constraints for the connectives derived from Introduction- and Elimination-Rules of Natural Deduction. The consequence relation $\vdash_P$ implicit in a Carnap 1952 function P coincides with logical consequence for L_P. As already indicated, this is not the case generally for relative probability functions for L_P, which are studied in Sections 3 and 4. In the final section analogous results are obtained for relative probability functions for L_ω.

Section 1. Degrees of Logical Consequence in the Single-Conclusion Sense

Relative probability functions, we hold, can be construed as generalising the relation of logical consequence. More precisely, it is the relation of single-conclusion consequence, a relation between a set X of premises and a statement A, that we have in mind. In order to bring out the affinity between logical consequence and relative probability functions it is preferable to consider relative probability functions which take sets of statements rather than statements as their second arguments, the sets in question conjunctively understood. Moreover, the first argument should also be a set of statements, said set again understood conjunctively, rather than a single statement, this for the following reason. Single-conclusion consequence can easily be viewed as a special case of a relation between two sets of statements, both conjunctively understood, because Y is a consequence of X in this sense precisely when every member of Y is a consequence of X in the single-conclusion sense. Single-conclusion consequence is thus a special case of consequence between two conjunctively understood sets, one that nevertheless contains all the information there is to be had about the full consequence relation between sets of statements. But this feature does not extend to relative probabilities: the probability of a conjunctively understood set of statements can usually not be determined from the probabilities of the statements that belong to it. Hence in pursuing the parallelism between single-conclusion consequence and relative probability functions we will in the first instance deal with binary functions that are defined on sets of statements, both sets understood conjunctively. Our discussion will as a result lead us to the relative probability functions introduced in Chapter 1, Section 3; but to begin with we will abstract from the specific features of L_P that are due to the truth-functional connectives and consider only the general constraints appropriate for a binary function on sets of statements that generalises single-conclusion consequence.

Characteristics of Single-Conclusion Consequence

The following three constraints for single-conclusion consequence, similar to those given in the previous chapter for multiple-conclusion consequence, correspond to structural rules implicit in Natural Deduction systems:

S 1 *If $A \in X$, then $X \vdash A$* (*Overlap*)

S 2 *If $X \vdash A$, then $Y \cup X \vdash A$* (*Thinning*)

S 3 *If $X \vdash A$ and $\{A\} \cup X \vdash B$, then $X \vdash B$* (*Cut*)[1]

In contrast with multiple-conclusion consequence, the general characterisation of single-conclusion consequence for an arbitrary language does not straightforwardly yield a definition of inconsistency.[2] The usual definitions of inconsistency rely on the presence of a negation connective or of a self-contradictory statement $\bot$. One may declare a set X of statements inconsistent when $X \vdash \bot$ or when both $X \vdash A$ and $X \vdash \neg A$ for some statement A. This kind of definition is of course not available when a language-independent definition of inconsistency is contemplated.

In classical logic inconsistent sets of statements are just those which have every statement as a consequence. While we don't think that this captures the essence of inconsistency, and while the identification of inconsistency with triviality is explicitly rejected in many non-classical logics on the ground that inconsistency of a set is not sufficient for it to imply every statement, this characterisation appears to be the only option for a language-independent definition of inconsistency. So, we adopt

X is inconsistent if and only if $X \vdash A$, for every statement A.

We must be aware that there are also reasons for holding that this characterisation of inconsistency is not strong enough; a consistent set of statements could imply every statement of the language. For one, in a language all of whose statements are logical truths every set of statements implies every statement. While this is true, it appears best to exclude such languages from consideration by stipulating that at least one statement is not a consequence of the empty set. This means, given our definition of inconsistency, that the empty set is not inconsistent. This is as it should be; so we add this fourth constraint to **S1–S3**:

S 4 *$\varnothing \nvdash A$ for at least one statement A*

Even with **S4** in place it remains true that the above definition of inconsistency will at times declare inconsistent sets of statements which are in fact consistent. In particular, the set of all statements of a language will always be inconsistent under the definition, even though one can conceive of a language whose statements can all be true simultaneously.

Kolmogorov Relative Probability Functions for Uninterpreted Languages

Turning now to the task of generalising single-conclusion consequence and of describing degrees of consequence, let P be a binary function defined on the *finite* sets of statements of a language L. Our aim is to formulate constraints for P which are themselves generalisations of **S1–S4**. $P(Y,X)$ is to be understood initially as the degree of implication of Y by X or, what amounts to the same, the degree of necessity of Y relative to X, and, for the time being, we take it that $P(Y,X) = 1$ amounts to logical implication of each of the members of Y by X. For fixed

X, $P(Y,X)$ is a function that assigns degrees of necessity and is consequently subject to constraints **AS1'** and **AS2** in Chapter 7. **AS2**, however, is not strong enough; in its place we need a constraint ensuring that the degree of implication by X of any subset of X is maximal, i.e. equals 1. Availing ourselves again of the notation P_X to denote the X-restriction of the binary function P, to wit

$$P_X(Y) = P(Y,X),$$

we have therefore

RS1 *If* $Y \subseteq X$, *then* $P(Y,X) = 1$

RS2' $0 \leq F_{P_X}(U,V)$

RS2' places boundaries on P that are required by the expectation that every X-restriction P_X of P is finitely additive. Added to these two constraints has to be the generalisation of the *Cut* rule, namely the *Multiplication Law,* in the form

RS3 $P(Z \cup Y,X) = P(Z,Y \cup X) \times P(Y,X)$

The constraint, first formulated in Chapter 1, Section 4, establishes the required connection between X-restrictions for different sets X of statements of L. Lastly, we adopt

RS4 $P(Z,\emptyset) \neq 1$ *for at least one set* Z *of statements*

RS1 corresponds to **S1** and **RS4** to **S4**. **RS3** corresponds to and generalises **S3**. **S2** should give rise to

$$\textit{If } P(Y,X) = 1, \textit{ then } P(Y,X \cup Z) = 1,$$

which is at present not listed as a proposed constraint, but will be considered shortly. **RS2'** does not mirror any of the constraints for single-conclusion consequence; it addresses specifically relations among degrees of consequence and is not required when the degree of consequence can be only either 0 or 1. It is clear that we have ended up with constraints that appear only insignificantly different from the **RS** constraints introduced in Chapter 1, Section 3. It should be noted, however, that **RS2'** is crucial for the language-independent characterisation of P; the connective-free constraints **RS1–RS4** among **RS1–RS6** are not sufficient. In order to appreciate the four constraints adopted here, we need to be aware of their consequences; Lemma 16 in Appendix I collects some of them.

Consequence and inconsistency were conceived to be limiting cases of degrees of consequence. Hence, it is natural, and in the spirit of Chapter 7, to postulate that X logically implies Y according to P (X and Y conjunctively understood sets of statements) just when $P(Y,X) = 1$, and that X is logically inconsistent according to P just when $P(X,\emptyset) = 0$. But as a characterisation of inconsistency the latter is not strong enough, since it does not entail

$$P(Z,X) = 1 \textit{ for every set } Z \textit{ of statements},$$

i.e. *X is P-abnormal,* which is the translation of the definition adopted above as the best way to describe inconsistency in terms of single-conclusion consequence. Hence, the envisaged definition of consequence and of inconsistency commits us to

If $P(X,\varnothing) = 0$, *then* $P(Z,X) = 1$ *for every set Z of statements*,

which is the constraint **RSK** of Chapter 1, Section 4. We also note that, in the presence of the other **RS** constraints, **RSK** implies the counterpart of **S2**:

If $P(Y,X) = 1$, *then* $P(Y,X \cup Z) = 1$.

For suppose that $P(Y,X) = 1$. Then $P(Y,X \cup Z) = 1$, if $P(Z,X) \neq 0$, by Lemma 16(h); and if $P(Z,X) = 0$, then $P(Y,X \cup Z) = 1$ by **RSK**. Consequently, given **RSK**, the **RS** counterpart of **S2** need not be formulated separately.

The generalisation of single-conclusion consequence encapsulated in the constraints **RS1**, **RS2'**, **RS3**, **RS4**, and **RSK**, though entirely feasible, adds nothing new to what we have already met with in Chapter 7. In fact, the binary functions on the finite sets of statements of L that meet these constraints can equally well be characterised by the definitional identity

$$(1)\quad P(Y,X) = \begin{cases} P(Y \cup X,\varnothing) \,/\, P(X,\varnothing) & \textit{if } P(X,\varnothing) \neq 0 \\ 1 & \textit{otherwise} \end{cases}$$

together with

$$(2)\quad P_{\varnothing}(\varnothing) = 1$$

and

$$(3)\quad 0 \leq F_{P_{\varnothing}}(U,V).$$

For proof assume first that P meets **RS1**, **RS2'**, **RS3**, **RS4**, and **RSK**. It is then immediate from **RS1** and **RS2'** that $P(\varnothing,\varnothing) = 1$ and $0 \leq F_{P_{\varnothing}}(U,V)$, while by **RS3** $P(Y,X) = P(Y \cup X,\varnothing) \,/\, P(X,\varnothing)$ when $P(X,\varnothing) \neq 0$, and by **RSK** $P(Y,X) = 1$ when $P(X,\varnothing) = 0$. Next, assume (1)–(3) and let $Y \subseteq X$. Then $P(Y,X)$ equals $P(X,\varnothing) \,/\, P(X,\varnothing)$ when $P(X,\varnothing) \neq 0$ and equals 1 when $P(X,\varnothing) = 0$. Hence, if $Y \subseteq X$, $P(Y,X)$ equals 1 whether or not $P(X,\varnothing) = 0$, i.e. P meets **RS1**. This done, we show by an inductive argument on the cardinality of V that

$$F_{P_X}(U,V) = \begin{cases} F_{P_{\varnothing}}(X \cup U,V) \,/\, P(X,\varnothing) & \textit{if } P(X,\varnothing) \neq 0 \\ 1 & \textit{if } P(X,\varnothing) = 0 \textit{ and } V = \varnothing \\ 0 & \textit{if } P(X,\varnothing) = 0 \textit{ and } V \neq \varnothing. \end{cases}$$

As regards the basis of the argument we note that $F_{P_X}(U,V)$ equals $P(U,X)$, which by (1) equals $P(X \cup U,\varnothing) \,/\, P(X,\varnothing)$, i.e. $F_{P_{\varnothing}}(X \cup U,\varnothing) \,/\, P(X,\varnothing)$, when $P(X,\varnothing) \neq 0$ and equals 1 when $P(X,\varnothing) = 0$. As to the inductive step, we recall that by the definition of F_{P_X}

$$F_{P_X}(U,V) = F_{P_X}(U,V-\{A\}) - F_{P_X}(\{A\} \cup U,V-\{A\}),$$

where A is the alphabetically latest member of V. So, by the inductive hypothesis, $F_{P_X}(U,V)$ equals

$$F_{P_{\varnothing}}(X \cup U,V-\{A\}) \,/\, P(X,\varnothing) - F_{P_{\varnothing}}(X \cup \{A\} \cup U,V-\{A\}) \,/\, P(X,\varnothing),$$

i.e. $F_{P_\varnothing}(X \cup U,V) / P(X,\varnothing)$, if $P(X,\varnothing) \neq 0$, and equals 0 if $P(X,\varnothing) = 0$. Given this result, it follows by (3) that $0 \leq F_{P_X}(U,V)$, i.e. **RS2'**. As for the proof of **RS3**, assume first that $P(Y \cup X,\varnothing) \neq 0$, in which case $P(X,\varnothing) \neq 0$ by Lemma 9(f), and so **RS3** by (1). Assume next that $P(Y \cup X,\varnothing) = 0$, but $P(X,\varnothing) \neq 0$. Then $P(Z \cup Y \cup X,\varnothing) = 0$ by Lemma 9(f), and hence **RS3** again by (1). Finally, assume that $P(Y \cup X,\varnothing) = P(X,\varnothing) = 0$, in which case (1) again yields **RS3**. And as to **RSK**, if $P(Y,X) = 0$, then $P(Y \cup X,\varnothing) = 0$ by the definition of $P(Y,X)$ and hence $P(Z,Y \cup X) = 1$ by that definition; i.e. P meets **RSK**.

The foregoing constraints (2) and (3) are of course the constraints **AS2** and **AS1'** for absolute probability functions on sets of statements formulated in Chapter 7. So the functions characterised by **RS1**, **RS2'**, **RS3**, **RS4**, and **RSK** are those relative probability functions that can be obtained from absolute probability functions P' via the definition

$$P(Y,X) = \begin{cases} P'(Y \cup X) / P'(X) & \text{if } P'(X) \neq 0 \\ 1 & \text{otherwise} \end{cases}$$

They are therefore what we have called *Kolmogorov functions*.

Relative Probability Functions for Uninterpreted Languages

Kolmogorov functions constitute but a subclass of all Popper functions, we pointed out in Chapter 3. Can Popper functions generally be interpreted as generalising the consequence relation? In order to do this, it turns out, we must identify

A is a logical consequence of X

not with

$$P(\{A\},X) = 1,$$

but with

$$P(\{A\},X \cup Z) = 1 \textit{ for every set } Z \textit{ of statements,}$$

which we abbreviate to

$$X \vdash_P A;$$

and, consequently,

X is an inconsistent set of statements

must be identified with

$$P(X,Z) = 0 \textit{ for every set } Z \textit{ such that } P(W,Z) \neq 1 \textit{ for at least one set } W,$$

i.e.

$$P(X,Z) = 0 \textit{ for every } P\textit{-normal } Z$$

(*X is P-inconsistent*, for short). For sets Y other than singletons

$$P(Y,X \cup Z) = 1 \textit{ for every } Z$$

then amounts to

$$X \vdash_P A \textit{ for every statement } A \textit{ in } Y,$$

this by virtue of Lemma 16(k).

S2 is now guaranteed by the very definition of $\vdash_P$, and P-inconsistency implies P-abnormality, given **RS1**. In fact

$$P(X,Z) = 0 \textit{ for every P-normal } Z$$

is stronger than

$$P(Z,X) = 1 \textit{ for every } Z,$$

and is in effect a more satisfactory rendering of inconsistency. It can distinguish between an inconsistent set of statements and one which entails every statement of the language without being inconsistent. Which means that the present explication of logical notions in terms of probability functions overcomes the problem of finding a satisfactory account of inconsistency in terms of single-conclusion consequence.

When $\vdash_P$ and P-inconsistency are adopted as characterisations of consequence and inconsistency, respectively, in terms of a function P, only the constraints **RS1**, **RS2'**, **RS3**, and **RS4** are needed to characterise that function. We have

Theorem 1. *Let P be a binary function on the finite sets of statements of L which meets constraints* **RS1**, **RS2'**, **RS3**, *and* **RS4**. *Then the relation* $\vdash_P$ *meets* **S1–S4**.

Proof: (a) If A is a member of X, A is a member of $X \cup Z$. Hence $P(\{A\},X \cup Z) = 1$ for every Z by **RS1**. So, **S1**. (b) **S2** is immediate from the definition of $\vdash_P$. (c) Assume $P(\{A\},X \cup Z) = 1$ for every Z and $P(\{B\},\{A\} \cup X \cup Z) = 1$ for every Z. Then $P(\{B,A\},X \cup Z) = P(\{B\},\{A\} \cup X \cup Z) \times P(\{A\},X \cup Z)$ by **RS3**. Hence $P(\{B,A\},X \cup Z) = 1$. But $P(\{B,A\},X \cup Z) \leq P(\{B\},X \cup Z) \leq 1$ by Lemmas 16(b) and 16(c). Hence **S3**. (d) **S4** by **RS1** and Lemma 16(k). ¤

However, it is now no longer convincing to call $P(\{A\},X)$ the degree to which the statements in X imply A, since consequence itself ($X \vdash_P A$) is not the same as $P(\{A\},X) = 1$ and is therefore only in a forced sense a limiting case of $P(\{A\},X)$. In the next section we will see that in the case where L is the language of the propositional calculus, $P(\{A\},X)$ is the degree to which the statements in X together with the statements in a certain set $S_P(X)$ have A as consequence, where $S_P(X)$ is defined thus:

$$S_P(X) = \{C : P(\{C\},Z) = 1 \textit{ for every set } Z \textit{ such that } P(Z,X) = 1\}.$$

Then $P(\{A\},X) = 1$ if and only if $X \cup S_P(X)$ logically implies A and $P(\{A\},X) = 0$ if and only if A is logically inconsistent with $X \cup S_P(X)$. As long as we are confining ourselves to language-independent constraints, we can only prove that $P(\{A\},X) = 1$ if $X \cup S_P(X) \vdash_P A$ and that $P(\{A\},X) = 0$ if A is P-inconsistent with $X \cup S_P(X)$. The interest of the result that $P(\{A\},X)$ is the degree to which $X \cup S_P(X)$ logically implies A will lie in showing that $S_P(X)$ can be given a persuasive interpretation and that it is not on account of $X \cup S_P(X)$ being inconsistent that A is implied by $X \cup S_P(X)$.

It would appear that the modified understanding of $P(\{A\},X)$, as the degree to which X

together with a certain set of statements implies A, undermines the strategy we adopted of formulating constraints for relative probability functions by generalising the constraints on single-conclusion consequence, because we operated with the understanding that $P(\{A\},X)$ is the degree to which X by itself implies A. In fact, this is not so; for **S1-S3** remain correct, when X is replaced by $X \cup S_P(X)$.

By identifying logical consequence for L with $\vdash_P$, allowing P to be any binary function on the finite sets of statements of L which meets **RS1, RS2', RS3**, and **RS4**, we in effect treat L as uninterpreted, i.e. we treat the statements of L as not having meanings that would determine their logical relations. Every function P interprets the statements of L to the extent that is required for them to stand in definite relations of logical implication, but to that extent only: the information contained in such a function does not specify truth-conditions for the statements in L.

Given any uninterpreted language L, the constraints **RS1**, **RS2'**, **RS3**, and **RS4** delineate the class of functions that express degrees of consequence in the single-conclusion sense. It is also the class of relative probability functions for L. No proof of this claim, beyond its intrinsic plausibility and the observation above that **RS1**, **RS2'**, **RS3**, **RS4** plus **RSK** single out relative probability functions of the Kolmogorov kind, is possible here, since probability functions for uninterpreted languages have not been previously considered. In our view, **RS1**, **RS2'**, **RS3**, and **RS4** constitute the language-independent constraints for relative probability functions on finite sets of statements. Further constraints will then be concerned with particular features of the syntax of the language in question, typically the sentential connectives. Usually relative probability functions take statements as arguments, rather than sets of statements, and it is such functions that we have studied for the most part. They can be obtained from the present ones by restricting their arguments to singleton sets.

Infinite sets of Statements

Allowing now infinite sets of statements as arguments in our functions, we add the constraint

RSω $P(Y,X) = \inf_{Y' \subseteq L} \{P(Y',X) : Y' \subseteq_f Y\}$.

RSω results from the by now familiar thought that the value of P for an infinite, conjunctively understood set Y is determined by its values for the finite subsets Y' of Y and, since $P(Z \cup Y,X) \leq P(Y,X)$ by Lemma 16(b), should be the infimum of those values. This idea cannot be extended to the second argument, since $P(Y,X \cup Z)$ may be either larger or smaller than $P(Y,X)$. The proof of Theorem 1 goes through without the restriction to finite sets: when P takes infinite sets of statements as well as finite ones as arguments, $\vdash_P$ still meets **S1-S4**. **RSω** has the character of a definition that plays a role in extending a function P' defined on finite sets of statements to infinite sets.

Theorem 2. *Let P' be a binary function defined on the finite sets of statements of L as far as the first argument is concerned, and on all sets of statements of L as far as the second argument is concerned, a function which meets constraints* **RS1, RS2', RS3,** *and* **RS4;** *and let P be the extension of P' to all sets of statements of L by means of the definition*

$$P(Y,X) = \begin{cases} P'(Y,X) & \text{if } Y \text{ is finite} \\ \inf_{Y'} \{P'(Y',X) : Y' \subseteq_f Y\} & \text{otherwise.} \end{cases}$$

Then P meets **RS1, RS2', RS3,** *and* **RS4.**

Proof:

RS1 RS2', and **RS4**: Obvious.

RS3: By Lemma 16(b) and the fact that P' meets **RS3.** ¤

Section 2. Carnap's 1952 Relative Probability Functions Interpreted as Generalising Logical Consequence

Having explored for an unspecified language L the view that relative probability functions are formally to be regarded as generalisations of the relation of (single-conclusion) consequence, we now turn our attention to the propositional language L_P. As relative probability functions for L_P have been thoroughly studied, our claim that they can be construed as generalisations of logical consequence is open to proof. We will in this section formulate constraints for the logical connectives which generalise the introduction and elimination rules employed in Natural Deduction versions of propositional logic, which, of course, also go back to Gentzen (1934–35). We list here constraints on the relation of logical consequence which correspond to a version of Natural Deduction rules that makes use of '$\bot$'. As before, we take $\bot$ to be $A \wedge \neg A$ for some fixed but arbitrary statement A. And '$\vdash$' now denotes the consequence relation for L_P.

S∧I *If* $X \vdash A$ and $X \vdash B$, *then* $X \vdash A \wedge B$

S∧E *If* $X \vdash A \wedge B$, *then* $X \vdash A$ *and* $X \vdash B$

S¬I *If* $\{A\} \cup X \vdash \bot$, *then* $X \vdash \neg A$

S¬E *If* $X \vdash \neg\neg A$, *then* $X \vdash A$

S⊥I *If* $X \vdash A$ *and* $X \vdash \neg A$, *then* $X \vdash \bot$

S⊥E *If* $X \vdash \bot$, *then* $X \vdash A$

These constraints, together with **S1–S4**, which were introduced in Section 1, give us an implicit characterisation of the (single-conclusion) consequence relation for L_P, on the understanding that the constraints other than **S4** are in effect rules of inference by means of which every instance of the consequence relation can be recursively obtained from basic instances, namely the instances of **S1**.

Constraints Generalising **S∧I, S∧E, S¬I, S¬E, S⊥I,** *and* **S⊥E**

Our first aim is to show that the functions P on finite sets of statements of L_P that meet **RS1, RS2', RS3, RS4**, and the constraints for connectives to be introduced next, generalise the relation of single-conclusion consequence for L_P and can be construed as assigning degrees of consequence. We prove, in particular, that

$$P(\{A\},X \cup Z) = 1 \textit{ for every set } Z \textit{ of statements,}$$

$X \vdash_P A$ for short, if and only if

$$X \vdash A.$$

We are also able to prove that

$$P(\{A\},X \cup Z) = 0 \textit{ for every set } Z \textit{ such that } X \cup Z \textit{ is logically consistent}$$

if and only if

$$\{A\} \cup X \textit{ is logically inconsistent.}$$

With '⊥' at hand we can express consistency and inconsistency in particularly simple ways. Namely,

$$P(\{\bot\},X) = 0$$

when X is logically consistent and

$$P(\{\bot\},X) = 1$$

when X is logically inconsistent. Hence X is P-normal if consistent, P-abnormal if inconsistent. The constraints for '∧', '¬', and '⊥' are these:

RS∧ $P(\{A \wedge B\},X) = P(\{A,B\},X)$

RS¬ $P(\{\neg A\},X) = 1 - P(\{A\},X) + P(\{\bot\},X)$

RS⊥ *If* $P(\{\bot\},X) \neq 0$, *then* $P(\{A\},X) = 1$

The observation made in the previous section that $P(\{A\},X)$ will turn out to be not the degree to which X by itself implies A, but the degree to which X together with a certain set $S_P(X)$ of statements implies A, does not affect the attempt to obtain constraints for degrees of consequence from Natural Deduction rules. For the latter are Introduction- and Elimination-Rules which affect statements on the conclusion side of the consequence relation, and replacing X by $X \cup S_P(X)$ does not affect their correctness.

To the constraints for connectives we have to add a further constraint that records the logical independence of the atomic statements. We have already pointed out in Chapter 7 that the independence of atoms is always assumed, often tacitly, in propositional logic and emerges in the practice that allows the assignment of any combination of truth-values to the atomic statements as a possible assignment.

IND *If* X *and* Y *are sets of atomic statements and* $X \cap Y = \varnothing$, *then* $P(\{\bot\}, X \cup Y^{\neg}) = 0$[3]

which amounts to saying that $X \cup Y^{\neg}$ is consistent, as long as $X \cap Y = \varnothing$.

By Lemmas 17(g), 17(h), and 17(i) the identities

$$P(\{\neg A\}, X) = 1 - P(\{A\}, X) + P(\{B \wedge \neg B\}, X)$$

and

$$\textit{If } P(\{B \wedge \neg B\}, X) \neq 0, \textit{ then } P(\{A\}, X) = 1$$

can be deduced from the constraints above. They show that the constraints for connectives are not in conflict with the stipulation that $\bot$ be $A \wedge \neg A$ for an arbitrary statement A. It is also instructive at this point to list the constraints for the defined connectives '$\vee$' and '$\supset$', for proof of which see Lemmas 17(o) and 17(p). They are:

RS$\supset$ $P(\{A \supset B\}, X) = 1 - P(\{A\}, X) + P(\{B\}, X \cup \{A\}) \times P(\{A\}, X)$

RS$\vee$ $P(\{A \vee B\}, X) = P(\{A\}, X) + P(\{B\}, X) - P(\{A, B\}, X)$

$\vdash_P$ *compared with* $\vdash$

In order to prove that the constraints **RS$\wedge$**, **RS$\neg$**, and **RS$\bot$** are indeed generalisations of the constraints on the consequence relation $\vdash$ listed above, we show in Theorem 3 that if P meets these constraints, then $\vdash_P$ meets the corresponding conditions placed on the consequence relation.

Theorem 3. *Let* P *be a function on the finite sets of statements of* L_P *that meets*

(i) **RS1**, **RS2'**, **RS3**, *and* **RS4**,

(ii) **RS$\wedge$**, **RS$\neg$**, *and* **RS$\bot$**.

Then $\vdash_P$ *meets* **S$\wedge$I**, **S$\wedge$E**, **S$\neg$I**, **S$\neg$E**, **S$\bot$I**, *and* **S$\bot$E**.

Proof: (a) $P(\{A\}, X) = P(\{B\}, X) = 1$ if and only if $P(\{A, B\}, X) = 1$ (by Lemma 16(k)), hence if and only if $P(\{A \wedge B\}, X) = 1$ (by **RS$\wedge$**). Hence **S$\wedge$I** and **S$\wedge$E**.

(b) By Lemma 17(d) constraint **RS$\neg$** can be given the equivalent form

$$P(\{\neg A\}, X) = 1 - P(\{A\}, X) + P(\{\bot\}, X \cup \{A\}) \times P(\{A\}, X).$$

So, on the assumption that $P(\{\bot\}, X \cup \{A\}) = 1$, $P(\{\neg A\}, X) = 1$. Hence **S$\neg$I**.

(c) Assume $P(\{\neg\neg A\}, X) = 1$. Then $P(\{A\}, X) = 1$ by Lemma 17(n). Hence **S$\neg$E**.

(d) Assume $P(\{A\}, X) = P(\{\neg A\}, X) = 1$. Then $P(\{\bot\}, X) = 1$ by **RS$\neg$**. Hence **S$\bot$I**.

(e) Assume $P(\{\bot\}, X) = 1$. Then $P(\{A\}, X) = 1$ by **RS$\bot$**. Hence **S$\bot$E**. ¤

We now show first that $X \vdash_P A$, if $X \vdash A$, and then, for functions meeting **IND**, that $X \vdash A$, if $X \vdash_P A$.

Theorem 4. *Let* P *be a function on the finite sets of statements of* L_P *that meets*

(i) **RS1**, **RS2'**, **RS3**, *and* **RS4**,

(ii) **RS$\wedge$**, **RS$\neg$**, *and* **RS$\bot$**.

Then

(a) *If* $X \vdash A$, *then* $P(\{A\},X \cup Z) = 1$ *for every set Z of statements of* L_P;

(b) *If* $P(\{\bot\},X) = 0$, *then X is logically consistent;*

(c) *If X is logically inconsistent, then* $P(Y,X) = 1$ *for every Y;*

(d) *If* $Y \cup X$ *is logically inconsistent, then* $P(Y,X \cup Z) = 0$ *for every Z such that* $P(\{\bot\},X \cup Z) = 0$.

(e) *If* $\{A\} \cup X$ *is logically inconsistent, then* $P(\{A\},X \cup Z) = 0$ *for every Z such that* $P(\{\bot\},X \cup Z) = 0$.

Proof: (a) By induction on the recursive specification of $\vdash$.

Basis: $X \vdash A$ by **S1**. Then $P(\{A\},X \cup Z) = 1$ for every Z by Theorem 1.

Inductive Step: $X \vdash A$ by one or another of the recursive constraints. If **SC2** or **SC3** is invoked, then $P(\{A\},X \cup Z) = 1$ for every Z by the inductive hypothesis and Theorem 1. If an introduction or elimination constraint is invoked, then $P(\{A\},X \cup Z) = 1$ for every Z by the inductive hypothesis and Theorem 3.

(b) Suppose X is logically inconsistent, i.e. $X \vdash \bot$. Then $P(\{\bot\},X) = 1 \neq 0$ by (a).

(c) Suppose X is logically inconsistent. Then $P(\{\bot\},X) = 1$ by (b), hence $P(\{\bot\} \cup Y,X) = 1$ by Lemma 17(c), and hence $P(Y,X) = 1$ by Lemma 16(f).

(d) Suppose $Y \cup X \vdash \bot$, and $P(\{\bot\},X \cup Z) = 0$. Then $P(\{\bot\},Y \cup X \cup Z) = 1$ by (a). Hence $P(Y,X \cup Z) = 0$ by Lemma 17(k).

(e) By (d). ¤

Theorem 5. *Let P be a binary function on the finite sets of statements of* L_P *which meets*

(i) **RS1**, **RS2'**, **RS3**, *and* **RS4**,

(ii) **IND**,

(iii) **RS**$\wedge$, **RS**$\neg$, *and* **RS**$\bot$.

Then:

(a) *If* $P(\{A\},X \cup Z) = 1$ *for every set Z of statements, then* $X \vdash A$;

(b) *If* $P(\{A\},Z) = 1$ *for every set Z of statements, then* $\vdash A$;

(c) *If X is logically consistent, then* $P(\{\bot\},X) = 0$;

(d) *If* $P(Y,X) = 1$ *for every Y, then X is logically inconsistent;*

(e) *If* $P(\{A\},X \cup Z) = 0$ *for every Z such that* $P(\{\bot\},X \cup Z) = 0$, *then* $\{A\} \cup X$ *is logically inconsistent.*

Proof: (a) Suppose $X \nvdash A$. Then by the truth-functional soundness and completeness of '$\vdash$' there exists an assignment α of truth-values to the atomic statements occurring in $\{A\} \cup X$ such that $\alpha(A) = 0$ and $\alpha(B) = 1$ for every statement B in X. Let U be the set of atomic statements that have truth-value 1, and V the set of those with truth-value 0 under that assignment. Then (i) $U \cup V^{\neg} \vdash B$ for every member B of X and hence $P(\{B\},U \cup V^{\neg}) = 1$ for every such B, this by Theorem 4(a), and so $P(X,U \cup V^{\neg}) = 1$ by Lemma 16(k); and (ii) $U \cup V^{\neg} \cup \{A\}$ is logically inconsistent and hence by **IND** and Theorem 4(e) $P(\{A\},U \cup V^{\neg}) = 0$. Hence by

RS3 $P(\{A\} \cup X, U \cup V^{\neg}) = 0$ and, since

$$P(\{A\} \cup X, U \cup V^{\neg}) = P(\{A\}, X \cup U \cup V^{\neg}) \times P(X, U \cup V^{\neg})$$

by **RS3**, $P(\{A\}, X \cup U \cup V^{\neg}) = 0$. Consequently, it is not the case that $P(\{A\}, X \cup Z) = 1$ for every set Z.

(b) By (a).

(c) Suppose $P(\{\bot\}, X) \neq 0$, i.e. $P(\{\bot\}, X) = 1$ by **RS⊥**. Then $P(\{\bot\}, X \cup Z) = 1$ for every Z by Lemma 17(j) and $X \vdash \bot$ by (a), i.e. X is logically inconsistent.

(d) By (b).

(e) Suppose $\{A\} \cup X$ is logically consistent. Then $P(\{\bot\}, \{A\} \cup X) = 0$ by (b) but $P(\{A\}, \{A\} \cup X) = 1$ by **RS1**. Hence there exists a set Z, namely $\{A\}$, such that $P(\{\bot\}, X \cup Z) = 0$ but $P(\{A\}, X \cup Z) \neq 0$. ¤

Carnap 1952 Functions

Before we further pursue the theme of degrees of consequence we need to make good our claim that the functions described here are relative probability functions. More specifically, the functions meeting **RS1**, **RS2'**, **RS3**, **RS4**, **RS∧**, **RS¬**, and **RS⊥** are the relative probability functions on sets of statements of L_P, while those meeting **IND** as well are the Carnap 1952 functions among them. We need only compare the present constraints with those originally formulated in Chapter 1, Section 3:

RS1	*If* $Y \subseteq X$, *then* $P(Y,X) = 1$
RS2	$0 \leq P(Y,X)$
RS3	$P(Z \cup Y, X) = P(Z, Y \cup X) \times P(Y,X)$
RS4	$\varnothing$ *is P-normal*
RS5	*If X is P-normal, then* $P(\{\neg A\},X) = 1 - P(\{A\},X)$
RS6	$P(\{A \wedge B\},X) = P(\{A,B\},X)$

as well as

RSC52 *If* $P(Y,X) = 1$ *for every set Y, then X is logically inconsistent*

for the Carnap 1952 functions. The main difference between the two sets of constraints is that firstly those among the present constraints that do not involve connectives provide a language-independent characterisation of probability functions; and, secondly, that the constraint **IND**, as opposed to **RSC52**, does not invoke semantic notions.[4]

Assume then that P is any function meeting **RS1**, **RS2'**, **RS3**, **RS4**, **RS∧**, **RS¬**, and **RS⊥**. Then P meets **RS2** by Lemma 16(c). As to **RS5**, if X is P-normal, then $P(\{\bot\},X) = 0$ by **RS⊥** and so $P(\{\neg A\},X) = 1 - P(\{A\},X) = 0$ by **RS¬**. And **RS6** is identical with **RS∧**. So P meets all of **RS1–RS6**.

On the other hand, assume that P meets **RS1–RS6**. Then:

RS2': That the X-restriction P_X of P meets **AS1, AS2, AS¬**, and **AS∧** has been proved in Chapter 5, Section 2. Hence $0 \leq F_{P_X}(Y,Z)$ by Lemma 10(j).

RS¬: Since $P(\{\bot\},X)$ equals 0 if X is P-normal, and equals 1 if X is P–abnormal, $P(\{\neg A\},X) = 1 - P(\{A\},X) + P(\{\bot\},X)$ in either case, by virtue of **RS5**.

RS⊥: $P(\{\bot\},X) = 0$ if X is P-normal by Lemma 15(h).

So, P meets all of **RS1, RS2', RS3, RS4, RS∧, RS¬**, and **RS⊥**.

We have established that **RS1, RS2', RS3, RS4, RS∧, RS¬**, and **RS⊥** characterise the relative probability functions on the finite sets of statements of L_P. To see that **IND** and **RSC52** single out the same subclass of relative probability functions, note first that, given **IND, RSC52** holds by Theorem 5(d). Secondly, if X and Y are sets of atomic statements and $X \cap Y = \emptyset$, then $X \cup Y^{\neg}$ is logically consistent. Hence $X \cup Y^{\neg}$ is P-normal by **RSC52**, and so $P(\{\bot\},X \cup Y^{\neg}) = 0$ by Lemma 15(h). Consequently, it is Carnap 1952 functions one obtains by generalising logical consequence to degree of consequence and assuming the independence of atomic statements. For the relations $\vdash$ and $\vdash_P$ coincide.

Section 3. The Assumption Sets of Relative Probability Functions

We have obtained the result that for Carnap 1952 functions the relation $\vdash_P$ ($X \vdash_P A$ defined as $P(\{A\},X \cup Z) = 1$ *for every* Z) coincides with the relation $\vdash$ of logical consequence for L_P (Theorems 4 and 5). Since the relative probability functions on the statements of L_P are interdefinable with the relative probability functions on the finite sets of statements of L_P, as we saw in Chapter 1, Section 3, we also have the result that, given a relative probability function P' for L_P meeting **RC52**,

$$P'(A,B \wedge C) = 1 \textit{ for every statement } C$$
$$\textit{if and only if}$$
$$B \textit{ logically implies } A.$$

Assumption Sets

But it is not plausible to regard $P(\{A\},X)$ as the degree to which X implies A, seeing that the highest degree — $P(\{A\},X) = 1$ — does not amount to implication (of A by X), nor the lowest — $P(\{A\},X) = 0$ — to incompatibility of A with X. When $P(\{A\},X) = 1$ even though $X \vdash_P A$ does not hold, and $P(\{B\},X) = 0$ even though B is not incompatible with X, it is not X by itself which implies A or is incompatible with B but rather X in conjunction with further statements, statements which depend on X. We introduce a function S_P which, for given P and X identifies these statements and define

$$S_P(X) = \{C : P(\{C\},Z) = 1 \textit{ for every set } Z \textit{ such that } P(Z,X) = 1\}.$$

We proceed to show that then $P(\{A\},X)$ can be understood as the degree to which A is a consequence of $X \cup S_P(X)$. Remarkably, this result obtains for all of Popper's relative probability functions defined on finite sets of statements, not only for the Carnap 1952 functions among them.

As we shall see, a special role is played by the set $S_P(\{\perp\})$. This set has been called *the assumption set* of the relative probability function P and has been studied in Leblanc and Morgan (1984). We now prefer to call all the sets $S_P(X)$ assumption sets. The terminology will be justified as we proceed.

For a better understanding of the function S_P, of the provenance of the assumption sets $S_P(X)$, and of their dependence on X, we recall the relation $\geq_P$ (Császár's relation) introduced in Chapter 4, Section 1. $\geq_P$ was defined as a relation among the statements of L_P:

$$A \geq_P B \textit{ if and only if } P(A, A \vee B) > 0,$$

P a relative probability function on the statements of L_P, and it was established that $\geq_P$ effects a linear order among those statements. $\geq_P$ is transitive by Theorem 5 of Chapter 4 and for any statements A and B either $A \geq_P B$ or $B \geq_P A$ (Theorem 3 of Chapter 4). By Theorem 7 of that chapter $\top \geq_P A \geq_P \perp$ for any statement A, and $A \sim_P \top$ if and only if $P(A,\top) > 0$. Since finite sets can be represented by statements, namely conjunctions of their members, Császár's relation is in effect defined for relative probability functions on finite sets of statements:

$$X \geq_P Y \textit{ if and only if } P(X, \{\wedge X \vee \wedge Y\}) > 0,$$ [5]

and, given the interdefinability of those functions and of relative probability functions on statements of L_P, we can freely draw on the results of Chapter 4 and on Lemma 4 in Appendix I, reading reference to statements as reference to their unit sets.

The first result is that the relation $\geq_P$ goes together with the inverse of set inclusion among the corresponding values of S_P.

Theorem 6. *Let P be a relative probability function on the finite sets of statements of L_P. Then $S_P(X) \supseteq S_P(Y)$, if $X \geq_P Y$.*

Proof: Suppose $X \geq_P Y$ and assume that $A \in S_P(Y)$. Further assume that $P(\{\wedge Z\},\{\wedge X\}) = 1$, which by Theorem 6 of Chapter 4 means that $Z \geq_P X$ by the transitivity of $\geq_P$. Then $Z \geq_P Y$, $P(\{\wedge Z \vee \wedge Y\},\{\wedge Y\}) = 1$ by Lemma 4(bb), and $P(\{A\},\{\wedge Z \vee \wedge Y\}) = 1$, since $A \in S_P(Y)$. Also, $P(\{A\} \cup \{\wedge Z\},\{\wedge Z \vee \wedge Y\}) = P(\{A\},\{\wedge Z\} \cup \{\wedge Z \vee \wedge Y\}) \times P(\{\wedge Z\},\{\wedge Z \vee \wedge Y\})$ by **RS3**. But $P(\{A\},\{\wedge Z\} \cup \{\wedge Z \vee \wedge Y\}) = P(\{A\},\{\wedge Z\})$ by Lemmas 4(x) and 4(bb) and $P(\{A\} \cup \{\wedge Z\},\{\wedge Z \vee \wedge Y\}) = P(\{A\},\{\wedge Z \vee \wedge Y\})$ by Lemmas 4(oo) and 4(bb). Hence, $P(\{A\},\{\wedge Z\}) = 1$, since $Z \geq_P Y$, i.e. $P(\{\wedge Z\},\{\wedge Z \vee \wedge Y\}) > 0$. So, if $P(Z,X) = 1$, $P(\{A\},Z) = 1$, i.e. $A \in S_P(X)$. Hence $S_P(X) \supseteq S_P(Y)$. ¤

Since either $X \geq_P Y$ or $Y \geq_P X$, the assumption sets $S_P(X)$ are ordered by set inclusion; and since $\{\top\} \geq_P X \geq_P \{\perp\}$, $S_P(\{\top\})$ (which equals $S_P(\varnothing)$, given that $\wedge\varnothing = \top$) is the largest and $S_P(\{\perp\})$ the smallest of them. The next theorem also identifies $S_P(\varnothing)$ and $S_P(\{\perp\})$.

Theorem 7. *Let P be a relative probability function on the finite sets of statements of* L_P. *Then:*

(a) *For any finite sets of statements X and Y,* $S_P(X) \subseteq S_P(Y)$ *or* $S_P(Y) \subseteq S_P(X)$;

(b) $S_P(\{\bot\}) \subseteq S_P(X) \subseteq S_P(\varnothing)$;

(c) $S_P(\varnothing) = \{C : P(\{C\},\varnothing) = 1\}$, *and* $S_P(X) = S_P(\varnothing)$ *if* $P(X,\varnothing) > 0$;

(d) $S_P(\{\bot\}) = \{C : P(\{C\},Z) = 1$ *for every set Z*$\}$ *and* $S_P(X) = S_P(\{\bot\})$ *if X is P-abnormal.*

Proof: (a) By Theorem 6 and Theorem 3 of Chapter 4.

(b) By Theorem 7 of Chapter 4.

(c) Suppose $A \in S_P(\varnothing)$. Then $P(\{A\},\varnothing) = 1$, since $P(\varnothing,\varnothing) = 1$. And since $\wedge X \geq_P \top$ if $P(\{\wedge X\},\{\wedge X \vee \top\}) > 0$, i.e. $P(\{\wedge X\},\top\}) > 0$, $S_P(X) = S_P(\{\bot\})$ if $P(X,\varnothing) > 0$.

(d) $P(Z,X) = 1$ for every set Z of statements if and only if X is P-abnormal, and $\{\bot\}$ is P-abnormal by Lemma 17(b). Hence, if X is P-abnormal, $A \in S_P(X)$ if and only if $P(\{A\},Z) = 1$ for every Z. ¤

For P-normal X, $S_P(X)$ is consistent with X:

Theorem 8. *Let P be a relative probability function on the finite sets of statements of* L_P. *Then:*

(a) $S_P(X)$ *is consistent for any X;*

(b) *If X is P-normal,* $S_P(X) \cup X$ *is consistent;*

(c) *If X is P-abnormal,* $S_P(X) \cup X$ *is inconsistent;*

(d) *X is P-normal if and only if* $X \cup S_P(\{\bot\})$ *is consistent;*

(e) *If* $S_P(X) \subset S_P(Y)$, *then* $X \cup S_P(Y)$ *is inconsistent.*

Proof: (a) As we have seen, the largest assumption set of P is $S_P(\varnothing) = \{A : P(\{A\},\varnothing) = 1\}$. Hence, for every finite subset Y of $S_P(\varnothing)$, $P(Y,\varnothing) = 1$ by Lemma 16(k) and $P(\{\bot\},Y) = P(\{\bot\},\varnothing)$ by Lemma 16(f). But $P(\{\bot\},\varnothing) = 0$ by **RS4** and Lemma 16(h). So, $P(\{\bot\},Y) = 0$ and Y is consistent by Theorem 4(b). But then $S_P(\varnothing)$ is consistent by the compactness of $\vdash$ on L_P. And, since $S_P(X) \subseteq S_P(\varnothing)$ for every set X of statements, $S_P(X)$ is consistent.

(b) Suppose X is P-normal, i.e. $P(\{\bot\},X) = 0$. $P(\{A\},X) = 1$ for every statement A in $S_P(X)$, since $P(X,X) = 1$ by **RS1**. Hence by Lemma 16(k) $P(Y,X) = 1$ for every finite subset Y of $S_P(X)$. $P(\{\bot\},X \cup Y) = P(\{\bot\},X)$ for every such Y by Lemma 16(f). Hence $P(\{\bot\},X \cup Y) = 0$ by the supposition on X, and $X \cup Y$ is consistent by Theorem 4(b). But then, in virtue of the compactness of $\vdash$, X is consistent with $S_P(X)$.

(c) Suppose X is P-abnormal. Assume that Z is P-normal. Then $P(\{\bot\} \cup X,Z) = 0$ by Lemma 15(h) and $P(\{\bot\},X \cup Z) = 1$ by Lemma 15(j); hence $P(X,Z) = 0$ by **RS3** and $P(\{\neg\wedge X\},Z) = 1$ by **RS**$\wedge$ and **RS**$\neg$. But $P(\{\neg\wedge X\},Z) = 1$ if Z is P-abnormal. Hence $P(\{\neg\wedge X\},Z) = 1$ for every Z and $\neg\wedge X \in S_P(X)$ by Theorem 7(d). Hence $X \cup S_P(X)$ is inconsistent.

(d) If X is P-normal, $S_P(\{\bot\}) \cup X$ is consistent by (b) and Theorem 7(b). If X is P-abnormal, $S_P(\{\bot\}) \cup X$ is inconsistent by (c) and Theorem 7(d).

(e) Suppose $S_P(X) \subset S_P(Y)$. Then $\wedge Y >_P \wedge X$ by Theorem 6. Now assume $P(Z,Y) = 1$ so that

$\wedge Z \geq_P \wedge Y$ by **RS$\wedge$**, Lemma 15(g), and Theorem 6 of Chapter 4. Then $\wedge Z >_P \wedge X$ by Theorem 5 of Chapter 4. Consequently $\wedge Z$ is P-normal by Theorem 7 of Chapter 4. Hence $P(\{\neg\wedge X\},\{\wedge Z\}) = 1 - P(\{\wedge X\},\{\wedge Z\})$. But $P(\{\wedge X\},\{\wedge Z\}) = 0$ by Theorem 6 of Chapter 4. So, $P(\{\neg\wedge X\},Z) = 1$ by Lemma 15(g) and $\neg\wedge X$ is a member of $S_P(Y)$. Hence X is inconsistent with $S_P(Y)$. ¤

Theorems 7 and 8 make clear the character of the function S_P. Two values of the function play special roles in understanding its significance, namely $S_P(\varnothing)$ and $S_P(\{\bot\})$. $S_P(\varnothing)$ consists of the statements whose probability is 1 relative to the empty set of hypotheses; such statements are true statements according to P, they are the *assumptions* made by P, *P-assumptions* for short. Every $S_P(X)$ is then a subset of the P-assumptions. These subsets, all of them deductively closed, as we will see later, are ordered by set inclusion, the smallest of them being $S_P(\{\bot\})$. For given X, $S_P(X)$ is the largest set in this ordering which is consistent with X, if there exists such a set; otherwise $S_P(X)$ equals $S_P(\{\bot\})$. We can therefore think of $S_P(X)$ as obtained by discarding from the set $S_P(\varnothing)$ of all P-assumptions just enough statements to make the resulting set $S_P(X)$ compatible with X. Some statements however cannot be given up in this way; the statements in $S_P(\{\bot\})$, the *core assumptions*, are not abandoned, even if X is inconsistent with them. Note that, in the light of Theorem 5(b), the set of core assumptions is the set of logical truths when P is a Carnap 1952 function.

Degrees of Logical Consequence

In the light of this interpretation we will call the family of sets $S_P(X)$ the *assumption sets* of P, and a particular $S_P(X)$ the *assumption set of P for X*; $S_P(\varnothing)$ is the set of P-assumptions, and $S_P(\{\bot\})$ that of core assumptions. We go on to justify further our talk of assumptions by showing that $P(\{A\},X)$ ought to be interpreted as the degree to which A is implied by X together with the assumptions $S_P(X)$ of P for X.

Theorem 9. *Let P be a relative probability function on the finite sets of statements of L_P. Then:*

(a) *If $X \cup S_P(X) \vdash A$, then $P(\{A\},X) = 1$;*

(b) *If X is P-normal and $\{A\} \cup X \cup S_P(X)$ is inconsistent, then $P(\{A\},X) = 0$;*

(c) *If $P(\{A\},X) = 0$, then $\{A\} \cup X \cup S_P(X)$ is inconsistent;*

(d) *If $P(\{A\},X) = 1$, then $X \cup S_P(X) \vdash A$.*

Proof: (a) Suppose $X \cup S_P(X) \vdash A$. By the compactness of the consequence relation for L_P there exists then a finite subset Y of $S_P(X)$ such that $X \cup Y \vdash A$, and hence $X \cup Y \vdash_P A$ by Theorem 4(a). Hence $P(A,X \cup Y) = 1$. But since $Y \subseteq S_P(X)$, $P(Y,X) = 1$ by Lemma 16(k). Hence $P(\{A\},X) = 1$ by Lemma 16(f).

(b) Suppose X P-normal, i.e. $P(\{\bot\},X) = 0$, and $\{A\} \cup X \cup S_P(X)$ inconsistent. Then $X \cup S_P(X) \vdash \neg A$. Hence $P(\{\neg A\},X) = 1$ by (a), and hence $P(\{A\},X) = 0$ by **RS$\neg$**.

(c) Suppose $P(\{A\},X) = 0$. Then $P(\{\bot\},X) = 0$ by **RS$\bot$**, X is consistent by Theorem 4(b),

and $P(\{\neg A\},X) = 1$ by **RS¬**. The statement $\neg(A \wedge \Lambda X)$, where ΛX is some conjunction of the members of X, is a member of $S_P(X)$. For, on the assumption that $P(Z,X) = 1$, $P(\neg(A \wedge \Lambda X),Z) = 1 - P(\{A \wedge \Lambda X\},Z)$ by **RS¬** and Lemma 17(l). But $P(\{A \wedge \Lambda X\},Z)$ equals $P(\{A\} \cup X,Z)$ by **RS∧**, which equals $P(X,Z) - P(\{\neg A\} \cup X,Z)$ by Lemmas 17(e) and 17(l), and hence equals 0 by Lemma 16(f). Consequently, $A \wedge \Lambda X$ is incompatible with a member of $S_P(X)$, and hence $\{A\} \cup X \cup S_P(X)$ is inconsistent.
(d) Suppose $P(\{A\},X) = 1$. If X is P-abnormal, $X \cup S_P(X)$ is inconsistent by Theorem 8(c) and so $X \cup S_P(X) \vdash A$. If X is P-normal, $P(\{\neg A\},X) = 0$ by **RS¬**. Then $\{\neg A\} \cup X \cup S_P(X)$ is inconsistent by (c), and hence $X \cup S_P(X) \vdash A$ again. ¤

According to Theorem 9, if X is consistent with the core assumptions, then $P(\{A\},X) = 1$ if and only if $X \cup S_P(X) \vdash A$, and $P(\{A\},X) = 0$ if and only if $X \cup S_P(X)$ is inconsistent with A, hence if and only if $X \cup S_P(X) \vdash \neg A$. For a set X which is consistent with the core assumption set $S_P(\{\perp\})$, $P(\{A\},X)$ can therefore be understood as the degree to which $X \cup S_P(X)$ implies A, i.e. to the degree to which the set of P-assumptions — revised, if necessary, so as to become compatible with X — together with the statements in X implies A. If X is inconsistent with the core assumptions, then, of course, X together with those core assumptions implies any statement A; and since in that case X is P-abnormal, $P(\{A\},X)$ equals 1 so that again $P(\{A\},X)$ is the degree to which X together with the core assumptions implies A. We conclude that when P is any relative probability function on the finite sets of statements of L_P, then $P(\{A\},X)$ is the degree to which X together with the set of P-assumptions, modified in the light of X, implies A.

We can now also reassess the significance of $X \vdash_P A$, i.e.

$$P(\{A\},X \cup Z) = 1 \textit{ for every finite set } Z \textit{ of statements,}$$

which we showed in Section 2 to amount to $X \vdash A$, when P is a Carnap 1952 function.

Theorem 10. *Let P be a relative probability function on the finite sets of statements of L_P. Then $P(\{A\},X \cup Z) = 1$ for every finite set Z if and only if $X \cup S_P(\{\perp\}) \vdash A$.*

Proof: (a) Suppose $P(\{A\},X \cup Z) = 1$ for every Z. Then, in particular, $P(\{A\},X \cup \{\neg A\}) = 1$. Hence, $X \cup \{\neg A\}$ is P-abnormal by Lemma 16(k) and $S_P(X \cup \{\neg A\}) = S_P(\{\perp\})$ by Theorem 7(d). Therefore $X \cup \{\neg A\} \cup S_P(\{\perp\}) \vdash A$ by Theorem 9(d), i.e. $X \cup S_P(\{\perp\}) \vdash A$.
(b) Suppose $X \cup S_P(\{\perp\}) \vdash A$. Then $X \cup Z \cup S_P(X \cup Z) \vdash A$, since $S_P(\{\perp\}) \subseteq S_P(X \cup Z)$ by Theorem 7(b). Hence $P(\{A\},X \cup Z) = 1$ by Theorem 9(a). ¤

So, if P is an arbitrary relative probability function on the finite sets of statements of L_P, then $X \vdash_P A$ precisely when A is a logical consequence of X together with the core assumptions of P. In the special case of a Carnap 1952 function this amounts to the result already obtained in Section 2, namely that $X \vdash_P A$ if and only if $X \vdash A$, because in this case the core assumptions are just the logical truths.

We can now also easily demonstrate that the assumption sets S_P are deductively closed, a not unexpected result.

Theorem 11. *Let P be a relative probability function on the finite sets of statements of L_P. Then $A \in S_P(X)$, if $S_P(X) \vdash A$.*

Proof: Suppose $S_P(X) \vdash A$. Then there must be a finite subset Y of $S_P(X)$ such that $Y \vdash A$ and hence $Y \cup S_P(\{\bot\}) \vdash A$. Hence by Theorem 10 $P(\{A\},Y \cup Z) = 1$ for every Z. Now assume $P(Z,X) = 1$. Then $P(Y,Z) = 1$ by Lemma 16(k), since $Y \subseteq S_P(X)$. Hence $P(\{A\},Z) = 1$ by Lemma 16(f), and so $A \in S_P(X)$. ¤

Section 4. "Probability Semantics"

Core Assumptions and Classes of Probability Functions

The logical truths of L_P are always among the core assumptions of relative probability functions on the finite sets of statements of L_P, and they constitute the core assumptions in the case of Carnap 1952 functions. So, the set of logical truths of L_P is the least comprehensive set of core assumptions. The most comprehensive sets of core assumptions, on the other hand, are the maximally consistent sets of statements of L_P.[6] Hence core assumption sets can be sorted as follows:

Group One, to consist of the set of all logical truths of L_P,

Group Two, to consist of the remaining sets of statements of L_P that are closed under logical consequence and consistent, but not maximally consistent, and

Group Three, to consist of all the maximally consistent sets of statements of L_P.

Corresponding to these groups of sets of statements are three mutually exclusive groups of relative probability functions. Group One consists of all Carnap 1952 functions, including the Carnap 1950 functions, for which every assumption set $S_P(X)$ consists of just the logical truths. Group Three consists of the two-valued Kolmogorov functions, i.e. those two-valued Popper functions meeting

RSK *If $P(X,\emptyset) = 0$, then X is P-abnormal,*

as will be proved in the next theorem, and Group Two of the remaining relative probability functions.

Theorem 12. *Let P be a relative probability function on the finite sets of statements of L_P. Then P is 2-valued and meets* **RSK** *if and only if $S_P(\{\bot\})$ is maximally consistent.*

Proof: (i) Suppose $S_P(\{\bot\})$ maximally consistent. Then $S_P(\{\bot\}) = S_P(X)$ for any X, since any assumption set $S_P(X)$ is consistent and deductively closed (Theorems 8 and 11) and since $S_P(\{\bot\}) = S_P(X)$ by Theorem 7. Let A be a statement and X and Y be finite sets of statements

of L_P.
Case 1: X is not a subset of $S_P(\{\bot\})$. As $S_P(\{\bot\})$ is maximally consistent, $X \cup S_P(\{\bot\})$ is inconsistent and by Theorem 8 X is P-abnormal. Hence $P(\{A\},X) = 1$.
Case 2: $X \subseteq S_P(\{\bot\})$ and $A \in S_P(\{\bot\})$, i.e. $A \in S_P(X)$. Then $X \cup S_P(X) \vdash A$ and hence $P(\{A\},X) = 1$ by Theorem 9(a).
Case 3: $X \subseteq S_P(\{\bot\})$ and $A \notin S_P(\{\bot\})$. Then X is P-normal by Theorem 8 and $\{A\} \cup S_P(\{\bot\})$ is inconsistent, since $S_P(\{\bot\})$ is maximally consistent. Hence $\{A\} \cup X \cup S_P(X)$ is inconsistent and so $P(\{A\},X) = 0$ by Theorem 9(b).
So, $P(\{A\},X)$ equals either 0 or 1. Hence $P(Y,X)$ equals 0 or 1 by Lemmas 16(d) and 16(k). Now suppose $P(X,\emptyset) = 0$. Then by Lemma 16(k) $P(\{A\},\emptyset) = 0$ for some member A of X, and so $\{A\} \cup S_P(\emptyset)$ is inconsistent by Theorem 9(c), Hence $\{A\} \cup S_P(\{\bot\})$ is inconsistent, and so $\{A\}$ is P-abnormal by Theorem 8(d) and X is P-abnormal by Lemma 15(j). So P meets **RSK**.
(ii) Suppose P a two-valued Kolmogorov function and assume that $\{A\} \cup S_P(\{\bot\})$ is consistent. Then $\{A\}$ is P-normal by Theorem 8(d). Hence $P(\{A\},\emptyset) \neq 0$ by **RSK** and hence $P(\{A\},\emptyset) = 1$, since P is two-valued. But then $P(\{A\},Z) = 1$ for every Z by Lemma 15(l), i.e. $A \in S_P(\{\bot\})$. Hence $S_P(\{\bot\})$ is maximally consistent. ¤

So the set of core assumptions of a two-valued Kolmogorov function is maximally consistent and is identical with $S_P(\emptyset)$, the set of P-assumptions. But any maximally consistent set S corresponds to a truth-value assignment α in such a way that S consists of the statements that are true under α, i.e. $S = \{A : \alpha(A) = 1\}$. Consequently, invoking the reasoning of the proof of Theorem 12, we have

$$P(Y,X) = \begin{cases} 0 & \textit{if } \alpha(X) = 1 \textit{ and } \alpha(Y) = 0 \\ 1 & \textit{otherwise.} \end{cases}$$

(Here the truth-value $\alpha(X)$ assigned to a set of statements X equals 1 if and only if $\alpha(B) = 1$ for every member B of X, and consequently $\alpha(\emptyset)$ equals 1.) Any truth-value assignment α determines a two-valued Kolmogorov function, and one might say that two-valued Kolmogorov functions are but truth-value functions in disguise, something we already indicated in Chapter 1, Section 2.[7]

Logical Consequence in the Probabilistic Sense

Now let A be any statement and X any set (finite or infinite) of statements of L_P. Suppose $X \nvdash A$, i.e. suppose there exists a truth-value assignment α with $\alpha(X) = 1$ and $\alpha(A) = 0$. Then, if P is the two-valued Kolmogorov function determined by α, $P(\{A\},\emptyset) = 0$ and $P(\{B\},\emptyset) = 1$ for every member B of X. Consequently, if $X \nvdash A$, then there exists a Popper function P (in fact a two-valued Kolmogorov function P) and a finite set Z of statements (the empty set) such that $P(\{B\},Z) = 1$ for every member B of X and $P(\{A\},Z) \neq 1$. On the other hand, if $X \vdash A$, P is any relative probability function, Z is any finite set of statements, and

$P(\{B\},Z) = 1$ for every member B of X, then $P(\{A\},Z) = 1$. For, by the compactness of $\vdash$ for L_P, $X' \vdash A$ for some finite subset X' of X; $Z \cup S_P(Z) \vdash B$ for every member B of X by Theorem 9(d); hence $Z \cup S_P(Z) \vdash A$ by *Cut*; hence $P(\{A\},Z) = 1$ by Theorem 9(a). Hence we have:

$X \vdash A$

if and only if,

for every relative probability function P and every finite set Z of statements,
$P(\{A\},Z) = 1$ *if* $P(\{B\},Z) = 1$ *for every member B of X.*

This result is often regarded as establishing the soundness and completeness of $\vdash$ relative to *logical consequence in the probabilistic sense*, defined as follows:

A is a logical consequence of X in the probabilistic sense

if and only if

for every relative probability function P and every finite set Z of statements,
$P(\{A\},Z) = 1$ *if* $P(\{B\},Z) = 1$ *for every member B of X.*

Or, using relative probability functions on the statements of L_P, which, as we have seen in Chapter 1, Section 3, correspond to the relative probability functions on the finite sets of statements of L_P, we arrive at the familiar account

A is a logical consequence of X in the probabilistic sense

if and only if

for every relative probability function P for L_P and every statement C,
$P(A,C) = 1$ *if* $P(B,C) = 1$ *for every statement B in X.*

The result

$X \vdash A$ *if and only if A is a logical consequence of X in the probabilistic sense,*

is frequently seen as a central result of *Probabilistic Semantics*. The definition of *logical consequence in the probabilistic sense* is formally analogous to the standard definition of logical consequence (A is a logical consequence of X if and only if A is true in every interpretation in which every member of X is true), and is commonly conceived of as a generalisation of the standard definition, with probability functions seen as generalised interpretations. This view receives support from the observation that the standard interpretations associated with L_P, namely truth-value assignments, figure among the probability functions for L_P in the shape of two-valued Kolmogorov functions, as we have just seen.

But the view that probability functions for L_P, characterised by constraints, are like interpretations of L_P is misconceived. What is conspicuously lacking is a plausible story explaining the constraints on probability functions as expressing the interpretation of the connectives. In our view, on the other hand, any one of Popper's relative probability functions generalises the notion of logical consequence in the sense that $P(\{A\},X)$ is the degree to which A is logically implied by X together with a set $S_P(X)$ which results from a revision of the set of P-assumptions, a revision which ensures that $S_P(X)$ is compatible with X. The constraints imposed on Popper functions can be explained as generalisations of constraints corresponding to the rules of inference in a Natural Deduction framework. In our view, to explain the result

If, for every relative probability function P and every finite set Z of statements,
$P(\{A\},Z) = 1$ *if* $P(\{B\},Z) = 1$ *for every member B of X,*
then $X \vdash A$.

i.e., in the light of Theorem 9,

If, for every relative probability function P and every finite set Z of statements,
$Z \cup S_P(Z) \vdash A$ *if* $Z \cup S_P(Z) \vdash B$ *for every member B of X,*
then $X \vdash A$.

one should concentrate on the families of assumption sets associated with probability functions, since they are all that matters here. According to Theorem 12 we find among them the families consisting of a single assumption set, that set a maximally consistent set of statements of L_P. So, assuming

For every relative probability function P and every finite set Z of statements,
$Z \cup S_P(Z) \vdash A$ *if* $Z \cup S_P(Z) \vdash B$ *for every member B of X,*

take Z to be $\varnothing$ and restrict attention to the families of sets just mentioned. Then we have

For every maximally consistent set S of statements,
$S \vdash A$ *if* $S \vdash B$ *for every member B of X,*

i.e.

For every maximally consistent set S of statements,
$A \in S$ *if* $X \subseteq S$,

from which it follows that

$$X \vdash A.$$

Section 5. Relative Probability Functions for L_ω

The result of the previous sections concern arguments with finitely many premisses. In order to be able to deal with implication by infinitely many statements we turn to probability functions for L_ω. In Chapter 2, Section 1, we formulated constraints that are appropriate for relative probability functions on sets of statements of L_ω. They are **RS1–RS6**, **RSω**, and

RS∧' $P(\{\wedge A\},X) = P(S(A),X)$,

where $S(A)$ is the set of elements of the sequence of statements $A = \langle A_1, A_2, \ldots, A_i, \ldots \rangle$. And we know from Section 2 that the constraints **RS1–RS6** are equivalent to **RS1, RS2', RS3, RS4, RS∧, RS¬,** and **RS⊥**. This enables us to draw on results obtained in the previous sections. We now prove results about the relationship between logical consequence and probability functions for L_ω, which extend those for L_P. First we introduce constraints for the consequence relation that deal with infinitary conjunctions.

S⋀I *If* $X \vdash A_i$ *for* $i = 1, 2, \ldots$, *then* $X \vdash \bigwedge A$

S⋀E *If* $X \vdash \bigwedge A$, *then* $X \vdash A_i$ *for* $i = 1, 2, \ldots$

S⋀I is also known as (a version of) the *ω-rule*. The constraints on the consequence relation for L_ω are consequently these: **S1–S4, S∧I, S∧E, S⋀I, S⋀E, S¬I, S¬E, S⊥I,** and **S⊥E**.

As before, we understand by '$\vdash_P$' the relation given by the identity

$$X \vdash_P A \text{ if and only if } P(\{A\}, X \cup Z) = 1 \text{ for every set } Z \text{ of statements,}$$

where P is a given relative probability function on the sets of statements of L_ω. The claim that **RS⋀'** generalises the constraints **S⋀I** and **S⋀E** is proved in the next theorem, which is followed by counterparts of Theorems 4 and 5.

Theorem 13. *Let P be a function on the countable sets of statements of* L_ω *that meets the constraints*

(i) **RS1, RS2', RS3, RS4,** *and* **RSω**,

(ii) **RS⋀'**.

Then $\vdash_P$ *meets* **S⋀I** *and* **S⋀E**.

Proof: By Lemma 16(k). ¤

Theorem 14. *Let P be a relative probability function on the countable sets of statements of* L_ω, *i.e. a function on the countable sets of statements of* L_ω *which meets*

(i) **RS1, RS2', RS3, RS4,** *and* **RSω**,

(ii) **RS∧, RS¬, RS⊥,** *and* **RS⋀'**.

Then:

(a) *If* $X \vdash A$, *then* $P(A, X \cup Z) = 1$ *for every set Z of statements;*

(b) *If* $P(\{\bot\}, X) = 0$, *then X is logically consistent;*

(c) *If X is logically inconsistent, then* $P(Y, X) = 1$ *for every Y;*

(d) *If* $X \cup Y$ *is logically inconsistent, then* $P(Y, X \cup Z) = 0$ *for every Z with* $P(\bot, X \cup Z) = 0$.

Proof is by transfinite induction on the recursive definition of ⊢ and follows the structure of the proof of Theorem 4. The additional constraints **S⋀I** and **S⋀E** which play a role in the inductive step guarantee that $P(A, X \cup Z) = 1$ for every Z by Theorem 13. ¤

The converse result can be obtained if P is assumed to meet the constraint **IND** from Section 2, formulated now for all sets of atomic statements.

IND *If X and Y are sets of atomic statements and* $X \cap Y = \varnothing$, *then* $P(\{\bot\}, X \cup Y^{\neg}) = 0$

Theorem 15. *Let P be a relative probability function on the countable sets of statements of* L_ω *which meets* **IND**. *Then:*
(a) *If* $P(\{A\}, X \cup Z) = 1$ *for every set Z of statements, then* $X \vdash A$;
(b) *If X is logically consistent, then* $P(\{\bot\}, X) = 0$;
(c) *If* $P(Y, X) = 1$ *for every Y, then X is logically inconsistent;*
(d) *If* $P(Y, X \cup Z) = 0$ *for every Z such that* $P(\{\bot\}, X \cup Z) = 0$, *then* $X \cup Y$ *is logically inconsistent.*

Proof: Like that of Theorem 5. ¤

Given a relative probability function for L_ω, assumption sets of P can be introduced as in Section 3, the assumption set of P for X being

$$S_P(X) = \{C : P(\{C\}, Z) = 1 \text{ for every set } Z \text{ such that } P(Z, X) = 1\}.$$

$S_P(X)$ is understood as being the result of so revising the set of P-assumptions $S_P(\varnothing)$ as to be compatible with X. In order to simplify notation we will use '$\wedge X$' in the following theorems for '$\wedge A$', where A is any sequence of the members of X, X a countable set of statements of L_ω.

Theorem 16. *Let P be a relative probability function on the countable sets of statements of* L_ω. *Then:*
(a) $S_P(X)$ *is consistent;*
(b) *If X is P-normal,* $S_P(X) \cup X$ *is consistent;*
(c) *If X is P-abnormal,* $S_P(X) \cup X$ *is inconsistent;*
(d) *X is P-normal if and only if* $X \cup S_P(\{\bot\})$ *is consistent;*
(e) *If* $S_P(X) \subset S_P(Y)$, *then* $X \cup S_P(Y)$ *is inconsistent.*

Proof: (a) $P(S_P(\varnothing), \varnothing) = 1$ by the definition of S_P and Lemma 16(k) and $P(\{\bot\}, S_P(\varnothing)) = P(\{\bot\}, \varnothing)$ by Lemma 16(f). But $P(\{\bot\}, \varnothing) = 0$ by **RS4** and Lemma 16(h). So, $P(\{\bot\}, S_P(\varnothing)) = 0$ and $S_P(\varnothing)$ is consistent by Theorem 14(b). And, since $S_P(X) \subseteq S_P(\varnothing)$ for every set X of statements, $S_P(X)$ is consistent.
(b) Suppose X is P-normal, i.e. $P(\{\bot\}, X) = 0$. $P(\{A\}, X) = 1$ for every statement A in $S_P(X)$, since $P(X, X) = 1$ by **RS1**. Hence by Lemma 16(k) $P(S_P(X), X) = 1$. $P(\{\bot\}, X \cup S_P(X)) = P(\{\bot\}, X)$ by Lemma 16(f). Hence $P(\{\bot\}, X \cup S_P(X)) = 0$ by the supposition on X, and $X \cup S_P(X)$ is consistent by Theorem 14(b).
(c) Suppose X is P-abnormal. Assume that Z is P-normal. Then $P(\{\bot\} \cup X, Z) = 0$ by Lemma 15(h) and $P(\{\bot\}, X \cup Z) = 1$ by Lemma 15(j); hence $P(X, Z) = 0$ by **RS3** and $P(\{\neg\wedge X\}, Z) = 1$ by **RS**$\wedge$ and **RS**$\neg$. But $P(\{\neg\wedge X\}, Z) = 1$ if Z is P-abnormal. Hence $P(\{\neg\wedge X\}, Z) = 1$ for every Z and $\neg\wedge X \in S_P(X)$ by Theorem 7(d). Hence $X \cup S_P(X)$ is inconsistent.
(d) If X is P-normal, $S_P(\{\bot\}) \cup X$ is consistent by (b) and Theorem 7(b). If X is P-abnormal,

$S_P(\{\bot\}) \cup X$ is inconsistent by (c) and Theorem 7(d).
(e) Suppose $S_P(X) \subset S_P(Y)$. Then $\Lambda Y >_P \Lambda X$ by Theorem 6. Now assume $P(Z,Y) = 1$ so that $\Lambda Z \geq_P \Lambda Y$ by **RS**$\wedge$, Lemma 15(g), and Theorem 6 of Chapter 4. Then $\Lambda Z >_P \Lambda X$ by Theorem 5 of Chapter 4. Consequently ΛZ is P-normal by Theorem 7 of Chapter 4. Hence $P(\{\neg\Lambda X\},\{\Lambda Z\}) = 1 - P(\{\Lambda X\},\{\Lambda Z\})$. But $P(\{\Lambda X\},\{\Lambda Z\}) = 0$ by Theorem 6 of Chapter 4. So, $P(\{\neg\Lambda X\},Z) = 1$ by Lemma 15(g) and $\neg\Lambda X$ is a member of $S_P(Y)$. Hence X is inconsistent with $S_P(Y)$. ¤

So, the central result

$$P(\{A\},X) = 1 \text{ if and only if } X \cup S_P(X) \vdash A$$

holds for countable statements X.

Theorem 17. *Let P be a relative probability function on the countable sets of statements of L_ω. Then:*
(a) *If $X \cup S_P(X) \vdash A$, then $P(\{A\},X) = 1$;*
(b) *If X is P-normal and $\{A\} \cup X \cup S_P(X)$ is inconsistent, then $P(\{A\},X) = 0$;*
(c) *If $P(\{A\},X) = 0$, then $\{A\} \cup X \cup S_P(X)$ is inconsistent;*
(d) *If $P(\{A\},X) = 1$, then $X \cup S_P(X) \vdash A$.*

Proof: (a) Suppose $X \cup S_P(X) \vdash A$. Hence $X \cup S_P(X) \vdash_P A$ by Theorem 14(a). Hence $P(A,X \cup S_P(X)) = 1$. But $P(S_P(X),X) = 1$ by Lemma 16(k). Hence $P(\{A\},X) = 1$ by Lemma 16(f).
(b) Suppose X P-normal - i.e. $P(\{\bot\},X) = 0$ - and $\{A\} \cup X \cup S_P(X)$ inconsistent. Then $X \cup S_P(X) \vdash \neg A$. Hence $P(\{\neg A\},X) = 1$ by (a), and hence $P(\{A\},X) = 0$ by **RS**$\neg$.
(c) Suppose $P(\{A\},X) = 0$. Then $P(\{\bot\},X) = 0$ by Lemma 17(h), X is consistent by Theorem 14(b), and $P(\{\neg A\},X) = 1$ by **RS**$\neg$. The statement $\neg(A \wedge \Lambda X)$ is a member of $S_P(X)$. For, on the assumption that $P(Z,X) = 1$, $P(\neg(A \wedge \Lambda X),Z) = 1 - P(\{A \wedge \Lambda X\},Z)$ by **RS**$\neg$ and Lemma 17(l). But $P(\{A \wedge \Lambda X\},Z)$ equals $P(\{A\} \cup X,Z)$ by **RS**Λ, which equals $P(X,Z) - P(\{\neg A\} \cup X,Z)$ by Lemmas 17(e) and 17(l), and hence equals 0 by Lemma 16(f). Consequently, $A \wedge \Lambda X$ is incompatible with a member of $S_P(X)$, and hence $\{A\} \cup X \cup S_P(X)$ is inconsistent.
(d) Suppose $P(\{A\},X) = 1$. If X is P-abnormal, $X \cup S_P(X)$ is inconsistent by Theorem 16(c) and so $X \cup S_P(X) \vdash A$. If X is P-normal, $P(\{\neg A\},X) = 0$ by **RS**$\neg$. Then $\{\neg A\} \cup X \cup S_P(X)$ is inconsistent by (c), and hence $X \cup S_P(X) \vdash A$ again. ¤

$P(\{A\},X)$ should therefore be interpreted as the degree to which $X \cup S_P(X)$ implies X.

Chapter 9

Absolute Probability Functions for Intuitionistic Logic

Although we introduced intuitionistic probability functions in Chapter 1, Section 4, we did not offer any justification there for the groups of constraints said to characterise those functions. Indeed the task of developing intuitionistic probability theory and of formulating constraints for intuitionistic probability functions has not been well defined. Most writers on the matter, working within the paradigm of "probabilistic semantics", seem to have assumed that the aim is, in the case of absolute probability functions, say, to describe a class I of unary functions for which it holds that

$P(A) = 1$ *for every function P in I if and only if A is a theorem of intuitionistic logic*[1].

But this condition does not lead to a single solution. The groups of constraints that have been proposed differ markedly from one another. An exception is Miller and Popper (1989), whose criterion of adequacy appears to be that the equivalence classes of P-indiscernible statements form a Heyting algebra, and who propose a system of constraints for relative intuitionistic probability functions which is equivalent to our **RI1–RI9**.[2]

Our own approach to intuitionistic probability theory is of course guided by the idea that relative probability functions generalise the semantic notion of implication and absolute ones that of logical necessity. In the present chapter we describe functions which generalise intuitionistic necessity, while in the next one we turn to functions which generalise intuitionistic consequence. In both chapters we will again take Gentzen type constraints for multiple-conclusion consequence and for single-conclusion consequence as our starting point.

Section 1. Intuitionistic Consistency Functions and Intuitionistic Absolute Probability Functions

The multiple-conclusion constraints for intuitionistic propositional logic are largely the same as those listed in Chapter 7 for classical logic. Constraints for '$\vee$' and '$\supset$' must be added, since these connectives are not definable in terms of '$\wedge$' and '$\neg$' in intuitionistic logic. And the two constraints concerning the introduction of '$\neg$' and '$\supset$' on the right differ from the classical ones in restricting Y to the empty set.

M1 *If $X \cap Y \neq \varnothing$, then $X \vdash Y$*

M2 *If $X \vdash Y$, then $Z \cup X \vdash Y$ and $X \vdash Y \cup Z$*

M3 *If $X \vdash Y \cup \{A\}$ and $\{A\} \cup X \vdash Y$, then $X \vdash Y$*

M$\wedge$R *If $X \vdash Y \cup \{A\}$ and $X \vdash Y \cup \{B\}$, then $X \vdash Y \cup \{A \wedge B\}$*

M∧L *If* $\{A,B\} \cup X \vdash Y$, *then* $\{A \wedge B\} \cup X \vdash Y$

M∨R *If* $X \vdash Y \cup \{A,B\}$, *then* $X \vdash Y \cup \{A \vee B\}$

M∨L *If* $\{A\} \cup X \vdash Y$ *and* $\{B\} \cup X \vdash Y$, *then* $\{A \vee B\} \cup X \vdash Y$

MI⊃R *If* $\{A\} \cup X \vdash \{B\}$, *then* $X \vdash \{A \supset B\}$

M⊃L *If* $X \vdash Y \cup \{A\}$ *and* $\{B\} \cup X \vdash Y$, *then* $\{A \supset B\} \cup X \vdash Y$

MI¬R *If* $\{A\} \cup X \vdash \varnothing$, *then* $X \vdash \{\neg A\}$

M¬L *If* $X \vdash Y \cup \{A\}$, *then* $\{\neg A\} \cup X \vdash Y$

These constraints characterise recursively the multiple-conclusion consequence relation for L_I, symbolised by '⊢' in this chapter.

Constraints for Intuitionistic Consistency Functions

We now generalise these constraints so that they suit intuitionistic Gentzen functions, i.e. binary functions $G(X,Y)$ on the finite sets of statements which generalise the multiple-conclusion consequence relation in the sense that they attribute degrees of consequence. We will show that these functions are interdefinable with the absolute probability functions for intuitionistic logic which were introduced in Chapter 1, Section 4. For greater simplicity the constraints that characterise them are formulated in terms of the consistency functions $F(X,Y)$, defined by

$$F(X,Y) = 1 - G(X,Y),$$

which were introduced in Chapter 7. The constraints themselves are largely borrowed from the characterisation of consistency functions for classical logic in Chapter 7.

F1 $0 \leq F(X,Y)$

F2 $F(\varnothing,\varnothing) = 1$

F3 $F(X,Y \cup \{A\}) + F(\{A\} \cup X,Y) = F(X,Y)$

F∧L $F(\{A \wedge B\} \cup X,Y) = F(\{A,B\} \cup X,Y)$

F∨R $F(X,Y \cup \{A \vee B\}) = F(X,Y \cup \{A,B\})$

The constraints for '¬' and '⊃' cannot of course be those for classical logic. In the case of '¬',

$$F(\{\neg A\} \cup X,Y) = F(X,Y \cup \{A\})$$

will not do, since, given the intuitionistic understanding of negation, the degree of consistency of the negation of A with a set of statements may be less than the degree of consistency of those statements with the failure of A to hold.

$$F(\{\neg A\} \cup X,Y) \leq F(X,Y \cup \{A\})$$

is, however, not sufficient as a constraint, since it does not take into account that $\neg A$ is the weakest statement incompatible with A, hence the one with the highest degree of consistency with the set in question of those statements that are incompatible with A. Since incompatibility of a set Z of statements with A amounts to $F(Z \cup \{A\},\varnothing) = 0$, we adopt

FI¬L $F(\{\neg A\} \cup X,Y) = \max_{Z \subseteq_f L} \{F(Z \cup X,Y) : F(Z \cup \{A\},\varnothing) = 0\}$

Z here ranges over all finite sets of statements of L_I. And a similar line of reasoning leads to

FI⊃L $F(\{A \supset B\} \cup X,Y) = \max_{Z \subseteq_f L} \{F(Z \cup X,Y) : F(Z \cup \{A\},\{B\}) = 0\}$

The following four constraints then follow by virtue of **F3**:

F∧R $F(X,Y \cup \{A \wedge B\}) = F(X,Y \cup \{A\}) + F(\{A\} \cup X,Y \cup \{B\})$

F∨L $F(\{A \vee B\} \cup X,Y) = F(\{A\} \cup X,Y) + F(\{B\} \cup X,Y \cup \{A\})$

FI⊃R $F(X,Y \cup \{A \supset B\}) = \min_{Z \subseteq_f L} \{F(X,Y \cup \{\Lambda Z\}) : F(Z \cup \{A\},\{B\}) = 0\}$

FI¬R $F(X,Y \cup \{\neg A\}) = \min_{Z \subseteq_f L} \{F(X,Y \cup \{\Lambda Z\}) : F(Z \cup \{A\},\varnothing) = 0\}$

The last two constraints make use of 'ΛX', for which we stipulate in accordance with Lemma 14(d):

(i) ΛX is any conjunction of the members of X;
(ii) $\Lambda\varnothing$ is $A \supset A$, A an arbitrary but fixed statement.

Later on in this section we will also employ 'V', using the similar convention, which accords with Lemma 14(e), that

(iii) VX is any disjunction of the members of X;
(iv) $V\varnothing$ is $A \wedge \neg A$, A an arbitrary but fixed statement.

It is worth noting that in contrast with the constraints for '∧' and '∨', which are identical to those used in Chapter 7, the four constraints for '⊃' and '¬' are not recursive. With respect to these constraints we also note that $\{A \supset B\}$ meets the condition on Z

$$F(Z \cup \{A\},\{B\}) = 0$$

that occurs in **F⊃L** and **F⊃R**, and that $\{A\}$ meets the condition on Z

$$F(Z \cup \{A\},\varnothing) = 0$$

that occurs in **F¬L** and **F¬R**. For

$$\begin{aligned} &F(\{A \supset B\} \cup \{A\},\{B\}) \\ &\quad = \max_{Z \subseteq_f L} \{F(Z \cup \{A\},\{B\}) : F(Z \cup \{A\},\{B\}) = 0\} \qquad \textbf{(FI⊃L)} \\ &\quad = 0 \end{aligned}$$

and

$$\begin{aligned} &F(\{\neg A\} \cup \{A\},\varnothing) \\ &\quad = \max_{Z \subseteq_f L} \{F(Z \cup \{A\},\varnothing) : F(Z \cup \{A\},\varnothing) = 0\} \qquad \textbf{(FI¬L)} \\ &\quad = 0 \end{aligned}$$

The constraints **F¬L**, **F¬R**, **F⊃L**, and **F⊃R** in Section 3 of Chapter 7 can be obtained from the present intuitionistic ones by adding this further constraint, reminiscent of the classical law of the excluded middle:

LEM $F(X,Y \cup \{A,\neg A\}) = 0$,

a constraint which follows from **F¬R** and **F1**, since

$$
\begin{aligned}
&F(\{\neg A\} \cup X,Y) \\
&\quad = F(\{\neg A\} \cup X,Y \cup \{A\}) && (\textbf{F3}\text{, Lemma 14(a)}) \\
&\quad = F(\{\neg A\} \cup X,Y \cup \{A\}) - F(X,Y \cup \{A,\neg A\}) && (\textbf{F3}) \\
&\quad = F(X,Y \cup \{A\}) && (\textbf{LEM})
\end{aligned}
$$

Thus F meets **F¬L**, and, by **F3**, **F¬R**. As far as the constraints for '⊃' are concerned, note first that $\{\neg A \vee B\}$ meets the condition on Z

$$F(Z \cup \{A\},\{B\}) = 0$$

that occurs in **F⊃R** and **F⊃L**, since $F(\{\neg A \vee B\} \cup \{A\},\{B\})$ equals $F(\{\neg A\} \cup \{A\},\{B\}) + F(\{B\} \cup \{A\},\{B\} \cup \{\neg A\})$ by **F∧L**, and hence equals 0 by Lemmas 14(a) and 12(a). Hence $F(X,Y \cup \{A \supset B\}) \leq F(X,Y \cup \{\neg A \vee B\})$ by **FI⊃R**. But $F(\{A \supset B\} \cup X,Y \cup \{\neg A \vee B\}) = F(\{A\} \cup \{A \supset B\} \cup X,Y \cup \{B\}) = 0$ by **F∨R**, **F¬R**, and the fact, already mentioned, that $\{A \supset B\}$ meets the condition on Z

$$F(Z \cup \{A\},\{B\}) = 0$$

in constraint **FI⊃R**. Consequently, $F(X,Y \cup \{\neg A \vee B\}) + F(\{\neg A \vee B\} \cup X,Y \cup \{A \supset B\}) = F(X,Y \cup \{A \supset B\})$ by **F3**, and hence $F(X,Y \cup \{\neg A \vee B\}) \leq F(X,Y \cup \{A \supset B\})$ by **F1**. So, $F(X,Y \cup \{\neg A \vee B\}) = F(X,Y \cup \{A \supset B\})$, and therefore $F(X,Y \cup \{A \supset B\}) = F(\{A\} \cup X,Y\ \{B\})$, i.e. **F⊃R**, by **F∨R** and **F¬R**.

Hence the class of consistency functions which meet the constraints **F1–F3**, **F∧L**, **F∨R**, **FI⊃L**, and **FI¬L**, introduced in this chapter, includes the consistency functions of Chapter 7, i.e. those which meet **F1–F3**, **F∧L**, **F∨R**, **F⊃R**, and **F¬R**. Thus the classical consistency functions form a subset of the intuitionistic ones. We proceed to show that the latter generalise indeed the relation of multiple-conclusion consequence for intuitionistic logic in the sense that the functions

$$G(X,Y) = 1 - F(X,Y)$$

attribute degrees of intuitionistic consequence, with $G(X,Y) = 1$ amounting to $X \vdash Y$. So, we define, for a given consistency function F, the relation $\vdash_F$ thus:

$$X \vdash_F Y \textit{ if and only if } F(X,Y) = 0.$$

We are now in a position to show that the constraints **F1–F3**, **F∧R**, **F∧L**, **F∨R**, **F∨L**, **FI⊃R**, **FI⊃L**, **FI¬R**, and **FI¬L** for consistency functions generalise the corresponding constraints **M1–M3**, **M∧R**, **M∧L**, **M∨R**, **M∨L**, **MI⊃R**, **M⊃L**, **MI¬R**, and **M¬L**.

Theorem 1. *Let F be a binary function on the finite sets of statements of L_I that meets*

(i) **F1–F3**,

(ii) **F∧L**, **F∨R**, **FI⊃L**, *and* **FI¬L**.

Then $\vdash_F$ meets

(a) **M1–M3**, *and*

(b) **M∧R, M∧L, M∨R, M∨L, MI⊃R, M⊃L, MI¬R,** *and* **M¬L.**

Proof: (a) Like the proof of Theorem 3 of Chapter 7.

(b) **M∧R**: Suppose $F(X,Y \cup \{A\}) = F(X,Y \cup \{B\}) = 0$. Then $F(\{A\} \cup X,Y \cup \{B\}) = 0$ by Lemma 12(f) and, by **F∧R**, $F(X,Y \cup \{A \wedge B\}) = F(X,Y \cup \{A\}) + F(\{A\} \cup X,Y \cup \{B\}) = 0$.

M∧L: Suppose $F(\{A,B\} \cup X,Y) = 0$. Then $F(\{A \wedge B\} \cup X,Y) = 0$ by **F∧L**.

M∨R: Suppose $F(X,Y \cup \{A,B\}) = 0$. Then $F(X,Y \cup \{A \vee B\}) = 0$ by **F∨R**.

M∨L: Suppose $F(\{A\} \cup X,Y) = F(\{B\} \cup X,Y) = 0$. Then $F(\{B\} \cup X,Y \cup \{A\}) = 0$ by Lemma 12(f) and, by **F∨L**, $F(\{A \vee B\} \cup X,Y) = F(\{A\} \cup X,Y) + F(\{B\} \cup X,Y \cup \{A\}) = 0$.

MI⊃R: Suppose $F(\{A\} \cup X,\{B\}) = 0$, i.e. $F(\{\Lambda X\} \cup \{A\},\{B\}) = 0$ by Lemma 14(d). Then $F(X,\{A \supset B\}) \leq F(X,\{\Lambda X\})$ by **F⊃R**. But $F(X,\{\Lambda X\}) = F(\{\Lambda X\},\{\Lambda X\}) = 0$ by Lemmas 14(d) and 12(a). So, $F(X,\{A \supset B\}) = 0$ by **F1**.

M⊃L: Suppose $F(X,Y \cup \{A\}) = F(\{B\} \cup X,Y) = 0$, and let Z be any set of statements with $F(Z \cup \{A\},\{B\}) = 0$. Then $F(Z \cup X,Y) = 0$ by Lemma 12(h) and hence $F(\{A \supset B\} \cup X,Y) = 0$ by **F⊃L**.

MI¬R: Suppose $F(\{A\} \cup X,\varnothing) = 0$, i.e. $F(\{\Lambda X\} \cup \{A\},\varnothing) = 0$ by Lemma 14(d). Then $F(X,\{\neg A\}) \leq F(X,\{\Lambda X\})$ by **F¬R**. But $F(X,\{\Lambda X\}) = F(\{\Lambda X\},\{\Lambda X\}) = 0$ by Lemmas 14(d) and 12(a). So, $F(X,\{\neg A\}) = 0$ by **F1**.

M¬L: Suppose $F(X,Y \cup \{A\}) = 0$. Let Z be any set of statements with $F(Z \cup \{A\},\varnothing) = 0$. Then $F(Z \cup X,Y) = 0$ by Lemma 12(h) and hence $F(\{\neg A\} \cup X,Y) = 0$ by **F¬L**. ¤

The theorem shows that, in the limit $F(X,Y) = 0$, the constraints for intuitionistic consistency functions become the constraints for the intuitionistic consequence relation. The proof also makes it plausible that the classical constraints

M⊃R *If* $\{A\} \cup X \vdash Y \cup \{B\}$, *then* $X \vdash Y \cup \{A \supset B\}$

and

M¬R *If* $\{A\} \cup X \vdash Y$, *then* $X \vdash Y \cup \{\neg A\}$

cannot be obtained. In the case of **M¬R**, e.g., it does not follow from

$$F(\{A\} \cup X,Y) = 0$$

that $\{\Lambda X\}$ satisfies the condition on Z

$$F(Z \cup \{A\},\varnothing) = 0$$

in **FI¬R**, and as a result the inference to

$$F(X,Y \cup \{\neg A\}) = 0$$

fails.

The next theorem establishes that if Y is an intuitionistic consequence of X, then, for every intuitionistic consistency function F, $F(X,Y) = 0$.

Theorem 2. *Let F be a binary function on the finite sets of statements of L_I which meets*

(i) **F1–F3**,

(ii) **F∧L, F∨R, FI⊃L,** *and* **FI¬L.**

Then $F(X,Y) = 0$, *and hence* $G(X,Y) = 1$, *if* $X \vdash Y$.

Proof: By induction on the recursive specification of ⊢ by means of the Gentzen constraints **M1–M3, M∧R, M∧L, M∨R, M∨L, MI⊃R, M⊃L, MI¬R,** *and* **M¬L.**
Basis: $X \vdash Y$ by **M1**. Then $F(X,Y) = 0$ by Theorem 1.
Inductive Step: $X \vdash Y$ by one or another of the remaining constraints. Then $F(X,Y) = 0$ by the inductive hypothesis and Theorem 1. ¤

Consistency Functions with $F(X,Y) \neq 0$ *if* $X \nvdash Y$

The converse result is not to be had. In general $F(X,Y)$ may equal 0, even if $X \nvdash Y$, which means that, in general, an intuitionistic Gentzen function cannot be regarded as a straightforward generalisation of intuitionistic consequence. We shall see in Section 3 that any such function generalises the relation of intuitionistic consequence relative to a set of assumptions. Those Gentzen functions which satisfy the additional condition

FC *If X and Y are finite sets of statements and* $G(X,Y) = 1$, *then* $X \vdash Y$

or, formulated in terms of the consistency function F,

FC *If X and Y are finite sets of statements and* $F(X,Y) = 0$, *then* $X \vdash Y$,

do of course generalise the consequence relation. However, they can apparently not be characterised autonomously, i.e. without making use of the notion of intuitionistic consequence. And at this point we have not even established that there are such functions. Before we do this we show that

If $X \nvdash Y$, *then, for some consistency function* F, $F(X,Y) \neq 0$,

relying in the process on well-known results in the semantics for intuitionistic propositional logic.[3] Of the various semantics that have been devised for intuitionistic logic the most useful one for our purposes employs topological spaces:

A topological space $\langle \Xi, \mathbf{I} \rangle$ is constituted by a non-empty set Ξ of points and a function **I** from subsets of Ξ to subsets of Ξ meeting

T1 $\mathbf{I}(\alpha \cap \beta) = \mathbf{I}(\alpha) \cap \mathbf{I}(\beta)$

T2 $\mathbf{I}(\alpha) \subseteq \alpha$

T3 $\mathbf{I}(\mathbf{I}(\alpha)) = \mathbf{I}(\alpha)$

T4 $\mathbf{I}(\Xi) = \Xi$

I is called the *interior operation*, and a subset α of Ξ is called *open*, if $\mathbf{I}(\alpha) = \alpha$, i.e. if it is identical with its interior. In the light of **T2** and **T3** the interior of a subset α of Ξ is the largest open set contained in α.

Further, understand by a valuation v in the open sets of a topological space $\langle\Xi,\mathbf{I}\rangle$ any function that associates with every statement in L_I an open set of $\langle\Xi,\mathbf{I}\rangle$ subject to these constraints:

V1 $v(A \wedge B) = v(A) \cap v(B)$

V2 $v(A \vee B) = v(A) \cup v(B)$

V3 $v(\neg A) = \mathbf{I}(\backslash v(A))$

V4 $v(A \supset B) = \mathbf{I}(\backslash v(A) \cup v(B))$,

where $\backslash\alpha$ is the complement of α in Ξ^4, and let X and Y be finite sets of statements of L_I such that $X \not\vdash Y$. Then there exists a valuation v on the open sets of a *finite* topological space $\langle\Xi,\mathbf{I}\rangle$ such that

$$v(\Lambda X) \not\subseteq v(\mathrm{V}Y),$$

i.e.

$$v(\Lambda X) \cap \backslash v(\mathrm{V}Y) \neq \varnothing.$$

Now let $\langle\Xi,\mathbf{I}\rangle$ be a finite topological space and v be a valuation in the open sets of this space. Define a binary function on the finite sets of statements of L_I as follows:

$$F(X,Y) = 1/\mathrm{card}\Xi \times \mathrm{card}[v(\Lambda X) \cap \backslash v(\mathrm{V}Y)],$$

where $\backslash\alpha$ is the complement of α in Ξ and $\mathrm{card}\alpha$ is the number of points in α. Then F is a consistency function for L_I. For:

F1: $0 \leq F(X,Y)$ is immediate from the definition of F.

F2: $F(\varnothing,\varnothing)$

$= 1/\mathrm{card}\Xi \times \mathrm{card}[v(\Lambda\varnothing) \cap \backslash v(\mathrm{V}\varnothing)]$

$= 1/\mathrm{card}\Xi \times \mathrm{card}[\Xi \cap \backslash\varnothing]$

$= 1$

F3: $1/\mathrm{card}\Xi \times \mathrm{card}[v(\Lambda X) \cap \backslash v(\mathrm{V}(Y \cup \{A\}))] + 1/\mathrm{card}\Xi \times \mathrm{card}[v(\Lambda(X \cup \{A\})) \cap \backslash v(\mathrm{V}Y)]$

$= 1/\mathrm{card}\Xi \times \mathrm{card}[v(\Lambda X) \cap \backslash(v(\mathrm{V}Y) \cup v(A))] + 1/\mathrm{card}\Xi \times \mathrm{card}[v(\Lambda X) \cap v(A) \cap \backslash v(\mathrm{V}Y)]$ **(V1, V2)**

$= 1/\mathrm{card}\Xi \times \mathrm{card}[v(\Lambda X) \cap \backslash v(\mathrm{V}Y) \cap \backslash v(A)] + 1/\mathrm{card}\Xi \times \mathrm{card}[v(\Lambda X) \cap v(A) \cap \backslash v(\mathrm{V}Y)]$

$= 1/\mathrm{card}\Xi \times \mathrm{card}[v(\Lambda X) \cap \backslash v(\mathrm{V}Y) \cap \Xi] + 1/\mathrm{card}\Xi \times \mathrm{card}[v(\Lambda X) \cap v(A) \cap \varnothing]$ (Lemma 19(e))

$= 1/\mathrm{card}\Xi \times \mathrm{card}[v(\Lambda X) \cap \backslash v(\mathrm{V}Y)]$

F$\wedge$L: $1/\mathrm{card}\Xi \times \mathrm{card}[v(\Lambda(X \cup \{A \wedge B\})) \cap \backslash v(\mathrm{V}(Y)]$

$= 1/\mathrm{card}\Xi \times \mathrm{card}[v(\Lambda X) \cap v(A \wedge B) \cap \backslash v(\mathrm{V}(Y)]$ **(V1)**

$= 1/\mathrm{card}\Xi \times \mathrm{card}[v(\Lambda X) \cap v(A) \cap v(B) \cap \backslash v(\mathrm{V}(Y)]$ **(V1)**

$= 1/\mathrm{card}\Xi \times \mathrm{card}[v(\Lambda(X \cup \{A,B\}) \cap \backslash v(\mathrm{V}(Y)]$ **(V1)**

F∨R: Analogous to **F∧L**.

FI⊃L: $1/\text{card}\Xi \times \text{card}[v(A \supset B) \cap v(A) \cap \backslash v(B)]$

$$= 1/\text{card}\Xi \times \text{card}[\mathbf{I}(\backslash v(A) \cup v(B)) \cap v(A) \cap \backslash v(B)] \qquad \textbf{(V4)}$$
$$= 1/\text{card}\Xi \times \text{card}[\mathbf{I}(\backslash v(A) \cup v(B)) \cap \backslash(\backslash v(A) \cup v(B))]$$
$$= 1/\text{card}\Xi \times \text{card}\varnothing \qquad \textbf{(T2)}$$
$$= 0$$

So, $F(\{A \supset B\} \cup \{A\},\{B\}) = 0$. Now suppose $F(\{Z \cup \{A\},\{B\}) = 0$, hence $v(\Lambda Z) \subseteq \mathbf{I}(\backslash v(A) \cup v(B))$. Then

$1/\text{card}\Xi \times \text{card}[v(\Lambda X) \cap v(A \supset B) \cap \backslash v(\text{V}Y)]$

$$= 1/\text{card}\Xi \times \text{card}[v(\Lambda X) \cap \mathbf{I}(\backslash v(A) \cup v(B)) \cap \backslash v(\text{V}Y)] \qquad \textbf{(V4)}$$
$$\geq 1/\text{card}\Xi \times \text{card}[v(\Lambda X) \cap v(\Lambda Z) \cap \backslash v(\text{V}Y)]$$

Hence $F(\{A \supset B\} \cup X,Y) \geq F(Z \cup X,Y)$.

FI¬L: Like **FI⊃L**.

Theorem 3. *Let* $\langle \Xi,\mathbf{I}\rangle$ *be a finite topological space, v a valuation in the open sets of* $\langle \Xi,\mathbf{I}\rangle$, *and F the function defined on the finite sets of statements of* L_I *by*

$$F(X,Y) = 1/\text{card}\Xi \times \text{card}[v(\Lambda X) \cap \backslash v(\text{V}Y)].$$

Then F meets the constraints **F1–F3, F∧L, F∨R, FI⊃L,** *and* **FI¬L.**

Theorem 4. *If* $X \nvdash Y$, *then there exists a consistency function F for the finite sets of statements of* L_I *for which* $F(X,Y) \neq 0$.

Proof: Suppose $X \nvdash Y$. Then there exists a valuation v on the open sets of a finite topological space $\langle \Xi,\mathbf{I}\rangle$ such that

$$v(\Lambda X) \cap \backslash v(\text{V}Y) \neq \varnothing.$$

Let F be the consistency function which exists by Theorem 3 and for which

$$F(X,Y) = 1/\text{card}\Xi \times \text{card}[v(\Lambda X) \cap \backslash v(\text{V}Y)].$$

Then $\text{card}[v(\Lambda X) \cap \backslash v(\text{V}Y)] \neq 0$ and hence $F(X,Y) \neq 0$. ¤

In order to generate a consistency function $F(X,Y)$ on the finite sets of statements of L_I which takes the value 0 only if $X \vdash Y$,

(i) assume that the pairs of sets X and Y with $X \nvdash Y$ are arranged in an ω-sequence $\langle X_1,Y_1\rangle$, $\langle X_2,Y_2\rangle$, ..., $\langle X_n,Y_n\rangle$, ... ;

(ii) let, for any $n \geq 1$, F_n be a consistency function for which $F_n(X_n,Y_n) \neq 0$;

(iii) let $a_1, a_2, ..., a_n, ...$ be a sequence of positive real numbers whose infinite sum equals 1;

(iv) let $F(X,Y)$ be $\Sigma_{n\geq 1}[a_n \times F_n(X,Y)]$.

F then meets all the constraints for intuitionistic consistency functions:

F1, F2, F∧L, and **F∨R** are obviously satisfied.

F3: $\Sigma_{n\geq 1}[a_n \times F_n(X,Y \cup \{A\})] + \Sigma_{n\geq 1}[a_n \times F_n(\{A\} \cup X,Y)]$

$= \Sigma_{n\geq 1}[a_n \times [F_n(X,Y \cup \{A\}) + F_n(\{A\} \cup X,Y)]]$
$= \Sigma_{n\geq 1}[a_n \times F_n(X,Y)]$

F∧R and **F∨L**: Analogous to **F3**.

FI⊃L: Since $F_n(\{A \supset B\} \cup \{A\},\{B\}) = 0$ for every $n \geq 1$, $F(\{A \supset B\} \cup \{A\},\{B\}) = 0$. Suppose $F(Z \cup \{A\},\{B\}) = 0$. Then, since $a_n > 0$, $F_n(Z \cup \{A\},\{B\}) = 0$ for every $n \geq 1$ by the definition of F and by **F1**. Hence $F_n(\{A \supset B\} \cup X,Y) \geq F_n(Z \cup X,Y)$ by **FI⊃L** and so $\Sigma_{n\geq 1}[a_n \times F_n(\{A \supset B\} \cup X,Y)] \geq \Sigma_{n\geq 1}[a_n \times F_n(Z \cup X,Y)]$. So, $F(\{A \supset B\} \cup X,Y) = \max_{Z\subseteq_f L}\{F(Z \cup X,Y) : F(Z \cup \{A\},\{B\}) = 0\}$.
FI¬L: Like **FI⊃L**.

Now let X and Y be any finite sets of statements of L_I such that $X \nvdash Y$. Then $\langle X,Y\rangle = \langle X_n,Y_n\rangle$ for some $n \geq 1$. Therefore $F_n(X_n,Y_n) > 0$, hence $a_n \times F_n(X_n,Y_n) > 0$, and consequently $F(X,Y) = F(X_n,Y_n) = \Sigma_{n\geq 1}[a_n \times F_n(X_n,Y_n)] > 0$.

So, given Theorem 2,

Theorem 5. *There exist Gentzen functions $G(X,Y)$ on the finite sets of statements of L_I for which*

$$G(X,Y) = 1 \textit{ if and only if } X \vdash Y.$$

In other words, there are Gentzen functions which meet the constraint **FC**, introduced above. These functions generalise the intuitionistic consequence relation in the sense that $G(X,Y)$ is the degree to which the statements in X, conjunctively understood, entail the statements in Y, disjunctively understood.

Intuitionistic Probability Functions

It is time now to turn the consistency functions for intuitionistic logic into probability functions. Given any function on the finite sets of statements of L_I that meets **F1–F3**, **F∧L**, **F∨R**, **FI⊃L**, and **FI¬L**, let P be defined thus:

$$P(X) = F(X,\varnothing).$$

Then P meets the constraints

AS1 $0 \leq P(X)$

AS2 $P(\varnothing) = 1$

ASI3 $P(X \cup Y) \leq P(X)$

AS∧ $P(X \cup \{A \wedge B\}) = P(X \cup \{A,B\})$

AS∨ $P(X \cup \{A \vee B\}) = P(X \cup \{A\}) + P(X \cup \{B\}) - P(X \cup \{A,B\})$

ASI⊃ $P(X \cup \{A \supset B\}) = \max_{Z\subseteq_f L}\{P(X \cup Z) : P(Z \cup \{A\}) = P(Z \cup \{A,B\})\}$

ASI¬ $P(X \cup \{\neg A\}) = \max_{Z\subseteq_f L}\{P(X \cup Z) : P(Z \cup \{A\}) = 0\}$

For:

AS1: $0 \leq F(X,\varnothing)$ by **F1**.

AS2: $F(\varnothing,\varnothing) = 1$ by **F2**.

ASI3: $F(X \cup \{A\},\varnothing) = F(X,\varnothing) - F(X,\{A\})$ by **F3**. Hence $F(X \cup \{A\},\varnothing) \leq F(X,\varnothing)$ by **F1**.

AS∧: $F(X \cup \{A \wedge B\},\varnothing) = F(X \cup \{A,B\},\varnothing)$ by **F∧L**.

AS∨: $F(X \cup \{A \vee B\},\varnothing)$ equals $F(X \cup \{A\},\varnothing) + F(\{B\} \cup X,\{A\})$ by **F∨L**, which equals $F(X \cup \{A\},\varnothing) + F(X \cup \{B\},\varnothing) - F(X \cup \{A,B\},\varnothing)$ by **F3**.

ASI⊃: $F(X \cup \{A \supset B\},\varnothing) = \max_{Z \subseteq_f L} \{F(X \cup Z,\varnothing) : F(Z \cup \{A\},\varnothing) = F(Z \cup \{A,B\},\varnothing)\}$ by **FI⊃L** and **F3**.

ASI¬: $F(X \cup \{\neg A\},\varnothing) = \max_{Z \subseteq_f L} \{F(X \cup Z,\varnothing) : F(Z \cup \{A\},\varnothing) = 0\}$ by **FI¬L**.

Conversely, let P be any function for the finite sets of statements of L_I which meets constraints **AS1**, **AS2**, **ASI3**, **AS∧**, **AS∨**, **ASI⊃**, and **ASI¬**, and let F be defined as follows:

$$F(X,Y) = P(X) - P(X \cup \{\mathrm{V}Y\}).$$

By Lemma 11(f) F is then the auxiliary function F_P defined, as in Chapter 5, Section 2, as

$$F_P(X,Y) = \begin{cases} P(X) & \textit{if } Y = \varnothing \\ F_P(X,Y-\{A\}) - F_P(\{A\} \cup X,Y-\{A\}) & \textit{otherwise,} \end{cases}$$

where A is the alphabetically latest statement in Y.

And $F(X,Y)$ meets the constraints for intuitionistic consistency functions:

F1: $0 \leq P(X) - P(X \cup \{\mathrm{V}Y\})$ by **ASI3**.

F2: $P(\varnothing) - P(\varnothing \cup \{\Lambda\varnothing\}) = P(\varnothing) - P(\{A \wedge \neg A\}) = 0$ by Lemma 11(d).

F3: $P(X) - P(X \cup \{\mathrm{V}(Y \cup \{A\})\}) + P(X \cup \{A\}) - P(X \cup \{A\} \cup \{\mathrm{V}Y\})$

$= P(X) - P(X \cup \{\mathrm{V}Y \vee A\}) + P(X \cup \{A\}) - P(X \cup \{\mathrm{V}Y,A\})$ (Lemmas 11(a), 11(b))

$= P(X) - P(X \cup \{\mathrm{V}Y\})$ (**AS∨**)

F∧L: By **AS∧**.

F∨R: $P(X) - P(X \cup \{\mathrm{V}(Y \cup \{A \vee B\})\})$

$= P(X) - P(X \cup \{\mathrm{V}(Y \cup \{A,B\})\})$ (Lemmas 11(a), 11(b))

FI⊃L: $F(\{A \supset B\} \cup X,Y) = P(\{A \supset B\} \cup X) - P(\{A \supset B\} \cup X \cup \{\mathrm{V}Y\})$.

Now $P(Z \cup X) - P(Z \cup X \cup \{\mathrm{V}Y\})$ takes on its maximal value for $Z = \{A \supset B\}$. For let Z be any finite set of statements which satisfies the condition

$$P(Z \cup \{A\}) = P(Z \cup \{A,B\}).$$

Then $P(Z \cup U) = P(\{A \supset B\} \cup Z \cup U)$ for every set U of statements, since

$P(\{A \supset B\} \cup Z \cup U)$

$= \max_{V \subseteq_f L} \{P(V \cup Z \cup U) : P(V \cup \{A\}) = P(V \cup \{A,B\})\}$ (**ASI⊃**)

$= P(Z \cup U)$ **(ASI3)**

and $P(Z \cup X) - P(Z \cup X \cup \{VY\})$

$= P(\{A \supset B\} \cup Z \cup X) - P(\{A \supset B\} \cup Z \cup X \cup \{VY\})$

$= P(\{A \supset B\} \cup X \cup Z \cup \{VY\}) - P(\{A \supset B\} \cup X \cup \{VY\})$ **(AS∨)**

$\leq P(\{A \supset B\} \cup X) - P(\{A \supset B\} \cup X \cup \{VY\})$ **(ASI3)**

Hence $P(\{A \supset B\} \cup X) - P(\{A \supset B\} \cup X \cup \{VY\})$

$= \max_{Z \subseteq_f L} \{P(Z \cup X) - P(Z \cup X \cup \{VY\}) : P(Z \cup \{A\}) = P(Z \cup \{A,B\})\}$

$= \max_{Z \subseteq_f L} \{F(Z \cup X, Y) : F(Z \cup \{A\}, \{B\}) = 0\}$

FI¬L: Analogously.

So, there exists a one-to-one correspondence between the functions F meeting **F1–F3**, **F∧L**, **F∨R**, **FI⊃L**, and **FI¬L** and the functions P meeting **AS1**, **AS2**, **ASI3**, **AS∧**, **AS∨**, **ASI⊃**, and **ASI¬**. Since they correspond to one another in the same way as the consistency functions and the absolute probability functions for L_P (see Chapter 7, Section 3), we declare the functions meeting **AS1**, **AS2**, **ASI3**, **AS∧**, **AS∨**, **ASI⊃**, and **ASI¬** to be absolute probability functions for L_I, thereby justifying the adoption of these constraints in Chapter 1, Section 4. Thereby justified are also the constraints **AI1–AI9** imposed on absolute probability functions on the statements of L_I, which were shown there to stand in one-to-one correspondence with those on the finite sets of statements of L_I.

Making use of these correspondences we obtain from Theorems 2 and 4 the following result, which, from the point of view of "probabilistic semantics", constitutes a soundness and completeness theorem of intuitionistic derivability relative to a semantics in terms of probability functions.

Theorem 6. *A statement A is a theorem of intuitionistic propositional logic if and only if for every absolute probability function P for L_I $P(A) = 1$.*

Section 2. Intuitionistic Probability Functions and the Meaning of Connectives

Given our understanding of probability functions, Theorem 6 reflects the fact that $P(A) = 1$ if and only if the assumption set S of P logically implies A. Then A is a theorem if A is implied by any assumption set whatsoever. This understanding of probability functions meshes in well with the usual explication of the significance of connectives in intuitionistic logic. For assertibility conditions are standardly specified in terms of epistemic support or, more precisely, in terms of conclusive evidence. Hence probabilities of intuitionistically understood statements, signifying degrees of assertibility, should be explicated in terms of, generally inconclusive, evidence, i.e. as degrees of confirmation. By taking degrees of confirmation to be degrees of consequence we have been able to obtain the constraints on intuitionistic probability functions, and in particular those concerning connectives, from intuitionistic rules of inference, appropriately generalised.

The available evidence, to the extent that it can be expressed in the language, is then contained in the assumption set S. As a result, we have on hand explications of the intuitionistic connectives in the spirit of the standard explications, but suiting situations of inconclusive evidence as much as situations in which conclusive evidence is available. So, the probability of a statement A, in general, is the degree of support it receives from the assumption set S. Or, since an implied statement has at least as much support as the implying one, the probability of statement A is the highest degree of support received by any statement which itself would conclusively establish A.

The probability of $\neg A$ is, in the same vein, the highest degree of support received by any set of statements which would conclusively refute A. The probability of $A \supset B$ is the highest degree of support received by any set of statements which, in conjunction with A, would conclusively establish B. And the probability of $A \vee B$ is, when A and B are incompatible, the degree of support for A plus the degree of support for B.

Section 3. Assumption Sets of Intuitionistic Probability and Consistency Functions

In this section we look at assumption sets for intuitionistic probability and consistency functions. The results obtained are similar to those of Chapter 7, Section 4. Since assumption sets are infinite sets, it is advantageous to extend our investigation to functions which take all sets, i.e. both finite and infinite sets, of statements as their arguments. For this purpose we add the following two constraints, familiar from Chapter 7 and appropriate in the light of Lemma 12(e), to our constraints for intuitionistic consistency functions:

FωR $\quad F(X,Y) = \inf_{Y'} \{F(X,Y') : Y' \subseteq_f Y\}$

FωL $\quad F(X,Y) = \inf_{X'} \{F(X',Y) : X' \subseteq_f X\}$

As one would expect, the intuitionistic consistency functions for the sets of statements are extensions of those for the finite sets of statements.

Theorem 7. *Any consistency function F' on the finite sets of statements of L_I extends to a consistency function defined on all sets of statements of L_I by means of the definition*

$$F(X) = \begin{cases} F'(X,Y) & \textit{if } X \textit{ and } Y \textit{ are finite} \\ \inf_{X'} \{\inf_{Y'} \{F'(X',Y') : Y' \subseteq_f Y\} : X' \subseteq_f X\} & \textit{if } X \textit{ or } Y \textit{ is infinite.} \end{cases}$$

Proof: Let F' be a function on the finite sets of statements of L_I which meets constraints **F1–F3**, **F∧L**, **F∨R**, **FI⊃L**, and **FI¬L** and let F be its extension to all sets of statements of L_I.

Then F meets these constraints and **FωR** and **FωL** as well:

F1 and **F2**: Obvious.

F3: $\inf_{X'}\{\inf_{Y'}\{F'(X',Y') : Y' \subseteq_f Y \cup \{A\}\} : X' \subseteq_f X\}$

$+ \inf_{X'}\{\inf_{Y'}\{F'(X',Y') : Y' \subseteq_f Y\} : X' \subseteq_f \{A\} \cup X\}$

$= \inf_{X'}\{\inf_{Y'}\{F'(X',Y' \cup \{A\}) : Y' \subseteq_f Y\} : X' \subseteq_f X\}$

$+ \inf_{X'}\{\inf_{Y'}\{F'(\{A\} \cup X',Y') : Y' \subseteq_f Y\} : X' \subseteq_f X\}$ (Lemma 12(e))

$= \inf_{X'}\{\inf_{Y'}\{F'(X',Y' \cup \{A\}) + \{F'(\{A\} \cup X',Y') : Y' \subseteq_f Y\} : X' \subseteq_f X\}$

$= \inf_{X'}\{\inf_{Y'}\{F'(X',Y') : Y' \subseteq_f Y\} : X' \subseteq_f X\}$ (**F3**)

F∧L and **F∨R**: Analogous proof.

FI⊃L: Since by **FI⊃L**, for every finite X' and Y', $F(\{A \supset B\} \cup X',Y') \geq F(Z \cup X',Y')$, if $F(Z \cup \{A\},\{B\}) = 0$, it follows that for every X and Y

$$\inf_{X'}\{\inf_{Y'}\{F'(\{A \supset B\} \cup X',Y') : Y' \subseteq_f Y\} : X' \subseteq_f X\}$$
$$\geq \inf_{X'}\{\inf_{Y'}\{F'(Z \cup X',Y') : Y' \subseteq_f Y\} : X' \subseteq_f X\},$$

if $F(Z \cup \{A\},\{B\}) = 0$, and hence that

$$\inf_{X'}\{\inf_{Y'}\{F'(X',Y') : Y' \subseteq_f Y\} : X' \subseteq_f \{A \supset B\} \cup X\}$$
$$\geq \inf_{X'}\{\inf_{Y'}\{F'(X',Y') : Y' \subseteq_f Y\} : X' \subseteq_f \{C\} \cup X\},$$

if $F(Z \cup \{A\},\{B\}) = 0$.

F¬L: Similarly. ¤

Since the absolute probability functions for finite sets of statements of L_I can also be extended to functions that are defined on all sets of statements of L_I and meet constraint

ASω $P(X) = \inf_{X'}\{P(X') : X' \subseteq X\}$,

the one-to-one correspondence between intuitionistic consistency functions and intuitionistic absolute probability functions extends to functions defined on all sets of statements of L_I. Since the intuitionistic consequence relation is compact, we also have the following extension of Theorem 2:

Theorem 8. *Let F be a binary function on the sets of statements of L_I which meets*

(i) **F1–F3**,

(ii) **F∧L**, **F∨R**, **FI⊃L**, *and* **FI¬L**,

(iii) **FωL** *and* **FωR**.

Then $F(X,Y) = 0$, *if* $X \vdash Y$.

Proof: Suppose $X \vdash Y$. Then by the compactness of intuitionistic logic there are finite subsets X' of X and Y' of Y with $X' \vdash Y'$. Hence $F(X',Y') = 0$ by Theorem 2 and so $F(X,Y) = 0$ by Lemma 12(f). ¤

We turn now to the assumption sets of those functions. As in Chapter 7, we define the assumption set S of an absolute probability function P to be the set

$$S = \{A : P(\{A\}) = 1\}.$$

If F is the corresponding consistency function, we then have

$$S = \{A : F(\varnothing,\{A\}) = 0\}.$$

Theorem 9. *Let F be a consistency function for the sets of statements of* L_I *and let S be its assumption set. Then*
(a) *If A is a member of S, then* $F(\{A\} \cup X,Y) = F(X,Y)$;
(b) *If* $S' \subseteq S$, $F(S' \cup X,Y) = F(X,Y)$.

Proof: (a) Assume $F(\varnothing,\{A\}) = 0$. Then $F(X,Y \cup \{A\}) = 0$ by **F1** and Lemma 12(e), and so $F(\{A\} \cup X,Y) = F(X,Y)$ by **F3**.
(b) By repeated use of (a) and by **FωL**. ¤

Theorem 10. *Let X and Y be finite sets of statements of* L_I *and F a consistency function for the sets of statements of* L_I. *Then* $F(X,Y) = 0$ *if and only if* $S \cup X \vdash Y$.

Proof: (a) Suppose $F(X,Y) = 0$, i.e. $F(\{\Lambda X\},\{\mathrm{V}Y\}) = 0$ by **F∧L** and **F∨R**. Then $F(\varnothing,\{\Lambda X \supset \mathrm{V}Y\}) = 0$ by Lemmas 14(b), 14(d), and 14(e). So, the statement $\Lambda X \supset \mathrm{V}Y$ is a member of the assumption set S of F. Therefore $S \vdash \{\Lambda X \supset \mathrm{V}Y\}$ by **M1** and so $S \cup X \vdash Y$ by **M⊃L**, **M3**, and repeated use of **M∧L** and **M∨R**.
(b) Assume $S \cup X \vdash Y$. Then $F(S \cup X,Y) = 0$ by Theorem 8. Hence $F(X,Y) = 0$ by Theorem 9(b). ¤

Assumption sets are consistent and closed under logical consequence, where both consistency and consequence are taken in their intuitionistic sense. For if $S \vdash \{A\}$, then $F(S,\{A\}) = 0$ by Theorem 10, hence $F(\varnothing,\{A\}) = 0$ by Theorem 9(b), i.e. $A \in S$. So, S is closed under intuitionistic consequence. And S is consistent. For suppose S inconsistent, i.e. $S \vdash \varnothing$. Then $F(S,\varnothing) = 0$ by Theorem 10, and hence $F(\varnothing,\varnothing) = 0$ by Theorem 9(b), contradicting **F2**.

Theorem 11. *Let X and Y be finite sets of statements of* L_P, *let F and* F_0 *be consistency functions for the sets of statements of* L_I, *let S be the assumption set of F, and let* F_0 *meet the condition* **FC**. *Then* $F(X,Y) = 0$ *if and only if* $F_0(S \cup X,Y) = 0$.

Proof: (a) Suppose $F(X,Y) = 0$. Then $S \cup X \vdash Y$ by Theorem 10 and hence $F_0(S \cup X,Y) = 0$ by Theorem 8.
(b) Suppose $F_0(S \cup X,Y) = 0$. Then $S \cup X \vdash Y$ by **FC** and hence $F(X,Y) = 0$ by Theorem 10. ¤

If S is any consistent set of statements of L_I which is closed under logical consequence, then there exists a consistency function for L_I which has S as its assumption set. For let F_0 be any consistency function for L_I meeting the condition **FC** and define F thus:

$$F(X,Y) = F_0(S \cup X,Y) \,/\, F_0(S,\varnothing).$$

Then F is a consistency function for L_I. For constraint **F2** is secured by the factor $1 / F_0(S,\varnothing)$ in the definition, while all other constraints can be simply verified by inspection. Suppose $F(\varnothing,\{A\}) = 0$. Then $F_0(S,\{A\}) = 0$ by Theorem 9(b) and $S \vdash \{A\}$ by **FC** so that $A \in S$, since S is closed under logical consequence. Now suppose $A \in S$. Then $F(\varnothing,\{A\}) = F_0(S,\{A\}) / F_0(S,\varnothing)$. But $F_0(S,\{A\}) = 0$ by **F1**. So, S is the assumption set of F.

Conversely, given any consistency function for the sets of statements of L_I, with assumption set S, there exists a consistency function F_0 which meets **FC** such that

$$F(X,Y) = F_0(S \cup X,Y) / F_0(S,\varnothing).$$

This is shown by the same argument that was employed in Chapter 7, Section 4: Let F' be any consistency function which meets **FC**, and let r be any real number larger than 0 but smaller than 1. Define the function F_0 by

$$F_0(X,Y) = r \times F(X,Y) + (1 - r) \times [F'(X,Y) - F'(S \cup X,Y)] / [1 - F'(S,\varnothing)].$$

That F_0 meets the constraints **F1–F3**, **F∧L**, **F∨R**, **FI⊃L**, **FI¬L**, **FωL** and **FωR** follows by Lemma 12(e) in the case of **F1** and by inspection in all other cases. F_0 meets **FC**. For let X and Y be finite sets of statements of L_I and assume that $F_0(X,Y) = 0$. Then $F(X,Y) = 0$ and $F'(X,Y) - F'(S \cup X,Y) = 0$. Hence $F'(S \cup X,Y) = 0$ by Theorem 11, and so $F'(X,Y) = 0$. So, $X \vdash Y$, which means that F_0 meets **FC**.

Now let X and Y be any sets of statements of L_I. Then

$$F_0(S \cup X,Y) = r \times F(S \cup X,Y)$$

but

$$F(S \cup X,Y) = F(X,Y)$$

by Theorem 9(b). So,

$$F(X,Y) = 1/r \times F_0(S \cup X,Y),$$

but $r = F_0(S,\varnothing)$ by Theorem 9(b) and **F2**, so that

$$F(X,Y) = F_0(S \cup X,Y) / F_0(S,\varnothing).$$

So, apart from a proportionality factor, the degrees of consistency attributed by F are the degrees of consistency *relative to the assumption set of* F attributed by some consistency function which meets **FC**.

Or, if we focus on the associated absolute probability function P for sets of statements of L_I, whose assumption set is S, we see that the degree of necessity assigned by P to $\{A\}$ is the degree of necessity of $\{A\}$ relative to S assigned by a function P' which meets the Carnap constraint

ASC *If* $P(X) = 0$, *then* X *is logically inconsistent,*

where of course logical inconsistency has to be understood intuitionistically.

Chapter 10

Relative Probability Functions for Intuitionistic Logic

Whereas logical consequence, classically understood, obtains when the truth of the premisses guarantees the truth of the conclusion, intuitionistically logical consequence is a matter of the premisses providing conclusive evidence for the conclusion; more precisely, logical consequence obtains when any conclusive evidence for the premisses constitutes conclusive evidence for the conclusion. Hence it is intuitionistically even more plausible than classically to regard relative probability functions, which are naturally taken to register evidential support for a statement relative to stated (and unstated) assumptions, as generalisations of the consequence relation.

Moreover, since the significance of a connective is typically explicated in terms of what counts as conclusive evidence for a complex statement governed by that connective, the constraints on intuitionistic relative probability functions that concern connectives can be seen as derived from the standard accounts and adapted to cover inconclusive support.

In Section 1 constraints for relative probability functions on sets of statements are formulated as generalisations of Natural Deduction rules. In the following sections assumptions sets, as introduced in Chapter 8, are defined for intuitionistic relative probability functions and the result

$$P(\{A\},X) = 1 \textit{ if and only if } X \cup S_P(X) \vdash A$$

is proved. Finally, the constraints appropriate for intuitionistic relative probability functions are derived.

Section 1. Binary Functions Generalising Intuitionistic Single-Conclusion Consequence

The single-conclusion constraints for intuitionistic propositional logic are the same as those for classical logic that we listed in Chapter 8, Section 2, except for the double negation constraint **S¬E**, which must be dropped, and the introduction- and elimination constraints for '∨' and '⊃', which must be added, since '∨' and '⊃' are not definable in terms of '∧' and '¬' in intuitionistic logic.

S 1 *If* $A \in X$, *then* $X \vdash A$

S 2 *If* $X \vdash A$, *then* $Y \cup X \vdash A$

S 3 *If* $X \vdash A$ *and* $\{A\} \cup X \vdash B$, *then* $X \vdash B$

S 4 $\varnothing \nvdash A$ *for at least one statement A*

S∧I *If* $X \vdash A$ *and* $X \vdash B$, *then* $X \vdash A \wedge B$

S∧E *If* $X \vdash A \wedge B$, *then* $X \vdash A$ *and* $X \vdash B$

S∨I *If* $X \vdash A$ *or* $X \vdash B$, *then* $X \vdash A \vee B$

S∨E *If* $X \cup \{A\} \vdash C$ *and* $X \cup \{B\} \vdash C$, *then* $X \cup \{A \vee B\} \vdash C$

S⊃I *If* $X \cup \{A\} \vdash B$, *then* $X \vdash A \supset B$

S⊃E *If* $X \vdash A \supset B$ *and* $X \vdash A$, *then* $X \vdash B$

S¬I *If* $\{A\} \cup X \vdash \bot$, *then* $X \vdash \neg A$

S⊥I *If* $X \vdash A$ *and* $X \vdash \neg A$, *then* $X \vdash \bot$

S⊥E *If* $X \vdash \bot$, *then* $X \vdash A$

These constraints characterise in a recursive manner the (single-conclusion) consequence relation for L_I, symbolised in this chapter by '$\vdash$'.

We go on to define the class of binary functions on finite sets of statements which, attributing degrees of consequence, generalise this consequence relation. These functions are our intuitionistic relative probability functions defined on finite sets of statements. Here are the constraints, generalisations of those above, which characterise them. The constraints are largely borrowed from the characterisation of relative probability functions for classical logic that was discussed in Chapter 8, Section 2. The new constraints **RSI⊃** and **RSI¬** are motivated by the same kind of consideration which prompted the formulation of **FI⊃L** and **FI¬L** in Chapter 9. We will later explain that the intuitionistic relative probability functions on finite sets of statements correspond one-to-one to the functions introduced in Chapter 1, Section 4, as relative probability functions for L_I.

RS1 *If* $Y \subseteq X$, *then* $P(Y,X) = 1$

RS2' $0 \leq F_{P_X}(Y,Z)$

RS3 $P(Z \cup Y,X) = P(Z,Y \cup X) \times P(Y,X')$

RS4 $P(Z,\varnothing) \neq 1$ *for at least one set Z of statements.*

RS∧ $P(\{A \wedge B\},X) = P(\{A,B\},X)$

RS∨ $P(\{A \vee B\},X) = P(\{A\},X) + P(\{B\},X) - P(\{A,B\},X)$

RSI⊃ $P(\{A \supset B\},X) = \max_{Y \subseteq_f L} \{P(Y,X) : P(\{A\} \cup Y,X) = P(\{A,B\} \cup Y,X)\}$

RSI¬ $P(\{\neg A\},X) = \max_{Y \subseteq_f L} \{P(Y,X) : P(\{A\} \cup Y,X) = P(\{\bot\},X)\}$[1]

RS⊥ *If* $P(\{\bot\},X) \neq 0$, *then* $P(\{A\},X) = 1$

Useful consequences of these constraints are assembled in Lemmas 16 and 18 in Appendix I. With respect to the constraints **RSI⊃** and **RSI¬** we note that $\{A \supset B\}$ meets the condition on Y

$$P(\{A\} \cup Y,X) = P(\{A,B\} \cup Y,X)\}$$

that occurs in **RSI⊃**, and that $\{\neg A\}$ meets the condition on Y

$$P(\{A\} \cup Y,X) = P(\{\bot\},X)\}$$

that occurs in **RSI¬**. For $P(\{A,A \supset B\},X) = P(\{A,B,A \supset B\},X)$ by Lemma 18(g) and $P(\{A,\neg A\},X) = P(\{\bot\},X)$ by Lemma 18(d). The two constraints **RSI⊃** and **RSI¬** are also the only ones in the list above that are different from those for the relative probability functions for the finite sets of statements of L_P that we discussed in Chapter 8. From them **RS⊃** and **RS¬** can be obtained if the following additional constraint, reminiscent of the law of excluded middle, is assumed:

LEM $\quad P(\{A \vee \neg A\},X) = 1$

For in the case of '¬' we have $P(\{A,\neg A\},X) = P(\{\bot\},X)$ by Lemma 18(d), $P(\{A \vee \neg A\},X) = P(\{A\},X) + P(\{\neg A\},X) - P(\{\bot\},X)$ by **RS∨** and, hence, by **LEM**, $P(\{\neg A\},X) = 1 - P(\{A\},X) + P(\{\bot\},X)$, i.e. **RS¬**. As to the constraints for '⊃', first note that $\{\neg A \vee B\}$ meets the condition on Y

$$P(\{A\} \cup Y,X) = P(\{A,B\} \cup Y,X)\}$$

in **RSI⊃**. For

$P(\{A,\neg A \vee B\},X)$

$= [P(\{\neg A\},\{A\} \cup X) + P(\{B\},\{A\} \cup X) - P(\{\neg A,B\},\{A\} \cup X)] \times P(\{A\},X)$ (**RS3**, **RS∨**)

$= [P(\{\bot\},\{A\} \cup X) + P(\{B\},\{A\} \cup X) - P(\{\bot,B\},\{A\} \cup X)] \times P(\{A\},X)$ (Lemma 17(m))

$= P(\{A,B\},X)$ (Lemma 17(c)),

and $P(\{A,B,\neg A \vee B\},X) = P(\{A,B\},X)$ by a similar argument. Hence $P(\{\neg A \vee B\},X) \leq P(\{A \supset B\},X)$ by **RSI⊃**. And $P(\{\neg A \vee B\},X) = 1 - P(\{A\},X) + P(\{A,B\},X)$ by **RS∨** and **RS¬**. Therefore $P(\{A \supset B,\neg A \vee B\},X)$ equals $P(\{A \supset B\},X) - P(\{A,A \supset B\},X) + P(\{A,B,A \supset B\},X)$ by **RS3**, and hence equals $P(\{A \supset B\},X)$, since $\{A \supset B\}$ meets the condition on Y in **RSI⊃**. So, $P(\{A \supset B\},X) \leq P(\{\neg A \vee B\},X)$ by Lemma 16(g). Consequently, $P(\{A \supset B\},X) = P(\{\neg A \vee B\},X)$, and hence $P(\{A \supset B\},X) = 1 - P(\{A\},X) + P(\{A,B\},X)$, i.e. **RS⊃**.

Hence the class of relative probability functions for the finite sets of statements of L_I which meet **RS1**, **RS2'**, **RS3**, **RS4**, **RS∧**, **RS∨**, **RSI⊃**, **RSI¬**, and **RS⊥** includes the relative probability functions of Chapter 8, i.e. those which meet **RS1**, **RS2'**, **RS3**, **RS4**, **RS∧**, **RS∨**, **RS⊃**, **RS¬**, and **RS⊥**. In other words, the classical relative probability functions on finite sets of statements constitute a subset of the intuitionistic ones.

It is our aim to show that the functions P on finite sets of statements of L_I that meet **RS1**, **RS2'**, **RS3**, **RS4**, **RS∧**, **RS∨**, **RSI⊃**, **RSI¬**, and **RS⊥** generalise the relation of single-conclusion consequence for intuitionistic logic. Guided by the considerations in Chapter 8, we define again, for a given function P, the relation $\vdash_P$ thus:

$X \vdash_P A$ *if and only if* $P(\{A\}, X \cup Z) = 1$ *for every set* Z *of statements of* L_I.

The first theorem corresponds to Theorems 1 and 3 in Chapter 8.

Theorem 1. *Let P be a relative probability function for the finite sets of statements of* L_I. *Then* $\vdash_P$ *meets* **S1–S4**, **S∧I**, **S∧E**, **S∨I**, **S∨E**, **S⊃I**, **S⊃E**, **S¬I**, **S⊥I**, *and* **S⊥E**.

Proof: (a) That $\vdash_P$ meets **S1–S4** is proved as in Theorem 1 of Chapter 8.

(b) That $\vdash_P$ meets **S∧I** and **S∧E** is proved as in Theorem 3 of Chapter 8.

(c) Suppose $P(\{A \vee B\}, X) = P(\{C\}, \{A\} \cup X) = P(\{C\}, \{B\} \cup X) = 1$. Then:

$P(\{A \vee B, C\}, X)$

$= P(\{A \vee B\}, \{C\} \cup X) \times P(\{C\}, X)$ (**RS3**)

$= [P(\{A\}, \{C\} \cup X) + P(\{B\}, \{C\} \cup X) - P(\{A,B\}, \{C\} \cup X)] \times P(\{C\}, X)$ (**RS∨**)

$= P(\{A,C\}, X) + P(\{B,C\}, X) - P(\{A,B,C\}, X)$ (**RS3**)

$= P(\{A\}, X) + P(\{B\}, X) - P(\{A,B\}, X)$ (**RS3**)

$= P(\{A \vee B\}, X)$ (**RS∨**)

$= 1$

Hence $P(\{C\}, X) = 1$ by Lemma 16(k). So, **S∨E**. **S∨I** is immediate by **RS∨**.

(d) Suppose $P(\{B\}, \{A\} \cup X) = 1$. Then $P(\{A,B\} \cup Y, X) = P(\{A\} \cup Y, X)$ by **RS3** and Lemma 16(f). Hence $P(\{A \supset B\}, X) = \max_{Y \subseteq_f L} \{P(Y, X)\} = 1$ by **RSI⊃** and Lemma 18(f). So **S⊃I**. Now suppose $P(\{A \supset B\}, X) = P(\{A\}, X) = 1$. Then $P(\{A, A \supset B\}, X) = 1$ by Lemma 16(k) and $P(\{A, B, A \supset B\}, X) = P(\{B\}, X)$ by Lemma 16(f). But $P(\{A, A \supset B\}, X) = P(\{A, B, A \supset B\}, X)$ by **RSI⊃**. Hence $P(\{B\}, X) = 1$. So **S⊃E**.

(e) Suppose $P(\{\bot\}, \{A\} \cup X) = 1$. Then $P(\{A\} \cup Y, X) = P(\{\bot\}, X)$ by Lemma 18(h). Hence $P(\{\neg A\}, X) = \max_{Y \subseteq_f L} \{P(Y, X)\} = 1$ by **RSI¬** and Lemma 18(f). So **S¬I**.

(f) Assume $P(\{A\}, X) = P(\{\neg A\}, X) = 1$ and hence $P(\{A, \neg A\}, X) = 1$ by Lemma 16(k). Then $P(\{\bot\}, X) = 1$, since $P(\{A, \neg A\}, X) = P(\{\bot\}, X)$ by **RSI¬**. Hence **S⊥I**. Assume $P(\{\bot\}, X) = 1$. Then $P(\{A\}, X) = 1$ by **RS⊥**. Hence **S⊥E**. ¤

Theorem 2. *Let P be a relative probability function for the finite sets of statements of* L_I. *Then :*

(a) $P(\{A\}, X \cup Z) = 1$ *for every set* Z *of statements, if* $X \vdash A$;

(b) *If* $P(\{\bot\}, X) = 0$, *then* X *is logically consistent.*

Proof: (a) By induction on the recursive definition of $\vdash$.

Basis: $X \vdash A$ by **S1**. Then $P(A, X \cup Z) = 1$ for every Z by Theorem 1.

Inductive Step: $X \vdash A$ by one or another of the recursive constraints. Then $P(A, X \cup Z) = 1$ for every Z by the inductive hypothesis and Theorem 1.

(b) Suppose X is logically inconsistent, i.e. $X \vdash \bot$. Then $P(\{\bot\}, X) = 1 \neq 0$ by (a). ¤

Section 2. Assumption Sets of Intuitionistic Relative Probability Functions

Assumption sets for relative probability functions on finite sets of statements of L_I are defined in the same way as for those on the finite sets of statements of L_P:

$$S_P(X) = \{C : P(\{C\},Z) = 1 \textit{ for every set } Z \textit{ such that } P(Z,X) = 1\}.$$

And the results concerning these sets are also the same as those obtained in Chapter 8, Section 3. Much of the reasoning there goes through for intuitionistic probability functions as well. So, the sets $S_P(X)$ are consistent, deductively closed, and linearly ordered by set inclusion with $S_P(\{\bot\}) \subseteq S_P(X) \subseteq S_P(\varnothing\}$. $S_P(\varnothing)$ equals $\{C : P(\{C\},\varnothing) = 1\}$ and if X is P-abnormal, $S_P(X) = S_P(\{\bot\}) = \{C : P(\{C\},Z) = 1 \textit{ for every set } Z\}$. The proofs of some of the results involve the constraint for **RS¬** and therefore require modification.

Theorem 3. *Let P be a relative probability function on the finite sets of statements of L_P. Then:*

(a) *If X is P-normal, $S_P(X) \cup X$ is consistent;*

(b) *If X is P-abnormal, $S_P(X) \cup X$ is inconsistent;*

(c) *X is P-normal if and only if $X \cup S_P(\{\bot\})$ is consistent.*

Proof: (a) Suppose X is P-normal, i.e. $P(\{\bot\},X) = 0$ by **RS⊥**. $P(\{A\},X) = 1$ for every statement A in $S_P(X)$, since $P(X,X) = 1$ by **RS1**. Hence by Lemma 16(k) $P(Y,X) = 1$ for every finite subset Y of $S_P(X)$. $P(\{\bot\},X \cup Y) = P(\{\bot\},X)$ for every such Y by Lemma 16(f). Hence $P(\{\bot\},X \cup Y) = 0$ by the supposition on X, and $X \cup Y$ is consistent by Theorem 2(b). But then, in virtue of the compactness of $\vdash$, X is consistent with $S_P(X)$.

(b) Suppose X P-abnormal and assume that Z is P-normal. Then $P(\{\bot\} \cup X,Z) = 0$ by Lemma 18(l) and $P(\{\bot\},X \cup Z) = 1$ by Lemma 18(m); hence $P(X,Z) = 0$ by **RS3** and $P(\{\neg\Lambda X\},Z) = 1$ by **RS∧** and Lemma 18(n). But $P(\{\neg\Lambda X\},Z) = 1$ if Z is P-abnormal. Hence $P(\{\neg\Lambda X\},Z) = 1$ for every Z and so $\neg\Lambda X \in S_P(X)$. Hence $X \cup S_P(X)$ is inconsistent.

(c) If X is P-normal, $S_P(\{\bot\}) \cup X$ is consistent by (a). If X is P-abnormal, $S_P(\{\bot\}) \cup X$ is inconsistent by (b). ¤

Consequently, for given X, $S_P(X)$ is the largest set in the family of assumption sets which is consistent with X, if there exists such a set; otherwise $S_P(X)$ equals $S_P(\{\bot\})$, the set of core assumptions.

Theorem 4. *Let P be a relative probability function on the finite sets of statements of L_I. Then:*

(a) *If $X \cup S_P(X) \vdash A$, then $P(\{A\},X) = 1$;*

(b) *If X is P-normal and $\{A\} \cup X \cup S_P(X)$ is inconsistent, then $P(\{A\},X) = 0$;*

(c) *If $P(\{A\},X) = 0$, then $\{A\} \cup X \cup S_P(X)$ is inconsistent;*

(d) *If $P(\{A\},X) = 1$, then $X \cup S_P(X) \vdash A$.*

Proof: (a) Suppose $X \cup S_P(X) \vdash A$. By the compactness of the consequence relation for L_I there exists then a finite subset Y of $S_P(X)$ such that $X \cup Y \vdash A$, and hence $X \cup Y \vdash_P A$ by Theorem 2(a). Hence $P(A,X \cup Y) = 1$. But since $Y \subseteq S_P(X)$, $P(Y,X) = 1$ by Lemma 16(k). Hence $P(\{A\},X) = 1$ by Lemma 16(f).
(b) Suppose X P-normal, i.e. $P(\{\bot\},X) = 0$, and $\{A\} \cup X \cup S_P(X)$ inconsistent. Then $X \cup S_P(X) \vdash \neg A$. Hence $P(\{\neg A\},X) = 1$ by (a), and hence $P(\{A\},X) = 0$ by Lemma 18(n).
(c) Suppose $P(\{A\},X) = 0$. Assuming $P(Z,X) = 1$, $P(\{A\} \cup X,Z) = 0$ by Lemma 16(f) and **RS3**. Hence $P(\{A \wedge \Lambda X\},Z) = 0$ by **RS**$\wedge$ and $P(\neg(A \wedge \Lambda X),Z) = 1$ by Lemma 18(n). So, $\neg(A \wedge \Lambda X)$ is a member of $S_P(X)$. Consequently, $A \wedge \Lambda X$ is incompatible with a member of $S_P(X)$, and hence $\{A\} \cup X \cup S_P(X)$ is inconsistent.
(d) Suppose $P(\{A\},X) = 1$. Let Z be any finite set of statements with $P(Z,X) = 1$. Then $P(\{\Lambda X \supset A\},Z) = 1$. For by Lemmas 16(j) and 16(f) $P(\{\Lambda X\} \cup Y,Z) = P(\{\Lambda X,A\} \cup Y,Z) = P(Y,Z)$. Hence by **RSI**$\supset$ $P(\{\Lambda X \supset A\},Z) = \max_{Y \subseteq_f L} \{P(Y,X)\}$, which equals 1 by Lemma 18(f). Therefore $\{\Lambda X \supset A\} \in S_P(X)$ and hence $X \cup S_P(X) \vdash A$. ¤

Our conclusion, analogous to that obtained in Chapter 8, is that when P is any relative probability function on the finite sets of statements of L_I, then $P(\{A\},X)$ is the degree to which X together with the set $S_P(X)$, i.e. the set of P-assumptions, modified in the light of X, implies A.

Section 3. Intuitionistic Relative Probability Functions and Logical Consequence

By Theorem 2

If $X \vdash A$ *, then* $P(\{A\},X \cup Z) = 1$ *for every set* Z *of statements*

holds for every relative probability functions for the finite sets of statements of L_I. Of particular interest among these functions are those for which the converse

If $P(\{A\},X \cup Z) = 1$ *for every set* Z *of statements, then* $X \vdash A$

also holds. However, we know of no autonomous constraint which would single out these functions. What we can prove is this:

If $X \nvdash A$*, then there exists a function* P *such that*
$P(\{A\},X \cup Z) \neq 1$ *for some set* Z *of statements of* L_I.

In order to obtain this, we make use of the analogous result for intuitionistic absolute probability functions which we proved in Chapter 9, Section 2. To begin with, we establish that to every intuitionistic absolute probability function there corresponds a relative one, where the relationship is of the kind that Chapter 3 was devoted to.

Theorem 5. *Let* P_1 *be an absolute probability function for the finite sets of statements of* L_I, *and let P be this binary function:*

$$P(Y,X) = \begin{cases} P_1(Y \cup X) \,/\, P_1(X) & \textit{if } P_1(X) \neq 0 \\ 1 & \textit{otherwise.} \end{cases}$$

Then P is a relative probability function for the finite sets of statements of L_I.

Proof:

RS1: Immediate.

RS2': That P_1 meets

$$0 \leq F_{P_1}(Y,Z)$$

was established in Chapter 9, Section 1.

RS3: $P(Z \cup Y,X) = P(Z,Y \cup X) \times P(Y,X)$ by definition when $P_1(X) = 0$, and hence, by Lemma 11(c), $P_1(Y \cup X) = 0$. So, suppose first that $P_1(X) \neq 0$ but $P_1(Y \cup X) = 0$ and therefore $P_1(Z \cup Y \cup X) = 0$, in which case **RS3** holds. Suppose next that $P_1(X) \neq 0$ and $P_1(Y \cup X) \neq 0$, in which case **RS3** holds, too. So P meets **RS3**.

RS4: $P(\{\bot\},\varnothing) \neq 1$ by Lemma 11(d) and **AS2**.

RS∧: By **AS∧**.

RS∨: By **AS∨**.

RSI⊃: By **ASI⊃**.

RSI¬: By **ASI¬**.

RS⊥: Suppose $P(\{\bot\},X) \neq 0$. Then $P_1(X) = 0$, since $P_1(X \cup \{\bot\}) = 0$ by Lemma 11(d). So, $P(\{A\},X) = 1$ by the definition of P.

Theorem 6. *If $X \nvdash A$, there then exists a relative probability function P for the finite sets of statements of L_I for which $P(\{A\},X) \neq 1$.*

Proof: Suppose $X \nvdash A$. By Theorem 4 of Chapter 9 there exists then a consistency function F such that $F(X,\{A\}) \neq 0$, hence there exists an intuitionistic absolute probability function P_1 such that $P_1(X \cup \{A\}) < P_1(X)$. Hence $P_1(X) \neq 0$ by Lemma 11(c). So, by Theorem 5 there exists an intuitionistic relative probability function P such that $P(\{A\},X) \neq 1$. ¤

Combining this with Theorem 2 we obtain:

Theorem 7. *$X \vdash A$ if and only if, for every relative probability function P on the finite sets of statements of L_I, $P(\{A\},X \cup Z) = 1$ for every set Z of statements of L_I.*

Intuitionistic relative probability functions on the statements of L_I were introduced in Chapter 1, Section 4, and are characterised by these constraints:

RI1 $0 \leq P(A,B)$

RI2 $P(A,A) = 1$

RI3 $P(A \wedge B,C) = P(A,B \wedge C) \times P(B,C)$

RI4 $P(A \wedge B,C) \leq P(B \wedge A,C)$

RI5 $P(A,B \wedge C) \leq P(A,C \wedge B)$

RI6 *At least one statement of L_I is P-normal*

RI7 $P(A,C) + P(B,C) = P(A \wedge B,C) + P(A \vee B,C)$

RI8 $P(A \supset B,C) = \max_{D \in L}\{P(D,C) : P((D \wedge A) \wedge B,C) = P(D \wedge A,C)\}$

RI9 *If C is P-normal, then* $P(\neg A,C) = \max_{D \in L}\{P(D,C) : P(A \wedge D,C) = 0\}$

These functions stand in one-to-one correspondence with the relative probability functions on the finite sets of statements of L_I. For, firstly, let P be a function on the finite sets of statements of L_I which meets **RS1, RS2', RS3, RS4, RS∧, RS∨, RSI⊃, RSI¬**, and **RS⊥**. Then the function P' defined by

$$P'(A,B) = P(\{A\},\{B\})$$

is sure to meet **RI1–RI9**:

RI1: $0 \leq P(\{A\},\{B\})$ by Lemma 16(c).

RI2: $P(\{A\},\{A\}) = 1$ by **RS1**.

RI3: $P(\{A \wedge B\},\{C\}) = P(\{A\},\{B \wedge C\}) \times P(\{B\},\{C\})$ by **RS3, RS∧**, and Lemma 18(c).

RI4: By **RS∧**.

RI5: By Lemma 18(c).

RI6: $P(\{\bot\},\{A \supset A\}) = 0$ by Lemma 18(k).

RI7: $P(\{A\},\{C\}) + P(\{B\},\{C\}) = P(\{A \wedge B\},\{C\}) + P(\{A \vee B\},\{C\})$ by **RS∧** and **RS∨**.

RI8: If $P(\{(\Lambda Y \wedge A) \wedge B\},\{C\}) = P(\{\Lambda Y \wedge A\},\{C\})$, then $P(\{\Lambda Y\},\{C\}) \leq P(\{A \supset B\},\{C\})$ by **RSI⊃** and **RS∧**.

RI9: If $P(\{\bot\},\{C\}) = 0$ and $P(\{A \wedge \Lambda Y\},\{C\}) = 0$, then $P(\{\Lambda Y\},\{C\}) \leq P(\{\neg A\},\{C\})$ by **RSI¬**.

Conversely, given any function P' for L_I which meets **RI1–RI9**, let the function P on the finite sets of statements of L_I be defined by

$$P(Y,X) = P'(\Lambda Y,\Lambda X),$$

where ΛX is an arbitrary conjunction of the members of X, when X is non-empty, and $\Lambda\varnothing$ is $A \supset A$. P will then meet **RS1, RS2', RS3, RS4, RS∧, RS∨, RSI⊃, RSI¬**, and **RS⊥**. Proof of the fact is omitted.

Consequently, Theorem 4 translates into the following theorem, which reveals the significance of relative probability functions for L_I. The function S_P is now defined thus:

$$S_P(C) = \{D : P(D,E) = 1 \textit{ for every statement } E \textit{ such that } P(E,C) = 1\}.$$

Theorem 8. *Let P be a relative probability function for L_I. Then*

(a) $P(A,C) = 1$ *if and only if* $\{C\} \cup S_P(C) \vdash A$;

(b) *$P(A,C) = 0$ if and only if $\{A\} \cup \{C\} \cup S_P(C)$ is inconsistent.*

We conclude that $P(A,C)$ is the degree to which the explicit assumption C in conjunction with the implicit assumptions in $S_P(C)$ implies A, i.e. constitutes evidence for A.

It is also worth noting that the constraints **RI7–RI9** reflect the standard intuitionistic characterisation of connectives. $P(\neg A,C)$ is the highest degree of support received from C and the implicit assumptions $S_P(C)$ by any statement which would conclusively refute A. $P(A \supset B,C)$ is the highest degree of support received from C together with $S_P(C)$ by any statement which, together with A, would conclusively establish B. And $P(A \vee B,C)$ is, when A and B are incompatible, the degree of support A derives from C together with $S_P(C)$ plus the degree of support B derives from C together with $S_P(C)$.

In the light of the one-to-one correspondence between functions on statements and functions on finite sets of statements Theorem 7 becomes:

Theorem 9. *Let A be a statement and X be a finite set of statements of L_I. Then $X \vdash A$ if and only if, for every relative probability function P on the statements of L_I and for every statement C, $P(A,C) = 1$ if $P(B,C) = 1$ for every statement B in X.*

Proof: Suppose $X \vdash A$, i.e., by Theorem 7, $P(\{A\},X \cup Z) = 1$ for every intuitionistic relative probability function on finite sets of statements and every set Z. Hence $P(\{A\},X \cup \{C\}) = 1$ and, if $P(\{B\},\{C\}) = 1$ for every $B \in X$, then $P(X,\{C\}) = 1$ by Lemma 16(k), and so $P(\{A\} \cup X,\{C\}) = 1$ by **RS3**. So, $P(\{A\},\{C\}) = 1$ by Lemma 16(k). Now suppose that $P(\{A\},\{C\}) = 1$ if $P(\{B\},\{C\}) = 1$ for every $B \in X$. Since $P(X,\{\Lambda(X \cup Z)\}) = 1$ by Lemma 18(c) and **RS1**, $P(\{B\},\{\Lambda(X \cup Z)\}) = 1$ for every $B \in X$ by Lemma 16(k). Hence $P(\{A\},\{\Lambda(X \cup Z)\}) = 1$ and, by Lemma 18(c) again, $P(\{A\},X \cup Z) = 1$. So, $X \vdash A$ by Theorem 7. ¤

From the point of view of "probabilistic semantics" Theorem 9 constitutes a soundness and completeness theorem; we have already argued that "probabilistic semantics" is not a genuine semantics. In our view Theorem 9, in the light of Theorem 8, amounts to

$$X \vdash A \text{ if and only if } Y \cup X \vdash A \text{ for every set } Y \text{ of statements of } L_I.$$

Appendix I

Absolute Probability Functions for L_P

AP1 $0 \leq P(A)$

AP2 $P(\neg(A \wedge \neg A)) = 1$

AP3 $P(A) = P(A \wedge B) + P(A \wedge \neg B)$

AP4 $P(A \wedge B) \leq P(B \wedge A)$

AP5 $P(A \wedge (B \wedge C)) \leq P((A \wedge B) \wedge C)$

AP6 $P(A) \leq P(A \wedge A)$

Lemma 1. *Let P be a measure function, i.e. a function meeting constraints* **AP1** *and* **AP3–AP6** *and let A, B, and C be any statements of* L_P. *Then:*

(a) $P(A \wedge B) \leq P(A)$ *and* $P(A \wedge B) \leq P(B)$;

(b) *If* $P(A) = 0$, *then* $P(A \wedge B) = P(B \wedge A) = 0$;

(c) $P(A \wedge A) = P(A)$;

(d) $P(A \wedge B) = P(B \wedge A)$;

(e) $P((A \wedge B) \wedge C) = P(A \wedge (B \wedge C))$;

(f) $P(A \wedge \neg A) = P(\neg A \wedge A) = 0$;

(g) $P((A \wedge \neg A) \wedge B) = P((\neg A \wedge A) \wedge B) = P(A \wedge (B \wedge \neg B)) = P(A \wedge (\neg B \wedge B)) = 0$;

(h) $P(A) = P(\neg(B \wedge \neg B) \wedge A) = P(A \wedge \neg(B \wedge \neg B))$;

(i) $P(A \wedge \neg\neg B) = P(A \wedge B)$;

(j) *If* $P(A) = P(A \wedge B)$, *then* $P(A \wedge C) = P((A \wedge B) \wedge C) = P((B \wedge A) \wedge C)$;

(k) *If* $P(A) = P(A \wedge B) = P(B)$, *then* $P(A \wedge C) = P(B \wedge C)$;

(l) $P((A \wedge A) \wedge B) = P(A \wedge B)$;

(m) $P(A \wedge B) = P((A \wedge B) \wedge (B \wedge A))$;

(n) $P((A \wedge B) \wedge C) = P((B \wedge A) \wedge C)$;

(o) $P((A \wedge (B \wedge C)) \wedge ((A \wedge B) \wedge C)) = P(A \wedge (B \wedge C))$;

(p) $P((A \wedge (B \wedge C)) \wedge D) = P(((A \wedge B) \wedge C) \wedge D)$;

(q) $P((A_1 \wedge \ldots \wedge A_i \wedge \ldots \wedge A_n) \wedge A_i) = P(A_1 \wedge \ldots \wedge A_i \wedge \ldots \wedge A_n)$;

(r) $P((A_1 \wedge \ldots \wedge A_i \wedge \ldots \wedge A_n) \wedge \neg A_i) = 0$;

(s) $P(A \vee B) = P(A) + P(B) - P(A \wedge B)$;

(t) $P(A) \leq P(A \vee B)$ *and* $P(B) \leq P(A \vee B)$;

(u) $P(A \vee B) = 0$ *if and only if* $P(A) = P(B) = 0$;

(v) $P((A \vee B) \wedge C) = P(A \wedge C) + P(B \wedge C) - P((A \wedge B) \wedge C)$;

(w) *If* $P(A \wedge C) = P(A' \wedge C)$ *for every statement C and* $P(B \wedge C) = P(B' \wedge C)$ *for every statement C, then* $P((A \wedge B) \wedge C) = P((A' \wedge B') \wedge C)$ *for every statement C;*

(x) *If* $P(A \wedge C) = P(A' \wedge C)$ *for every statement C, then* $P(\neg A \wedge C) = P(\neg A' \wedge C)$ *for every statement C;*

(y) *If* $P(A_i \wedge B) = 0$ *for* $i = 1, \dots, n$, *then* $P((A_1 \vee \dots \vee A_n) \wedge B) = 0$.

Proof:

Note: IE is not used in the proofs of (a) through (r), nor do the lemmas from (s) on play any role in the justification of IE in Chapter 1, Section 1.

(a) $P(A)$

$= P(A \wedge B) + P(A \wedge \neg B)$ **(AP3)**

$\geq P(A \wedge B)$ **(AP1)**

Hence $P(A \wedge B) \leq P(B)$ by **AP4**.

(b) By (a) and **AP1**.

(c) By (a) and **AP6**.

(d) By **AP4**.

(e) $P((A \wedge B) \wedge C)$

$= P(C \wedge (A \wedge B))$ ((d))

$\leq P((C \wedge A) \wedge B)$ **(AP5)**

$\leq P(B \wedge (C \wedge A))$ ((d))

$\leq P((B \wedge C) \wedge A)$ **(AP5)**

$\leq P(A \wedge (B \wedge C))$ ((d))

$= P(A \wedge (B \wedge C))$ **(AP5)**

(f) $P(A) = P(A \wedge A) + P(A \wedge \neg A)$ by **AP3.** So, $P(A \wedge \neg A) = 0$ by (c). So (f) by (d).

(g) By (f) and (b).

(h) $P(A)$

$= P(A \wedge (B \wedge \neg B)) + P(A \wedge \neg(B \wedge \neg B))$ **(AP3)**

$= P(A \wedge \neg(B \wedge \neg B))$ ((g))

Hence (h) by (d).

(i) $P(A)$

$= P(A \wedge \neg B) + P(A \wedge \neg\neg B)$ **(AP3)**

$= P(A \wedge \neg B) + P(A \wedge B)$ **(AP3)**

So, $P(A \wedge \neg\neg B) = P(A \wedge B)$.

(j) Assume $P(A) = P(A \wedge B)$. Then $P(A \wedge \neg B) = 0$ by **AP3**, and both $P((C \wedge A) \wedge \neg B) = 0$ and $P((A \wedge C) \wedge \neg B) = 0$ by (b), (d), and (e). Hence $P(C \wedge A) = P((C \wedge A) \wedge B)$ and $P(A \wedge C) = P((A \wedge C) \wedge B)$ by **AP3**; and hence $P(A \wedge C) = P((A \wedge B) \wedge C) = P((B \wedge A) \wedge C)$ by (d) and (e).

(k) By (j) and (d).

(l) By (a) and (j).

(m) By (d), (e), and (a) $P(((A \wedge B) \wedge B) \wedge \neg A) \leq P(\neg A \wedge (A \wedge B))$. Hence $P(((A \wedge B) \wedge B) \wedge \neg A) = 0$ by (e) and (g). So,

$P(A \wedge B)$

$= P((A \wedge B) \wedge B) + P((A \wedge B) \wedge \neg B)$ (**AP3**)

$= P((A \wedge B) \wedge B)$ ((e), (g))

$= P(((A \wedge B) \wedge B) \wedge A) + P(((A \wedge B) \wedge B) \wedge \neg A)$ (**AP3**)

$= P(((A \wedge B) \wedge B) \wedge A)$

$= P((A \wedge B) \wedge (B \wedge A))$ ((e))

(n) By (k) and (m).

(o) By (a), (d), (e), and (g)

(1) $P(((A \wedge (B \wedge C)) \wedge A) \wedge \neg B) = 0$ and

(2) $P((((A \wedge (B \wedge C)) \wedge (A \wedge B)) \wedge \neg C) = 0$. So,

$P(A \wedge (B \wedge C))$

$= P((A \wedge (B \wedge C)) \wedge A) + P((A \wedge (B \wedge C)) \wedge \neg A)$ (**AP3**)

$= P((A \wedge (B \wedge C)) \wedge A)$ ((d), (e), (g))

$= P(((A \wedge (B \wedge C)) \wedge A) \wedge B)$ (**AP3** and (1))

$= P((A \wedge (B \wedge C)) \wedge (A \wedge B))$ ((e))

$= P(((A \wedge (B \wedge C)) \wedge (A \wedge B)) \wedge C)$ (**AP3,** (2))

$= P((A \wedge (B \wedge C)) \wedge ((A \wedge B) \wedge C))$ ((e))

(p) By (k) and (o).

(q) By (d), (l), (n), and (p).

(r) By (g), (n), and (p).

(s) $P(A \vee B)$

$= P((A \vee B) \wedge A) + P((A \vee B) \wedge \neg A)$ (**AP3**)

$= P(A) + P(B) - P(A \wedge B)$ (IE, **AP3,** (d))

(t) By (s) and (a).

(u) By (s), (t), and **AP1**.

(v) By (s) and IE.

(w) Assume $P(A \wedge C) = P(A' \wedge C)$ for every statement C and $P(B \wedge C) = P(B' \wedge C)$ for every statement C. Then

$P((A \wedge B) \wedge C)$

$= P(A \wedge (B \wedge C))$ ((e))

$= P(A' \wedge (B \wedge C))$ (Ass. on A)

$= P(B \wedge (A' \wedge C))$ (IE)

$= P(B' \wedge (A' \wedge C))$ (Ass. on B)

$= P((A' \wedge B') \wedge C))$ (IE)

(x) Assume $P(A \wedge C) = P(A' \wedge C)$ for every statement C. Then

$P(\neg A \wedge C)$

$= P(C) - P(A \wedge C)$ ((d), **AP3**)

$= P(C) - P(A' \wedge C)$ (Assumption)
$= P(\neg A' \wedge C)$ ((d), **AP3**)

(y) By induction on n. $P((A_1 \vee ... \vee A_n) \wedge B) = 0$ by the hypothesis of the induction and $P(A_{n+1} \wedge B) = 0$ by assumption. Hence $P(((A_1 \vee ... \vee A_n) \vee A_{n+1}) \wedge B) = 0$ by **AP4**, IE, and (u).

Lemma 2. *Let P be an absolute probability function of Popper's, i.e. a function meeting constraints* **AP1–AP6**, *and let A and B be any statements of* L_P. *Then:*
(a) $P(\neg A) = 1 - P(A)$;
(b) $P(A) \leq 1$;
(c) *There exist statements A and B of* L_P *such that* $P(A) \neq P(B)$;
(d) *If* $P(A) = 1$, *then* $P(A \wedge B) = P(B)$;
(e) *If* $P(A) = P(B) = 1$, *then* $P(A \wedge B) = 1$;
(f) *If* $P(A) = 1$, *then* $P(A \vee B) = 1$;
(g) *For every statement A of* L_P *there exists a statement B of* L_P *such that* $P(A) \leq P(B)$ *and* $P(A \wedge B) = P(A) \times P(B)$.[1]

Proof:
(a) $P(\neg A)$
$= P(\neg(A \wedge \neg A) \wedge \neg A)$ (Lemma 1(h))
$= P(\neg(A \wedge \neg A)) - P(\neg(A \wedge \neg A) \wedge A)$ (**AP3**)
$= 1 - P(\neg(A \wedge \neg A) \wedge A)$ (**AP2**)
$= 1 - P(A)$ (Lemma 1(h))

(b) By **AP1** and (a).

(c) By (b) and **AP2**.

(d) $P(A \wedge B) = P(B) - P(B \wedge \neg A)$ by Lemma 1(d) and **AP3**. But $P(\neg A) = 0$ by (a) and hence $P(B \wedge \neg A) = 0$ by Lemma 1(b). So, $P(A \wedge B) = P(B)$.

(e) By (d).

(f) By (d) and Lemma 1(s).

(g) Let A be an arbitrary statement of L_P. Then

$$P(A) \leq P(\neg(A \wedge \neg A))$$

by (b) and **AP2**. But

$$P(A \wedge \neg(A \wedge \neg A)) = P(A)$$

by Lemma 1(h) and **AP4**. So, by **AP2**

$$P(A \wedge \neg(A \wedge \neg A)) = P(A) \times P(\neg(A \wedge \neg A)).$$

So, for every statement A of L_P there exists a statement B of L_P such that

$$P(A) \leq P(B) \text{ and } P(A \wedge B) = P(A) \times P(B).$$

Lemma 3. *Let P be a measure function, i.e. a function meeting constraints* **AP1** *and* **AP3–**

AP6 *and assume that* $P(A) = P(A \wedge B) = P(B)$. *Then:*
(a) $P(D \wedge C) = P(D[A//B] \wedge C)$ *for every statement C, where* $D[A//B]$ *is the result of substituting A for none, or one, or several occurrences of B in D;*
(b) $P(D) = P(D[A//B])$.

Proof: (a) By induction on the complexity of D, using Lemmas 1(w) and 1(x).
(b) $P(\neg(C \wedge \neg C) \wedge D) = P(\neg(C \wedge \neg C) \wedge D[A//B])$ by (a). Hence, $P(D) = P(D[A//B])$ by Lemma 1(h).

Relative Probability Functions for L_P

RP1 $0 \leq P(A,B)$

RP2 $P(A,A) = 1$

RP3 *If B is P-normal,* $P(A,B) + P(\neg A,B) = 1$

RP4 $P(A \wedge B,C) = P(A,B \wedge C) \times P(B,C)$

RP5 $P(A \wedge B,C) \leq P(B \wedge A,C)$

RP6 $P(A,B \wedge C) \leq P(A,C \wedge B)$

RP7 *At least one statement of* L_P *is P-normal*

Lemma 4. *Let P be a relative probability function of Popper's i.e. a function meeting constraints* **RP1-RP7**. *Let A, B, and C be any statements of* L_P. *Then:*
(a) $P(A,B) \leq 1$;
(b) $P(A,A \wedge B) = P(A,B \wedge A) = 1$;
(c) $P(A \wedge A,B) = P(A,B)$;
(d) $P(A \wedge B,C) \leq P(A,C)$ *and* $P(A \wedge B,C) \leq P(B,C)$;
(e) *If* $P(A,C) = 0$, *then* $P(A \wedge B,C) = P(B \wedge A,C) = 0$;
(f) $P(A,\bot) = 1$, *i.e.* $\bot$ *is P-abnormal;*
(g) *If A is P-normal, then* $P(\neg A,A) = 0$;
(h) *If B is P-normal, then* $P(A \wedge \neg B,B) = P(\neg B \wedge A,B) = 0$;
(i) *If C is P-normal, then* $P(A,C) = P(A \wedge B,C) + P(A \wedge \neg B,C)$;
(j) $P(A \wedge B,C) = 1$ *if and only if* $P(A,C) = P(B,C) = 1$;
(k) *If B is P-normal, then* $P(\bot,B) = 0$;
(l) *If B is P-normal, then* $P(A \wedge \bot,B) = 0$;
(m) $P(\top,B) = P(\neg(A \wedge \neg A),B) = 1$;
(n) $P(A \wedge (B \wedge C),(A \wedge B) \wedge C) = P((A \wedge B) \wedge C,A \wedge (B \wedge C)) = 1$;
(o) $P(C,A) = P(C,B)$ *for every statement C if and only if* $P(A,B) = P(B,A) = 1$;
(p) $P(D,A \wedge (B \wedge C)) = P(D,(A \wedge B) \wedge C)$;
(q) $P(A \wedge (B \wedge C),D) = P((A \wedge B) \wedge C,D)$;
(r) *If* $P(A,C) = P(B,C)$ *for every statement C, then* $P(C,A) = P(C,B)$ *for every statement C;*
(s) $P_\top(A \wedge B) = P(A,B) \times P_\top(B)$;

(t) ⊤ *is P-normal;*

(u) $P(A \wedge B,B) = P(A,B)$;

(v) *If A is P-abnormal, then so is* $A \wedge B$;

(w) *If A is P-abnormal and B P-normal then* $P(A,B) = 0$;

(x) *If* $P(C,B) = 1$, *then* $P(A,B) = P(A,B \wedge C)$;

(y) $P(A \vee B,C) = P(A,C) + P(B,C) - P(A \wedge B,C)$;

(z) *If* $P(A \wedge B,C) = 0$, *then* $P(A \vee B,C) = P(A,C) + P(B,C)$;

(aa) $P(A,C) \leq P(A \vee B,C)$;

(bb) $P(A \vee B,A) = 1$;

(cc) $P((A \vee B) \wedge C,D) = P(A \wedge C,D) + P(B \wedge C,D) - P(A \wedge B) \wedge C,D)$;

(dd) *If* $P(A \wedge B,D) = 0$, *then* $P((A \vee B) \wedge C,D) = P(A \wedge C,D) + P(B \wedge C,D)$;

(ee) *If* $P(A,C) = P(A',C)$ *for every statement C and* $P(B,C) = P(B',C)$ *for every statement C, then* $P(A \wedge B,C) = P(A' \wedge B',C)$ *for every statement C;*

(ff) *If* $P(A,C) = P(A',C)$ *for every statement C, then* $P(\neg A,C) = P(\neg A',C)$ *for every statement C;*

(gg) $P(A,A \vee B) + P(B,A \vee B) - P(A \wedge B,A \vee B) = 1$;

(hh) *If* $P(A,A \vee B) = 0$, *then* $P(B,A \vee B) = 1$;

(ii) *If* $P(A,B) > 0$, *then* $P(A \vee C,B \vee C) > 0$;

(jj) $P(A,A \vee (B \vee C)) = P(A,A \vee B) \times P(A \vee B,A \vee (B \vee C))$;

(kk) $P(A,A \vee B) + (1 - P(A,B)) \times P(B,A \vee B) = 1$;

(ll) $A \vee B$ *is P-abnormal if and only if both A and B are P-abnormal;*

(mm) *A is P-abnormal if and only if* $P(\neg A,B) = 1$ *for every statement B;*

(nn) $A \sim_P \bot$ *if and only if A is P-abnormal;*

(oo) *If* $P(C,B) = 1$, *then* $P(A,B) = P(A \wedge C,B)$;

(pp) *If* $P(A,B) = 1$, *then* $P(A,A \vee B) = 1$;

(qq) *If* $P(B,C) = 1$, *then* $P(A,B) = P(A,B \vee C)$;

(rr) *If* $P(B,A) = 1$, *then* $P(A,B \vee C) = P(A \wedge B,B \vee C)$;

(ss) *If* $P(B,A) = 1$, *then* $P(A,B \vee C) = P(A,B) \times P(B,B \vee C)$;

(tt) *If* $P(B,A) = 1$, *then* $P(A,B \vee C) \leq P(A,B)$;

(uu) *If* $P(A,A \vee B) = P(B,A \vee B)$ *and* $P(B,B \vee C) = P(C,B \vee C)$, *then* $P(A,A \vee C) = P(C,A \vee C)$;

(vv) *If C is P-normal, then* $P(A \equiv B,C) = 1$ *if and only if* $P(A \wedge \neg B,C) = P(\neg A \wedge B,C) = 0$;

(ww) $P(A \equiv B,C) = 1$ *if and only if* $P(A,C) = P(A \wedge B,C) = P(B,C)$;

(xx) $P(A \equiv B,C) = 1$ *for every C if and only if* $P(A,C) = P(B,C)$ *for every C;*

(yy) *If* $P(A \equiv B,\top) = P(B \equiv C,\top) = 1$, *then* $P(A \equiv C,\top) = 1$;

(zz) *If* $P(A,B) = P(B,A) = 1$, *then* $P(A \equiv B,\top) = 1$;

(ab) *If* $P(A \equiv B,\top) = 1$ *and* $P(A,\top) > 0$, *then* $P(A,B) = P(B,A) = 1$;

(ac) $P(A \equiv B,\top) = 1$ *if and only if* $P(A,B) = P(B,A) = 1$ *or* $P(\neg A,\neg B) = P(\neg B,\neg A) = 1$;

(ad) *If* $A \succeq_P B$ *and A is P-abnormal, then so is B;*

(ae) *If $A \sim_P B$, then A is P-normal if and only if B is.*

Proof:

Note: IE is not used in the proofs of (a) through (r) nor do the lemmas from (s) on play any role in the justification of IE in Chapter 1, Section 2.

(a) $P(A,B) = 1$ when B is P-abnormal. So, suppose B is P-normal. Then $P(A,B) = 1 - P(\neg A,B)$ by **RP3**, and hence $P(A,B) \leq 1$ by **RP1**.

(b) $P(B \wedge A,A \wedge B) = 1$ by **RP2** and **RP5**. Hence $P(B,A \wedge (A \wedge B)) \times P(A,A \wedge B) = 1$ by **RP4**. Hence $P(A,A \wedge B) = 1$ by **RP1** and (a); and $P(A,B \wedge A) = 1$ by **RP6**.

(c) By **RP4** and (b).

(d) $P(A \wedge B,C) = P(A,B \wedge C) \times P(B,C)$ by **RP4**. But $0 \leq P(A,B \wedge C) \leq 1$ and $0 \leq P(B,C) \leq 1$ by **RP1** and (a). Hence $P(A \wedge B,C) \leq P(B,C)$, and hence $P(A \wedge B,C) \leq P(A)$ by **RP5**.

(e) By (d) and **RP1**.

(f) $P(B,B \wedge \neg B) = P(\neg B,B \wedge \neg B) = 1$ by (b). Hence $B \wedge \neg B$ is P-abnormal by **RP3**, and hence $P(A,B \wedge \neg B) = 1$ for every A in L_P.

(g) By **RP2** and **RP3**.

(h) Suppose B is P-normal. Then $P(A \wedge \neg B,B) = P(A,\neg B \wedge B) \times P(\neg B,B)$ by **RP4**, and hence $P(A \wedge \neg B,B) = 0$ by (g). So, $P(\neg B \wedge A,B) = 0$ by **RP5**.

(i) Suppose C is P-normal. Then, by definition in case $A \wedge C$ is P-abnormal and otherwise by **RP3**, $P(B,A \wedge C) + P(\neg B,A \wedge C) = P(C,A \wedge C) + P(\neg C,A \wedge C)$. Hence by (b) $P(B,A \wedge C) \times P(A,C) + P(\neg B,A \wedge C) \times P(A,C) = P(A,C) + P(\neg C,A \wedge C) \times P(A,C)$; hence by **RP4**, (h), and the hypothesis on C, $P(B \wedge A,C) + P(\neg B \wedge A,C) = P(A,C)$. So, (i) by **RP5**.

(j) If $P(A \wedge B,C) = 1$, then $P(A,C) = P(B,C) = 1$ by (d) and (a). Conversely, suppose that $P(A,C) = P(B,C) = 1$, and let C be P-normal. Then $P(\neg B,C) = 0$ by **RP3**, hence $P(A \wedge \neg B,C) = 0$ by (f), hence $P(A \wedge B,C) = P(A,C)$ by (i), and hence $P(A \wedge B,C) = 1$. But $P(A \wedge B,C) = 1$ by definition when C is P-abnormal. So, $P(A \wedge B,C) = 1$ in either case.

(k) By (i) and (c).

(l) $P(A \wedge (C \wedge \neg C),B) = P(A,(C \wedge \neg C) \wedge B) \times P(C \wedge \neg C,B)$ by **RP4**. Hence (l) by (k).

(m) By definition when B is P-abnormal; otherwise, by **RP3** and (k).

(n) $P((A \wedge B) \wedge C,(B \wedge C) \wedge ((A \wedge B) \wedge C)) = 1$ by (b)); hence $P(A,(B \wedge C) \wedge ((A \wedge B) \wedge C)) = 1$ by (j); and hence $P(A \wedge (B \wedge C),(A \wedge B) \wedge C) = P(B \wedge C,(A \wedge B) \wedge C)$ by **RP4**. But $P(C,(A \wedge B) \wedge C) = 1$ (b) and $P(B,(A \wedge B) \wedge C) = 1$ by (b) and (j). So, $P(B \wedge C,(A \wedge B) \wedge C) = 1$ by (j), and so $P(A \wedge (B \wedge C),(A \wedge B) \wedge C) = 1$. Similarly, $P((A \wedge B) \wedge C,A \wedge (B \wedge C)) = 1$.

(o) Suppose $P(C,A) = P(C,B)$ for every statement C. Then in particular $P(B,A) = P(B,B)$ and $P(A,B) = P(A,A)$. Hence $P(A,B) = P(B,A) = 1$ by **RP2**. Now suppose $P(A,B) = P(B,A) = 1$. Then $P(C \wedge A,B) = P(C,A \wedge B) \times P(A,B)$ and $P(C \wedge B,A) = P(C,B \wedge A) \times P(B,A)$ by

RP4. Hence $P(C \wedge A,B) = P(C \wedge B,A)$ by **RP6**. But since $P(C \wedge \neg A,B) = 0$ by **RP4** and **RP3**, $P(C \wedge A,B) = P(C,B)$ by (i) if B is P-normal, while $P(C \wedge A,B) = P(C,B)$ by definition if B is P-abnormal. Similarly, $P(C \wedge B,A) = P(C,A)$. So $P(C,A) = P(C,B)$ for every statement C.

(p) By (n) and (o).

(q) $P(A \wedge (B \wedge C),D)$

$$
\begin{aligned}
&= P(A,(B \wedge C) \wedge D) \times P(B,C \wedge D) \times P(C,D) && (\textbf{RP4})\\
&= P(A,B \wedge (C \wedge D)) \times P(B,C \wedge D) \times P(C,D) && ((\text{p}))\\
&= P(A \wedge B,C \wedge D) \times P(C,D) && (\textbf{RP4})\\
&= P((A \wedge B) \wedge C,D) && (\textbf{RP4})
\end{aligned}
$$

(r) Suppose that $P(A,C) = P(B,C)$ for every statement C, and let C be an arbitrary statement. Then

$P(C,A)$

$$
\begin{aligned}
&= P(A,C \wedge A) \times P(C,A) && ((\text{b}))\\
&= P(B,C \wedge A) \times P(C,A) && (\text{Assumption})\\
&= P(B \wedge C,A) && (\textbf{RP4})\\
&= P(C \wedge B,A) && (\textbf{RP5})\\
&= P(C,B \wedge A) \times P(B,A) && (\textbf{RP4})\\
&= P(C,B \wedge A) \times P(A,A) && (\text{Assumption})\\
&= P(C,B \wedge A) && (\textbf{RP2}).
\end{aligned}
$$

But $P(C,B) = P(C,A \wedge B)$ by simply interchanging A and B throughout. Hence $P(C,A) = P(C,B)$ by **RP6**.

(s) By **RP4** and IE.

(t) Suppose $\neg(A \wedge \neg A)$ is P-abnormal, and let B and C be arbitrary statements. Then $P(B \wedge C,\neg(A \wedge \neg A)) = 1$, hence $P(B,C \wedge \neg(A \wedge \neg A)) \times P(C,\neg(A \wedge \neg A)) = 1$ by **RP4**, and hence $P(B,C) = 1$ by **RP1**, (a), and (m). So, contrary to **RP7**, $P(B,C) = 1$ for any statements B and C. Hence (t).

(u) By **RP4** and **RP2**.

(v) Suppose A is P-abnormal, and let C be an arbitrary statement of L_P. Then, $P(C \wedge B,A) = 1$, hence $P(C,A \wedge B) \times P(B,A) = 1$ by **RP4** and **RP6**, hence $P(C,A \wedge B) = 1$, and hence $A \wedge B$ is P-abnormal.

(w) $P(A \wedge \neg A,B) = 0$ by (k). So $P(\neg A,A \wedge B) \times P(A,B) = 0$ by **RP4** and **RP5**. But $P(\neg A,A \wedge B) = 1$ by (v). So, $P(A,B) = 0$.

(x) When B is P-abnormal, so is $B \wedge C$ by (v), and hence $P(A,B) = P(A,B \wedge C)$. So, suppose B is P-normal. Then, $P(A,B) = P(A,B \wedge C) \times P(C,B) + P(A,B \wedge \neg C) \times P(\neg C,B)$ by (i) and **RP4**. So, if $P(C,B) = 1$, in which case $P(\neg C,B) = 0$ by **RP3**, then $P(A,B) = P(A,B \wedge C)$.

(y) By definition when C is P-abnormal. So, suppose C is P-normal. Then $P(A \vee B,C) = P((A \vee B) \wedge A,C) + P((A \vee B) \wedge -A),C)$ by (i), hence $P(A \vee B,C) = P(A,C) + P(A,C) +$

$P(B \wedge \neg A, C)$ by IE.

(z) By (y).

(aa) By (z) and **RP1**.

(bb) By (aa) and **RP2**.

(cc) By **RP4** and (y).

(dd) By (cc) and (e).

(ee) Assume $P(A,C) = P(A',C)$ for every statement C and $P(B,C) = P(B',C)$ for every statement C. Then:

$P(A \wedge B,C)$

$= P(A, B \wedge C) \times P(B,C)$	(**RP4**)
$= P(A', B \wedge C) \times P(B,C)$	(Ass. on A)
$= P(A' \wedge B, C)$	(**RP4**)
$= P(B \wedge A', C)$	(**RP5**)
$= P(B, A' \wedge C) \times P(A',C)$	(**RP4**)
$= P(B', A' \wedge C) \times P(A',C)$	(Ass. on B)
$= P(B' \wedge A', C)$	(**RP4**)
$= P(A' \wedge B', C)$	(**RP5**)

(ff) $P(\neg A,C) = P(\neg A',C)$ by definition if C is P-abnormal. So, assume that C is P-normal. Then: $P(\neg A,C)$

$= 1 - P(A,C)$	(**RP3**)
$= 1 - P(A',C)$	(Assumption)
$= P(\neg A',C)$	(**RP3**)

(gg) By (y) and **RP2**.

(hh) If $P(A, A \vee B) = 0$, then $P(A \wedge B, A \vee B) = 0$ by (e), and hence $P(B, B \vee A) = 1$ by (gg).

(ii) If $P(C, B \vee C) > 0$, then $P(A \vee C, B \vee C) > 0$ by (aa) and IE. So, suppose $P(C, B \vee C) \not> 0$, and hence $P(C, B \vee C) = 0$ by **RP1**. Then $P(B, B \vee C) = 1$ by (hh). But $P((A \vee C) \wedge B, B \vee C) = P(A \vee C, B) \times P(B, B \vee C)$ by **RP4** and IE. $P((A \vee C) \wedge B, B \vee C) = P(A \vee C, B)$, hence $P(A,B) \leq P((A \vee C) \wedge B, B \vee C)$ by (aa) again, and hence $P(A,B) \leq P(A \vee C, B \vee C)$ by (d). So, if $P(A,B) > 0$, then $P(A \vee C, B \vee C) > 0$.

(jj) $P(A, A \vee (B \vee C))$

$= P(A \wedge (A \vee B), (A \vee B) \vee C)$	(IE)
$= P(A, A \vee B) \times P(A \vee B, (A \vee B) \vee C)$	(**RP4**, IE)
$= P(A, A \vee B) \times P(A \vee B, A \vee (B \vee C))$	(IE)

(kk) By (gg), **RP4**, and IE.

(ll) Suppose $A \vee B$ is P-abnormal. Then by definition $P(C \wedge A, A \vee B) = P(A, A \vee B) = 1$ for every statement C, hence by **RP4** $P(C, A \wedge (A \vee B)) = 1$, hence by IE and (r) A is P-abnormal. And by a similar reasoning B is P-abnormal. Suppose now that $A \vee B$ is P-normal, but A and B are P-abnormal. Then $P(A, A \vee B) = P(B, A \vee B) = 0$ by (w). Hence $P(A \wedge B, A \vee B) = 0$ by

(e) and hence $P(A \vee B, A \vee B) = 0$ by (z), q.e.a..

(mm) Suppose first that A is P-abnormal and let B be P-normal. Then $P(A,B) = 0$ by (w) and hence $P(\neg A,B) = 1$ by **RP3**. But $P(\neg A,B) = 1$ by definition if B is P-abnormal. Suppose next that $P(\neg A,B) = 1$ for every statement B. Then $P(\neg A,A) = 1$ and hence A is P-abnormal by (g).

(nn) By (f) and (ll) A is P-abnormal if and only if $A \vee \bot$ is P-abnormal. And by (k) $A \vee \bot$ is P-abnormal if and only if $P(A,A \vee \bot) > 0$ and $P(\bot,A \vee \bot) > 0$.

(oo) By (y) and (bb).

(pp) By (kk).

(qq) Suppose $P(B,C) = 1$. Then $P(A \wedge B,B \vee C)$ equals $P(A,B)$ by IE and (pp). But since $P(B,B \vee C) = 1$ by (pp), $P(A \wedge B,B \vee C) = P(A,B \vee C)$ by (oo). So $P(A,B) = P(A,B \vee C)$.

(rr) Suppose $P(B,A) = 1$. Then $P(B \wedge (B \vee C),A) = 1$ by IE, hence $P(B,(B \vee C) \wedge A) \times P(B \vee C,A) = 1$ by **RP4**, and hence $P(B,A \wedge (B \vee C)) = 1$ by **RP1**, (a) and **RP6**. But $P(B \wedge A,B \vee C) = P(B,A \wedge (B \vee C)) \times P(A,B \vee C)$ by **RP4**. So $P(A,B \vee C) = P(A \wedge B,B \vee C)$ by **RP5**.

(ss) By (rr) and **RP4**.

(tt) By (ss) and **RP1**.

(uu) Suppose $P(A,A \vee B) = P(B,A \vee B)$ and $P(B,B \vee C) = P(C,B \vee C)$. By (hh) $P(B,A \vee B) > 0$ and $P(B,B \vee C) > 0$, and hence by IE and (ii) $P(A \vee B,A \vee (B \vee C)) > 0$ and $P(B \vee C,A \vee (B \vee C)) > 0$. Since $P(A,A \vee (B \vee C)) = P(A,A \vee B) \times P(A \vee B,A \vee (B \vee C))$ and $P(B,A \vee (B \vee C)) = P(B,A \vee B) \times P(A \vee B,A \vee (B \vee C))$ by IE and (jj), $P(A,A \vee (B \vee C)) = P(B,A \vee (B \vee C)) > 0$. Similarly, $P(B,A \vee (B \vee C)) = P(C,A \vee (B \vee C)) > 0$, and hence $P(A,A \vee (B \vee C)) = P(C,A \vee (B \vee C)) > 0$. Therefore $P(A,A \vee C) \times P(A \vee C,A \vee (B \vee C)) = P(C,A \vee C) \times P(A \vee C,A \vee (B \vee C)) > 0$ by IE and (jj). Hence $P(A,A \vee C) = P(C,A \vee C)$.

(vv) Suppose C P-normal. Then $P(A \equiv B,C) = P(\neg(A \wedge \neg B) \wedge \neg(\neg A \wedge B),C)$ by the definition of '$\equiv$'. And $P(\neg(A \wedge \neg B) \wedge \neg(\neg A \wedge B),C) = 1$ if and only if $P(\neg(A \wedge \neg B),C) = P(\neg(\neg A \wedge B),C) = 1$ by (j), if and only if $P(A \wedge \neg B,C) = P(\neg A \wedge B,C) = 0$ by **RP3**.

(ww) Trivially, $P(A \equiv B,C) = 1$ if and only if $P(A,C) = P(A \wedge B,C) = P(B,C)$ if C is P-abnormal. If C is P-normal, $P(A \equiv B,C) = 1$ if and only if $P(A \wedge \neg B,C) = P(\neg A \wedge B,C) = 0$ by (vv). But $P(A \wedge \neg B,C) = P(\neg A \wedge B,C) = 0$ if and only if $P(A,C) = P(A \wedge B,C) = P(B,C)$ by (i).

(xx) If $P(A \equiv B,C) = 1$ for every statement C, then $P(A,C) = P(B,C)$ by (ww). Now suppose $P(A,C) = P(B,C)$ for every statement C. Then

$P(A \wedge B,C)$

$= P(A,B \wedge C) \times P(B,C)$ (**RP4**)

$= P(B,B \wedge C) \times P(B,C)$

$= P(B,C)$ ((b))

(yy) Suppose $P(A \equiv B,\mathrm{T}) = P(B \equiv C,\mathrm{T}) = 1$. Then

$$P(A \wedge \neg C,\mathrm{T})$$
$$= P((A \wedge \neg C) \wedge B,\mathrm{T}) + P((A \wedge \neg C) \wedge \neg B,\mathrm{T}) \quad \text{((i))}$$
$$= P(A \wedge (B \wedge \neg C),\mathrm{T}) + P((A \wedge \neg B) \wedge \neg C,\mathrm{T}) \quad \text{(IE)}$$
$$= 0 \quad \text{((vv), (e))}$$

Similarly, $P(\neg A \wedge C,\mathrm{T}) = 0$, and hence $P(A \equiv C,\mathrm{T}) = 1$ by (vv).

(zz) Suppose $P(A,B) = P(B,A) = 1$. Then $P(A \wedge B,\mathrm{T}) = P(B,\mathrm{T})$ and $P(B \wedge A,\mathrm{T}) = P(A,\mathrm{T})$ by **RP4** and IE. Hence $P(A \equiv B,\mathrm{T}) = 1$ by (ww).

(ab) Suppose $P(A \equiv B,\mathrm{T}) = 1$ and $P(A,\mathrm{T}) > 0$. Then by (b) $P(A,\mathrm{T})$, $P(B,\mathrm{T})$, and $P(A \wedge B,\mathrm{T})$ are identical and larger than 0. Hence $P(A,B) = P(B,A) = 1$ by (s) and **RP5**.

(ac) Suppose first that $P(A \equiv B,\mathrm{T}) = 1$. By **RP3** and (t) either $P(A,\mathrm{T}) > 0$ or $P(\neg A,\mathrm{T}) > 0$. Hence either $P(A,B) = P(B,A) = 1$ or $P(\neg A,\neg B) = P(\neg B,\neg A) = 1$ by (ab) and IE. Suppose next that either $P(A,B) = P(B,A) = 1$ or $P(\neg A,\neg B) = P(\neg B,\neg A) = 1$. If $P(A,B) = P(B,A) = 1$, then $P(A \equiv B,\mathrm{T}) = 1$ by (zz); and if $P(\neg A,\neg B) = P(\neg B,\neg A) = 1$, then $P(\neg A \equiv \neg B,\mathrm{T}) = 1$ by (zz), i.e. by IE $P(A \equiv B,\mathrm{T}) = 1$ in either case.

(ad) Suppose $A \succeq_P B$ and A is P-abnormal. If B is P-normal, then so is $A \vee B$ by (ll), and hence by (w) $P(A,A \vee B) = 0$, contrary to the supposition. So B is P-abnormal.

(ae) By (ad).

Lemma 5 *Let P be a binary function that meets constraints* **RP1**-**RP7**, *and let* **A** = $\langle A_1, A_2, \dots, A_n, \dots\rangle$ *be any sequence of statements, and B and C any statements of* L_P. *Then:*
(a) *If* $P(A_i \wedge B,C) = 0$ for $i = 1, \dots, n$, *then* $P((A_1 \vee \dots \vee A_n) \wedge B,C) = 0$;
(b) *If* $m \leq m + k \leq n$, *then*

$$P(A_1 \wedge \dots \wedge A_n,A_1 \wedge \dots \wedge A_m) \leq P(A_1 \wedge \dots \wedge A_n,A_1 \wedge \dots \wedge A_{m+k});$$

(c) $\lim_{m\to\infty} \lim_{n\to\infty} P(B_1 \vee \dots \vee B_m,B_1 \vee \dots \vee B_n)$ *equals either* 0 *or* 1;
(d) $\lim_{m\to\infty} \lim_{n\to\infty} P(B_1 \wedge \dots \wedge B_n,B_1 \wedge \dots \wedge B_m)$ *equals either* 0 *or* 1.

Proof: (a) Assume $P(A_i \wedge B,C) = 0$ for $i = 1, \dots, n$. Proof is then by induction on n. $P(((A_1 \vee \dots \vee A_n) \vee A_{n+1}) \wedge B,C) = P((A_1 \vee \dots \vee A_n) \wedge B,C) + P(A_{n+1} \wedge B,C)$ by (dd). But $P((A_1 \vee \dots \vee A_n) \wedge B,C)$ by the inductive hypothesis and $P(A_{n+1} \wedge B,C) = 0$ by assumption.

(b) $P(A_1 \wedge \dots \wedge A_n,A_1 \wedge \dots \wedge A_m)$

$$= P((A_1 \wedge \dots \wedge A_n) \wedge (A_1 \wedge \dots \wedge A_{m+k}),A_1 \wedge \dots \wedge A_m) \quad \text{(IE)}$$
$$= P(A_1 \wedge \dots \wedge A_n,A_1 \wedge \dots \wedge A_{m+k}) \times P(A_1 \wedge \dots \wedge A_{m+k},A_1 \wedge \dots \wedge A_m) \quad \text{(\textbf{RP4}, IE)}$$
$$\leq P(A_1 \wedge \dots \wedge A_n,A_1 \wedge \dots \wedge A_{m+k}) \quad \text{(Lemma 4(a))}$$

(c) $\lim_{m\to\infty} \lim_{n\to\infty} P(A_1 \vee \dots \vee A_n,A_1 \vee \dots \vee A_m)$

$$= \lim_{m\to\infty} \lim_{k\to\infty} \lim_{n\to\infty} [P(A_1 \vee \dots \vee A_n,A_1 \vee \dots \vee A_k) \times P(A_1 \vee \dots \vee A_k,A_1 \vee \dots \vee A_m)] \quad \text{(\textbf{RP4})}$$
$$= [\lim_{m\to\infty} \lim_{n\to\infty} P(A_1 \vee \dots \vee A_n,A_1 \vee \dots \vee A_m)]^2.$$

(d) Analogously.

Lemma 6. *Let P be a binary function meeting constraints* **RP1–RP7**, *suppose that* $P(A,C) = P(B,C)$ *for every statement C, and let D[A//B] be the result of substituting A for none, one, or several occurrences of B in D. Then:*

(a) $P(D,C) = P(D[A//B],C)$ *for any statements C and D;*

(b) $P(C,D) = P(C,D[A//B])$ *for any statements C and D.*

Proof: (a) By induction on the complexity of D, using Lemmas 4(ee) and 4(ff).

(b) By (a) and Lemma 4(r).

Intuitionistic Absolute Probability Functions

AI1 $0 \leq P(A))$

AI2 $\max_{A \in L}\{P(A)\} = 1$

AI3 $\max_{B \in L}\{P(A \wedge B)\} = P(A)$

AI4 $P(A \wedge B) \leq P((A \wedge B) \wedge B)$

AI5 $P(A \wedge (B \wedge C)) = P(A \wedge (C \wedge B))$

AI6 $P(A \wedge (B \wedge (C \wedge D))) = P(A \wedge ((B \wedge C) \wedge D))$

AI7 $P(A \wedge (B \vee C)) = P(A \wedge B) + P(A \wedge C) - P(A \wedge (B \wedge C))$

AI8 $P((A \supset B) \wedge C) = \max_{D \in L}\{P(C \wedge D) : P((D \wedge A) \wedge B) = P(D \wedge A)\}$

AI9 $P(A \wedge \neg B) = \max_{C \in L}\{P(A \wedge C) : P(C \wedge B) = 0\}$

Lemma 7. *Let P be a unary function on the statements of* L_I *meeting constraints* **AI1–AI9**. *Then:*

(a) $P(A \wedge B) \leq P(A)$;

(b) $P(A \wedge B) = P((A \wedge B) \wedge B)$;

(c) $P(A) = P((B \supset B) \wedge A)$;

(d) $P(A \wedge B) = P(B \wedge A)$;

(e) $P(A \wedge (B \wedge C)) = P((A \wedge B) \wedge C)$;

(f) *Let* $C_1, \ldots, C_n$ *be distinct statements of L and let A and B be pure conjunctive statements, i.e. statements whose only connective is '*$\wedge$*'. Let A' be the result of uniformly substituting for every atomic statement in A one of* $C_1, \ldots, C_n$ *and be such that every one of* $C_1, \ldots, C_n$ *is substituted for at least one atomic statement in A. Similarly, let B' be the result of uniformly substituting for every atomic statement in B one of* $C_1, \ldots, C_n$ *and be such that every one of* $C_1, \ldots, C_n$ *is substituted for at least one atomic statement in B. Then* $P(A') = P(B')$;

(g) $P(A \vee B) = P(A) + P(B) - P(A \wedge B)$;

(h) $P(\neg A) = \max_{B \in L}\{P(B) : P(B \wedge A) = 0\}$;

(i) $P(A) = P(A \wedge A)$;

(j) $P(A \supset B) = \max_{C \in L}\{P(C) : P((C \wedge A) \wedge B) = P(C \wedge A)\}$;

(k) $P(((A \supset B) \wedge A) \wedge B) = P((A \supset B) \wedge A) = P(A \wedge B)$;

(l) $P(A \wedge \neg A) = 0$;

(m) $P(A \supset A) = 1$;

(n) $P(A) \leq P(A \vee B)$;

(o) *If* $P(A) = 1$, *then* $P(A \vee B) = 1$;

(p) *If* $P(A) = 1$, *then* $P(A \wedge B) = P(B)$;

(q) *If* $P(A) = 1$ *and* $P(A \supset B) = 1$, *then* $P(B) = 1$;

(r) *If* $P(A) = 0$, *then* $P(A \wedge B) = 0$ *for every statement* B;

(s) *If* $P(A) = 0$, *then* $P(\neg A) = 1$;

(t) $P(\neg(A \wedge \neg A)) = 1$.

Proof: (a) By **AI3**.

(b) By **AI4** and (a).

(c) $P((B \supset B) \wedge A)$

$$= \max_{C \in L}\{P(A \wedge D) : P(((A \wedge D) \wedge B) \wedge B) = P((A \wedge D) \wedge B)\} \qquad \textbf{(AI8)}$$
$$= \max_{C \in L}\{P(A \wedge D)\} \qquad \text{((b))}$$
$$= P(A) \qquad \textbf{(AI3)}$$

(d) By (c) and **AI5**.

(e) By (c) and **AI6**.

(f) By **AI5**, **AI6**, (b), (d), and (e).

(g) By (c) and **AI7**.

(h) By (c) and **AI9**.

(i) By (b), (e), and (c).

(j) $P(A \supset B)$

$$= P((A \supset B) \wedge (C \supset C)) \qquad \text{((c), (d))}$$
$$= \max_{C \in L}\{P((C \supset C) \wedge D) : P((((C \supset C) \wedge D) \wedge A) \wedge B) = P(((C \supset C) \wedge D) \wedge A)\} \qquad \textbf{(AI8)}$$
$$= \max_{C \in L}\{P(D) : P((D \wedge A) \wedge B) = P(D \wedge A)\} \qquad \text{((f), (c))}$$

(k) (i) $P(((A \supset B) \wedge A) \wedge B)$

$$= P((A \supset B) \wedge (A \wedge B)) \qquad \text{((e))}$$
$$= \max_{C \in L}\{P((A \wedge B) \wedge D) : P((((A \wedge B) \wedge D) \wedge A) \wedge B) = P(((A \wedge B) \wedge D) \wedge A)\} \qquad \textbf{(AI8)}$$
$$= \max_{C \in L}\{P((A \wedge B) \wedge D)\} \qquad \text{((f))}$$
$$= P(A \wedge B) \qquad \textbf{(AI3)}$$

(ii) $P((A \supset B) \wedge A) = P(A \wedge B)$ by a similar reasoning.

(l) By **AI9** and (f).

(n) By (g), (a), and (d).

(o) By (n) and **AI2**.

(p) By (a) and (g).

(q) Suppose $P(A) = P(A \supset B) = 1$. Then $P((A \supset B) \wedge A) = 1$ by (p); so $P(((A \supset B) \wedge A) \wedge B) = P(B)$, again by (p). Hence $P(B) = 1$ by (k).

(r) By (a) and **AI1**.

(s) By (h), (r) and **AI2**.

(t) By (s) and (l).

The Auxiliary Function F_P

Definition of F_P.

(i)

$$F_P(X,Y) = \begin{cases} P(X) & \textit{if } Y = \varnothing \\ F_P(X,Y-\{A\}) - F_P(\{A\} \cup X, Y-\{A\}) & \textit{otherwise,} \end{cases}$$

when Y is finite and A is the alphabetically latest statement in Y;

(ii) $\quad F_P(X,Y) = \inf_{Y'} \{F_P(X,Y') : Y' \subseteq_f Y\}$,

when Y is infinite.

Lemma 8. *Let P be a unary function defined on the finite sets of statements of L, and let X and Y be two such sets. Then:*

(a) *If* $X \cap Y \neq \varnothing$, *then* $F_P(X,Y) = 0$;

(b) $F_P(X,Y) = F_P(X,Y \cup \{A\}) + F_P(\{A\} \cup X,Y)$;

(c) $F(X,Y) - F_P(\{A_1,...,A_{n-1}\} \cup X,Y) =$

$$F_P(X,Y \cup \{A_1\}) + F_P(\{A_1\} \cup X,Y \cup \{A_2\}) + ... + F_P(\{A_1,...,A_{n-1}\} \cup X,Y \cup \{A_n\}).$$

Proof:

(a) Assume that $A \in X$ and $A \in Y$. Proof is by induction on the cardinality of Y.

Basis: card(Y) = 1, i.e. $Y = \{A\}$. Then $F_P(X,\{A\}) = 0$ by the definition of F_P.

Inductive Step: card(Y) > 1. Let B be the alphabetically latest statement in Y.

Case 1: B is A. Then $F_P(X,Y) = F_P(X,Y-\{A\}) - F_P(X,Y-\{A\}) = 0$.

Case 2: B is a statement other than A. Then

$F_P(X,Y)$

$= F_P(X,Y-\{B\}) - F_P(\{B\} \cup X,Y-\{B\})$ (Def. F_P)

$= 0$ (Ind. Hyp.)

(b) By (a) if $A \in Y$. So, assume that $A \notin Y$. Then proof is by induction on the cardinality of Y.

Basis: card(Y) = 0, i.e. $Y = \varnothing$. Then $F_P(X,\{A\}) = F_P(X,\varnothing) - F_P(\{A\} \cup X,\varnothing)$ by the definition of F_P.

Inductive Step: card(Y) > 0. Let B be the alphabetically latest statement in $Y \cup \{A\}$.

Case 1: B is A. Then $F_P(X,Y \cup \{A\}) = F_P(X,Y) - F_P(\{A\} \cup X,Y)$ by the definition of F_P.

Case 2: B *is* a statement other than Y. Then

$$
\begin{aligned}
&F_P(X,Y \cup \{A\}\\
&\quad = F_P(X,(Y-\{B\}) \cup \{A\}) - F_P(\{B\} \cup X,(Y-\{B\}) \cup \{A\}) && \text{(Def. } F_P\text{)}\\
&\quad = F_P(X,Y-\{B\}) - F_P(\{A\} \cup X,Y-\{B\}) - F_P(\{B\} \cup X,Y-\{B\})\\
&\qquad\qquad + F_P(\{A,B\} \cup X,Y-\{B\}) && \text{(Ind. Hyp.)}\\
&\quad = F_P(X,Y) - F_P(\{A\} \cup X,Y) && \text{(Def. } F_P\text{)}
\end{aligned}
$$

(c) By repeated use of (b).

Absolute Probability Functions for Sets of Statements (Language-Independent Constraints)

AS1' $0 \leq F_P(X,Y)$

AS2 $P(\varnothing) = 1$

Lemma 9. *Let P be a unary function defined on the finite sets of statements of L which satisfies* **AS1'** *and* **AS2**; *and let X, Y, and Z be finite sets of statements of L. Then:*

(a) $0 \leq P(X)$;

(b) $F_P(Z \cup X,Y) \leq F_P(X,Y)$;

(c) $F_P(X,Y \cup Z) \leq F_P(X,Y)$;

(d) $P(Z \cup X) \leq P(X)$;

(e) $F_P(X,Y) \leq P(X)$;

(f) *If* $P(X) = 0$, *then* $P(Z \cup X) = 0$;

(g) $0 \leq 1 - P(\{A\}) - P(X) + P(\{A\} \cup X)$.

Proof: (a) By the Basic Clause in the definition of F_P and the assumption that F_P meets **AS1'**.

(b) By induction on the cardinality of Z.

Basis: card(Z) = 0, i.e. $Z = \varnothing$. Then trivially $F_P(Z \cup X,Y) \leq F_P(X,Y)$.

Inductive Step: card(Z) > 0. Let A be any member of Z. Then

$$
\begin{aligned}
&F_P(Z \cup X,Y)\\
&\quad = F_P((Z - \{A\}) \cup X,Y) - F_P((Z - \{A\}) \cup X,Y \cup \{A\}) && \text{(Lemma 8(b))}\\
&\quad \leq F_P((Z - \{A\}) \cup X,Y) && \textbf{(AS1')}\\
&\quad \leq F_P(X,Y) && \text{(Ind. Hyp.)}
\end{aligned}
$$

(c) Proof like that of (b), but with the induction run on the second argument.

(d) By (b) and the Basic Clause in the definition of F_P.

(e) By (c) and the Basic Clause in the definition of F_P.

(f) By (a) and (d).

(g) By induction on the cardinality of X.

Basis: card(X) = 0, i.e. $X = \varnothing$. Then $0 = P(\varnothing) - P(\{A\}) - P(X) + P(\{A\} \cup X)$ and hence $0 \leq 1 - P(\{A\}) - P(X) + P(\{A\} \cup X)$ by **AS2**.

Inductive Step: card(X) > 0. Let B be any member of X. Then $0 \leq 1 - P(\{A\}) - P(X - \{B\}) + P(\{A\} \cup (X - \{B\}))$ by the inductive hypothesis and

$$0 = F_P(X-\{B\},\varnothing) - F_P(X,\varnothing) - F_P(\{A\} \cup (X-\{B\}),\{B\}) - F_P(X-\{B\},\{A,B\})$$

(Lemma 8(b))

$\leq F_P(X-\{B\},\varnothing) - F_P(X,\varnothing) - F_P(\{A\} \cup (X-\{B\}),\{B\})$ **(AS1')**

$\leq F_P(X-\{B\},\varnothing) - F_P(X,\varnothing) - F_P(\{A\} \cup (X-\{B\}),\varnothing) + F_P(\{A\} \cup X,\varnothing)$

(Lemma 8(b))

I.e. $0 \leq P(X - \{B\}) - P(\{A\} \cup (X - \{B\})) - P(X) + P(\{A\} \cup X)$. Hence $0 \leq 1 - P(\{A\}) - P(X) + P(\{A\} \cup X)$.

Absolute Probability Functions for Sets of Statements

AS1	$0 \leq P(X)$
AS2	$P(\varnothing) = 1$
AS¬	$P(X \cup \{\neg A\}) = P(X) - P(X \cup \{A\})$
AS∧	$P(X \cup \{A \wedge B\}) = P(X \cup \{A,B\})$
ASω	$P(X) = \inf_{X'} \{P(X') : X' \subseteq_f X\}$
ASΛ	$P(X \cup \{\Lambda\mathbf{A}\}) = \lim_{n\to\infty} P(X \cup \{A_1,\ldots,A_n\})$
ASΛ	$P(X \cup \{\Lambda\mathbf{A}\}) = \inf_{Z'} \{P(X \cup Z') : Z' \subseteq_f \mathrm{S}(\mathbf{A})\}$
AS∀	$P(X \cup \{(\forall x)A\}) = \lim_{n\to\infty} P(X \cup \{A(\mathrm{t}_1/x),\ldots,A(\mathrm{t}_n/x)\})$
AS∀	$P(X \cup \{(\forall x)A\}) = \inf_{Z'} \{P(X \cup Z') : Z' \subseteq_f \mathrm{I}((\forall x)A,\mathcal{D}_0 \cup \mathcal{D}_\omega)\}$

Lemma 10. *Let P be a function defined on the finite sets of statements of L_P, L_Q, or L_ω that meets constraints* **AS1**, **AS2**, **AS¬**, *and* **AS∧**, *or a function on the sets of statements of L_P, L_Q, or L_ω that meets those constraints and* **ASω**; *let $Y^\neg$ be the set whose members are the negations of the members of Y; and let* **A** *be the ω-sequence of statements $\langle A_1,\ldots,A_n,\ldots\rangle$. Then:*

(a) $P(X \cup \{A\}) \leq P(X)$;
(b) $P(X \cup Y) \leq P(X)$;
(c) *If* $P(X) = 0$, $P(X \cup Y) = 0$;
(d) $P(X) \leq 1$;
(e) $P(\{A\}) = P(\{B\}) = 1$ *if and only if* $P(\{A,B\}) = 1$;
(f) $P(X \cup \{A \wedge \neg A\}) = 0$;
(g) $P(X \cup \{A \vee B\}) = P(X \cup \{A\}) + P(X \cup \{B\}) - P(X \cup \{A,B\})$;
(h) $F_P(X,Y) = F_P(\{A\} \cup X,Y) + F_P(\{\neg A\} \cup X,Y)$;
(i) $F_P(X,Y) = P(X \cup Y^\neg)$;
(j) $0 \leq F_P(X,Y)$;
(k) $F_P(\varnothing,\varnothing) = 1$;
(l) $F_P(X,Y) = F_P(X,Y \cup \{A\}) + F_P(\{A\} \cup X,Y)$;
(m) $F_P(X,Y \cup \{\neg A\}) = F_P(\{A\} \cup X,Y)$;
(n) $F_P(\{A \wedge B\} \cup X,Y) = F_P(\{A,B\} \cup X,Y)$;
(o) *If P meets* **ASΛ**, *then* $F_P(\{\Lambda\mathbf{A}\} \cup X,Y) = \lim_{n\to\infty} F_P(\{A_1,\ldots,A_n\} \cup X,Y)$;

(p) $\inf_{Z'} \{P(X \cup Z') : Z' \subseteq_f S(A)\} = \lim_{n \to \infty} P(X \cup \{A_1,...,A_n\})$;
(q) $\inf_{Z'} \{P(X \cup Z') : Z' \subseteq_f I((\forall x)A, \mathcal{D}_0 \cup \mathcal{D}_\omega)\} = \lim_{n \to \infty} P(X \cup \{A(t_1/x),...,A(t_n/x)\})$;
(r) $F_P(X,Y) = P(X) - P(X \cup \{\vee Y\})$.

Proof: (a) By **AS1** and **AS¬**.

(b) By (a).

(c) By (b) and **AS1**, in case Y is finite. By (b), **AS1**, and **ASω**, if Y is infinite.

(d) By (b) and **AS2**.

(e) Suppose $P(\{A\}) = P(\{B\}) = 1$. Then $P(\{\neg A\}) = 0$ by **AS2** and **AS¬** and $P(\{\neg A,B\}) = 0$ by (c). Hence $P(\{A,B\}) = 1$ by **AS¬**. Now suppose $P(\{A,B\}) = 1$. Then $P(\{A\}) = P(\{B\}) = 1$ by (a) and (d).

(f) By **AS$\wedge$** and **AS¬**.

(g) By the definition of '$\vee$' and **AS¬**.

(h) By induction on the cardinality of Y, when Y is finite.
Basis: card(Y) = 0, i.e. $Y = \varnothing$. Then $F_P(X,Y) = F_P(\{A\} \cup X,Y) + F_P(\{\neg A\} \cup X,Y)$ by **AS¬**.
Inductive Step: card(Y) > 0. Then, by the definition of F_P, $F_P(X,Y) = F_P(X,Y-\{B\}) - F_P(\{B\} \cup X,Y-\{B\})$, where B is the alphabetically latest statement in Y. Hence $F_P(X,Y) = F_P(\{A\} \cup X,Y-\{B\}) + F_P(\{\neg A\} \cup X,Y-\{B\}) - F_P(\{A\} \cup \{B\} \cup X,Y-\{B\}) - F_P(\{\neg A\} \cup \{B\} \cup X,Y-\{B\})$ by the inductive hypothesis and hence $F_P(X,Y) = F_P(\{A\} \cup X,Y) + F_P(\{\neg A\} \cup X,Y)$ by the definition of F_P.
If Y is infinite, the second part of the definition of F_P and Lemma 9(c) need to be invoked as well.

(i) By induction on the cardinality of Y, if Y is finite.
Basis: $Y = \varnothing$. Then $F_P(X,Y) = P(X)$ by the definition of F_P. Hence $F_P(X,Y) = P(X \cup Y^{\neg})$.
Inductive Step: card(Y) > 0. Let A be the alphabetically latest member of Y and let Y' be $Y-\{A\}$. Then $F_P(X,Y' \cup \{A\})$

$= F_P(X,Y') - F_P(\{A\} \cup X,Y')$ (Def. of F_P)
$= P(X \cup (Y')^{\neg}) - P(\{A\} \cup X \cup (Y')^{\neg})$ (Ind. Hyp.)
$= P(\{\neg A\} \cup X \cup (Y')^{\neg})$ (**AS¬**)
$= P(X \cup Y^{\neg})$

If Y is infinite, **ASω** and the second part of the definition of F_P need to be invoked as well.

(j) By induction on the cardinality of Y, if Y is finite.
Basis: card(Y) = 0, i.e. $Y = \varnothing$. Then $0 \le F_P(X,Y)$ by **AS1**.
Inductive Step: card(Y) > 0. Then, by the definition of F_P, $F_P(X,Y) = F_P(X,Y-\{A\}) - F_P(\{A\} \cup X,Y-\{A\})$, where A is the alphabetically latest statement in Y. Hence $F_P(X,Y) = F_P(\{\neg A\} \cup X,Y-\{A\})$ by (h), and so $0 \le F_P(X,Y)$ by the inductive hypothesis.
If Y is infinite, by the second part of the definition of F_P.

(k) By **AS2** and the definition of F_P.

(l) For finite Y', $F_P(X,Y') = F_P(X,Y' \cup \{A\}) + F_P(\{A\} \cup X,Y')$ by Lemma 8(b). Hence

$$\inf_{Y'} \{F_P(X,Y') : Y' \subseteq_f Y\}$$
$$= \inf_{Y'} \{F_P(X,Y' \cup \{A\}) + F_P(\{A\} \cup X,Y') : Y' \subseteq_f Y\}$$
$$= \inf_{Y'} \{F_P(X,Y' \cup \{A\}) : Y' \subseteq_f Y\} + \inf_{Y'} \{F_P(\{A\} \cup X,Y') : Y' \subseteq_f Y\} \quad ((b))$$
$$= \inf_{Y'} \{F_P(X,Y') : Y' \subseteq_f Y \cup \{A\}\} + \inf_{Y'} \{F_P(\{A\} \cup X,Y') : Y' \subseteq_f Y\} \quad ((b))$$

(m) By (g) and Lemma 8(b).

(n) Like (h), but using **AS∧** instead of **AS¬**.

(o) Like (h), but using **ASΛ** instead of **AS¬**.

(p) Let n be an arbitrary positive number. Then $\{A_1,...,A_n\}$ is a finite subset of S(A). Hence $\inf_{Z'} \{P(X \cup Z') : Z' \subseteq_f \mathrm{S}(A)\} \leq \lim_{n\to\infty} P(X \cup \{A_1,...,A_n\})$ by (b). Now let Z' be an arbitrary finite subset of S(A). Then there exists a positive n such that $Z' \subseteq \{A_1,...,A_n\}$. Hence $\lim_{n\to\infty} P(X \cup \{A_1,...,A_n\}) \leq \inf_{Z'} \{P(X \cup Z') : Z' \subseteq_f \mathrm{S}(A)\}$ by (b) again.

(q) Like (p).

(r) By (c) if $Y = \varnothing$. Otherwise by induction on the cardinality of Y:

$$F_P(X,Y \cup \{A\})$$
$$= F_P(X,Y) - F_P(\{A\} \cup X,Y) \quad ((l))$$
$$= P(X) - P(X \cup \{\bigvee Y\}) - P(X \cup \{A\}) \neq P(X \cup \{\bigvee Y,A\}) \quad (\text{Ind. Hyp.})$$
$$= P(X) - P(X \cup \{\bigvee Y\}) \quad ((g))$$

Intuitionistic Absolute Probability Functions for Sets of Statements

AS1 $0 \leq P(X)$

AS2 $P(\varnothing) = 1$

ASI3 $P(X \cup Y) \leq P(X)$

AS∧ $P(X \cup \{A \wedge B\}) = P(X \cup \{A,B\})$

AS∨ $P(X \cup \{A \vee B\}) = P(X \cup \{A\}) + P(X \cup \{B\}) - P(X \cup \{A,B\})$

ASI⊃ $P(X \cup \{A \supset B\}) = \max_{Y \subseteq_f L} \{P(X \cup Y) : P(Y \cup \{A\}) = P(Y \cup \{A,B\})\}$

ASI¬ $P(X \cup \{\neg A\}) = \max_{Y \subseteq_f L} \{P(X \cup Y) : P(Y \cup \{A\}) = 0\}$

ASω $P(X) = \inf_{X'} \{P(X') : X' \subseteq_f X\}$.

Lemma 11. *Let P be a function defined on the finite sets of statements of L_I that meets constraints* **AS1**, **AS2**, **ASI3**, **AS∧**, **AS∨**, **ASI⊃**, *and* **ASI¬**, *or a function on the sets of statements of L_I that meets those constraints and* **ASω**. *Then:*

(a) $P(X \cup \{A \vee B\}) = P(X \cup \{B \vee A\})$;

(b) $P(X \cup \{A \vee (B \vee C)\}) = P(X \cup \{(A \vee B) \vee C\})$;

(c) *If* $P(X) = 0$, *then* $P(X \cup Y) = 0$;

(d) $P(X \cup \{A \wedge \neg A\}) = 0$;

(e) $P(X \cup \{A \supset A\}) = P(X)$;

(f) $F_P(X,Y) = P(X) - P(X \cup \{\bigvee Y\})$.

Proof: (a) By **AS∨**.

(b) By **AS∨**.

(c) By **ASI3** and **AS1**.

(d) $P(X \cup \{A \wedge \neg A\})$

$$= P(X \cup \{A, \neg A\}) \qquad (\mathbf{AS}\wedge)$$
$$= \max_{Y \subseteq_f L} \{P(X \cup \{A\} \cup Y) : P(Y \cup \{A\}) = 0\} \qquad (\mathbf{AIS}\neg)$$
$$= 0 \qquad ((c))$$

(e) $P(X \cup \{A \supset A\})$

$$= \max_{Y \subseteq_f L} \{P(X \cup Y) : P(Y \cup \{A\}) = P(Y \cup \{A, A\})\} \qquad (\mathbf{AIS}\supset)$$
$$= \max_{Y \subseteq_f L} \{P(X \cup Y)\}$$
$$= P(X) \qquad (\mathbf{ASI3})$$

(f) By induction on the cardinality of Y.

Basis: $Y = \varnothing$. Then $\bigvee Y$ is $A \wedge \neg A$ and $F_P(X) = P(X) - P(X \cup \{\bigvee Y\})$ by (d).

Inductive Step: $\text{card}Y > 0$. Then

$F_P(X, Y-\{A\}) - F_P(\{A\} \cup X, Y-\{A\})$

$$= P(X) - P(X \cup \{\bigvee(Y-\{A\})\}) - P(X \cup \{A\}) + P(X \cup \{\bigvee(Y-\{A\})\} \cup \{A\})$$

(Ind. Hyp.)

$$= P(X) - P(X \cup \{\bigvee Y\}) \qquad (\mathbf{AS}\vee)$$

Consistency Functions

F1 $0 \leq F(X,Y)$

F2 $F(\varnothing,\varnothing) = 1$

F3 $F(X, Y \cup \{A\}) + F(\{A\} \cup X, Y) = F(X,Y)$

FωR $F(X,Y) = \inf_{Y'} \{F(X,Y') : Y' \subseteq_f Y\}$

FωL $F(X,Y) = \inf_{X'} \{F(X',Y) : X' \subseteq_f X\}$

F∧R $F(X, Y \cup \{A \wedge B\}) = F(X, Y \cup \{A\}) + F(\{A\} \cup X, Y \cup \{B\})$

F∧L $F(\{A \wedge B\} \cup X, Y) = F(\{A,B\} \cup X, Y)$

F¬R $F(X, Y \cup \{\neg A\}) = F(\{A\} \cup X, Y)$

F¬L $F(\{\neg A\} \cup X, Y) = F(X, Y \cup \{A\})$

Lemma 12. *Let F be a binary function on the finite sets of statements of L that meets constraints* **F1–F3**, *or a function on the sets of statements of L that meets* **F1–F3** *and* **FωL** *and* **FωR**. *Then:*

(a) *If* $X \cap Y \neq \varnothing$, *then* $F(X,Y) = 0$;
(b) $F(X,Y) = \Sigma_{Z'}\{F(Z' \cup X, Y \cup (Z - Z')) : Z' \subseteq Z\}$;
(c) $F(X,Y) = \Sigma_{W \subseteq L}\{F(W, L-W) : X \subseteq W$ *and* $Y \subseteq L-W\}$;
(d) $F(\{A\} \cup X, Y) \leq F(X,Y)$ *and* $F(X, Y \cup \{A\}) \leq F(X,Y)$;
(e) $F(Z \cup X, Y) \leq F(X,Y)$ *and* $F(X, Y \cup Z) \leq F(X,Y)$;
(f) *If* $F(X,Y) = 0$, *then* $F(X' \cup X, Y \cup Y') = 0$;
(g) $F(X,Y) \leq 1$;
(h) *If* $F(X, Y \cup \{A\}) = 0$ *and* $F(\{A\} \cup X', Y') = 0$, *then* $F(X \cup X', Y \cup Y') = 0$;
(i) $F(X,\varnothing) = 1$ *if and only if* $F(\{A\},\varnothing) = 1$ *for every* $A \in X$;
(j) $F(X,Y) - F(\{A_1,\ldots,A_n\} \cup X, Y)$
$$= F(X, Y \cup \{A_1\}) + F(\{A_1\} \cup X, Y \cup \{A_2\}) + \ldots + F(\{A_1,\ldots,A_{n-1}\} \cup X, Y \cup \{A_n\});$$
(k) $F(\mathrm{S}(A) \cup X, Y) = \lim_{n\to\infty} F(\{A_1,\ldots,A_n\} \cup X, Y)$;
(l) $F(\mathrm{I}((\forall x)A) \cup X, Y) = \lim_{n\to\infty} F(\{A(\mathrm{t}_1/x),\ldots,A(\mathrm{t}_n/x)\} \cup X, Y)$.

Proof: (a) Suppose $X \cap Y \neq \varnothing$, and let A belong to both X and Y. Then $X = \{A\} \cup X$ and $Y = Y \cup \{A\}$. Hence $F(X,Y) + F(X,Y) = F(X,Y)$ by **F3**. Hence $F(X,Y) = 0$.

(b) Proof by induction on the cardinality n of Z.

Basis: $n = 0$. Then (b) holds trivially.

Inductive Step: $n > 0$. Let A be a member of Z and Z* be Z–$\{A\}$. Then
$$F(X,Y) = F(X, Y \cup \{A\}) + F(\{A\} \cup X, Y)$$
by **F3**. But
$$F(X, Y \cup \{A\}) = \Sigma_{Z'}\{F(Z' \cup X, Y \cup \{A\} \cup (Z^* - Z')) : Z' \subseteq Z^*\}$$
and
$$F(\{A\} \cup X, Y) = \Sigma_{Z'}\{F(Z' \cup \{A\} \cup X, Y \cup (Z^* - Z')) : Z' \subseteq Z^*\}$$
by the inductive hypothesis. So (a).

(c) Let Z in (b) be L. Then $F(X,Y) = \Sigma_{W \subseteq L}\{F(W \cup X, Y \cup (L-W))\}$. But
$$(W \cup X) \cap (Y \cup (L-W)) \neq \varnothing$$
if and only if $X \subseteq W$ and $Y \subseteq (L-W)$. Hence (c) by (a).

(d) $F(\{A\} \cup X, Y) \leq F(X,Y)$ and $F(X, Y \cup \{A\}) \leq F(X,Y)$ by **F3** and **F1**.

(e) By repeated use of (d) and, in case Z is infinite, by **FωL** and **FωR** as well.

(f) By (e) and **F1**.

(g) $F(\varnothing,\varnothing) = 1$ by **F2**. Hence $F(X,Y) \leq 1$ by (e).

(h) By (f) and **F3**.

(i) Suppose $F(X,\varnothing) = 1$. Then $F(\{A\},\varnothing) = 1$ by (e) and (g). Now suppose $F(\{A\},\varnothing) = 1$ for every $A \in X$. For finite X we use induction on the cardinality of X.

Basis: card(X) = 1, i.e. $X = \{A\}$. Then of course $F(X,\varnothing) = 1$.

Inductive Step: card(X) > 1. Let A be any member of X. Then $F(X-\{A\},\varnothing) = 1$ by the inductive hypothesis. Hence $F(X,\varnothing) = 1 - F(X-\{A\},\{A\})$. But $F(\varnothing,\{A\}) = 0$ by **F2** and **F3**, and hence $F(X-\{A\},\{A\}) = 0$ by (f). So, $F(X,\varnothing) = 1$.

(j) By induction on n.
Basis: $n = 1$. Then $F(X,Y) - F(\{A_1\} \cup X,Y) = F(X,Y \cup \{A_1\})$ by **F3**.
Inductive Step: $n > 1$. Then $F(X,Y) - F(\{A_1,...,A_{n-1}\} \cup X,Y)$

$$= F(X,Y \cup \{A_1\}) + F(\{A_1\} \cup X,Y \cup \{A_2\}) + ... + F(\{A_1,...,A_{n-2}\} \cup X,Y \cup \{A_{n-1}\})$$

by the inductive hypothesis. But

$$F(\{A_1,...,A_{n-1}\} \cup X,Y) = F(\{A_1,...,A_n\} \cup X,Y) + F(\{A_1,...,A_{n-1}\} \cup X,Y \cup \{A_n\})$$

by **F3**. Hence (j).
(k) $F(S(A) \cup X,Y)$

$$= \inf_{Z'} \{F(Z',Y) : Z' \subseteq_f S(A) \cup X\} \qquad \text{(F}\omega\text{L)}.$$

$$= \inf_{Z'} \{F(Z' \cup X,Y) : Z' \subseteq_f S(A)\} \qquad \text{((e))}$$

Let n be an arbitrary positive number. Then $\{A_1,...,A_n\}$ is a finite subset of $S(A)$. Hence $\inf_{Z'} \{F(Z' \cup X,Y) : Z' \subseteq_f S(A)\} \leq \lim_{n\to\infty} F(\{A_1,...,A_n\} \cup X,Y)$ by (e). Now let Z' be an arbitrary finite subset of $S(A)$. Then there exists a positive n such that $Z' \subseteq \{A_1,...,A_n\}$. Hence $\lim_{n\to\infty} F(\{A_1,...,A_n\} \cup X,Y) \leq \inf_{Z'} \{F(Z' \cup X,Y) : Z' \subseteq_f S(A)\}$ by (e) again.
(l) Like (k).

Lemma 13. *Let F be a function on the sets of statements of* L_P *that meets*
(i) **F1–F3**, **FωR** *and* **FωL**,
(ii) **F∧R**, **F∧L**, **F¬R**, *and* **F¬L**;
and let $S = \{A : F(\{A\},\emptyset) = 1\}$ *be the assumption set of F. Then:*
(a) *If* $F(A,\emptyset) = 1$ *then* $F(\{A\} \cup X,Y) = F(X,Y)$;
(b) *If* $S' \subseteq S$, *then* $F(S' \cup X,Y) = F(X,Y)$;
(c) $F(X,Y \cup \{A \wedge \neg A\}) = F(X,Y)$;
(d) $F(\{\neg(A \wedge \neg A)\} \cup X,Y) = F(X,Y)$;
(e) $F(X,Y \cup \{A \vee B\}) = F(X,Y \cup \{A,B\})$;
(f) $F(X,Y \cup \{A \supset B\}) = F(\{A\} \cup X,Y \cup \{B\})$;
(g) *If X is finite, then* $F(X,Y) = F(\{\bigwedge X\},Y)$, *where* $\bigwedge X$ *is any conjunction of the members of X, when X is not empty, and* $\bigwedge X = \neg(A \wedge \neg B)$, *when* $X = \emptyset$;
(h) *If Y is finite, then* $F(X,Y) = F(X,\{\bigvee Y\})$, *where* $\bigvee Y$ *is any disjunction of the members of Y, when Y is not empty, and* $\bigvee Y = A \wedge \neg B$, *when* $Y = \emptyset$;
(i) *If X and Y are finite,* $F(X,Y) = F(\emptyset,\{\bigwedge X \supset \bigvee Y\})$.

Proof: (a) Assume $F(\{A\},\emptyset) = 1$. Then $F(\emptyset,\{A\}) = 0$ by **F2** and **F3**. Hence $F(X,Y \cup \{A\}) = 0$ by Lemma 12(f), and so $F(\{A\} \cup X,Y) = F(X,Y)$ by **F3**.

(b) $F(S' \cup X,Y) = F(X,Y)$ by repeated use of (a) when S' is finite. Hence $F(S' \cup X,Y) = F(X,Y)$ by **FωL** when S' is infinite.

(c) $F(X,Y \cup \{A \wedge \neg A\})$

$$= F(X,Y \cup \{A\}) + F(\{A\} \cup X,Y \cup \{\neg A\}) \qquad \text{(F}\wedge\text{R)}$$

$$= F(X,Y \cup \{A\}) + F(\{A\} \cup X,Y) \qquad \text{(F}\neg\text{R)}$$

$$= F(X,Y) \qquad \text{(F3)}$$

(d) By **F¬L** and (c).

(e) $F(X,Y \cup \{A \vee B\})$

$= F(X,Y \cup \{\neg(\neg A \wedge \neg B)\})$ (Def. '$\vee$')

$= F(\{\neg A \wedge \neg B\} \cup X,Y)$ (**F¬R**)

$= F(\{\neg A,\neg B\} \cup X,Y)$ (**F∧L**)

$= F(X,Y \cup \{A,B\})$ (**F¬L**)

(f) $F(X,Y \cup \{A \supset B\})$

$= F(X,Y \cup \{\neg(A \wedge \neg B\})$ (Def. '$\supset$')

$= F(\{A,\neg B\} \cup X,Y)$ (**F¬R**, **F∧L**)

$= F(\{A\} \cup X,Y \cup \{B\})$ (**F¬L**)

(g) By **F∧L**, when X is not empty, and by (d), when $X = \varnothing$.

(h) By (e), when Y is not empty, and by (c), when $X = \varnothing$.

(i) By (g), (h), and (f).

Intuitionistic Consistency Functions

F1 $0 \le F(X,Y)$

F2 $F(\varnothing,\varnothing) = 1$

F3 $F(X,Y \cup \{A\}) + F(\{A\} \cup X,Y) = F(X,Y)$

F∧R $F(X,Y \cup \{A \wedge B\}) = F(X,Y \cup \{A\}) + F(\{A\} \cup X,Y \cup \{B\})$

F∧L $F(\{A \wedge B\} \cup X,Y) = F(\{A,B\} \cup X,Y)$

F∨R $F(X,Y \cup \{A \vee B\}) = F(X,Y \cup \{A,B\})$

F∨L $F(\{A \vee B\} \cup X,Y) = F(\{A\} \cup X,Y) + F(\{B\} \cup X,Y \cup \{A\})$

FI⊃R $F(X,Y \cup \{A \supset B\}) = \min_{Z \subseteq_f L} \{F(X,Y \cup Z) : F(Z \cup \{A\},\{B\}) = 0\}$

FI⊃L $F(\{A \supset B\} \cup X,Y) = \max_{Z \subseteq_f L} \{F(Z \cup X,Y) : F(Z \cup \{A\},\{B\}) = 0\}$

FI¬R $F(X,Y \cup \{\neg A\}) = \min_{Z \subseteq_f L} \{F(X,Y \cup Z) : F(Z \cup \{A\},\varnothing) = 0\}$

FI¬L $F(\{\neg A\} \cup X,Y) = \max_{Z \subseteq_f L} \{F(Z \cup X,Y) : F(Z \cup \{A\},\varnothing) = 0\}$

Lemma 14. *Let F be a function on the finite sets of statements of L_I that meets*

(i) **F1–F3**,

(ii) **F∧R**, **F∧L**, **F∨R**, **F∨L**, **FI⊃R**, **FI⊃L**, **FI¬R**, *and* **FI¬L**;

or a function on the sets of statements of L_I which meets those constraints and

(iii) **FωR** *and* **FωL**.

Then:

(a) $F(\{A,\neg A\} \cup X,Y) = 0$;

(b) $F(\{A\},\{B\}) = 0$ *if and only if* $F(\varnothing,\{A \supset B\}) = 0$;

(c) $F(X,Y \cup \{A \supset A\}) = 0$;

(d) *If Z is finite, then* $F(Z \cup X,Y) = F(\{\wedge Z\} \cup X,Y)$;

(e) *If Z is finite, then* $F(X,Y \cup Z) = F(X,Y \cup \{\bigvee Z\})$;

Proof: (a) $F(\{A,\neg A\} \cup X,Y)$

$$= \max_{Z \subseteq_f L} \{F(Z \cup \{A\} \cup X,Y) : F(Z \cup \{A\},\varnothing) = 0\} \qquad \textbf{(FI}\neg\textbf{L)}$$
$$= 0 \qquad \text{(Lemma 12(f))}$$

(b) Suppose $F(\{A\},\{B\}) = 0$. Then $F(\{A \supset B\},\varnothing) = 1$ by **F⊃L** and **F1**. Hence $F(\varnothing,\{A \supset B\}) = 0$ by **F3** and **F1**. Now assume $F(\varnothing,\{A \supset B\}) = 0$, hence, by Lemma 12(f), $F(\{A\},\{B\} \cup \{A \supset B\}) = 0$. Then $F(\{A \supset B\} \cup \{A\},\{B\}) = 0$ by **F⊃L**, and hence $F(\{A\},\{B\}) = 0$ by Lemma 12(f).

(c) By (b), **F1**, and Lemma 12(f).

(d) By **F∧L** if $Z \neq \varnothing$ and by (a) and **F∧L** if $Z = \varnothing$.

(e) By **F∨R** if $Z \neq \varnothing$ and by (c) and **F∨R** if $Z = \varnothing$.

Relative Probability Functions for Sets of Statements of L_P

RS1	*If* $Y \subseteq X$, $P(Y,X) = 1$
RS2	$0 \leq P(Y,X)$
RS3	$P(Z \cup Y,X) = P(Z,Y \cup X) \times P(Y,X)$
RS4	$\varnothing$ *is P-normal*
RS5	*If X is P-normal, then* $P(\{\neg A\},X) = 1 - P(\{A\},X)$
RS6	$P(\{A \wedge B\},X) = P(\{A,B\},X)$
RSω	$P(Y,X) = \inf_{Y' \subseteq L} \{P(Y',X) : Y' \subseteq_f Y\}$
RSK	*If* $P(X,\varnothing) = 0$, *then X is P-abnormal*
RSC52	*If* $P(Y,X) = 1$ *for every Y, then X is logically inconsistent*
RSC50	*If* $P(X,\varnothing) = 0$, *then X is logically inconsistent*

Lemma 15. *Let P be a relative probability function defined on the finite sets of statements which meets constraints* **RS1–RS6**, *or a function on the sets of statements which meets those constraints and* **RSω**. *Then:*

(a) $P(Z \cup Y,X) \leq P(Y,X)$;

(b) $P(Y,X) \leq 1$;

(c) $P(Z \cup Y,X \cup Z) = P(Y,X \cup Z)$;

(d) *If X is P-normal, then* $P(Y \cup \{\neg A\},X) = P(Y,X) - P(Y \cup \{A\},X)$;

(e) $P(Y \cup \{A \wedge B\},X) = P(Y \cup \{A,B\},X)$;

(f) $P(Y,X \cup \{A \wedge B\}) = P(Y,X \cup \{A,B\})$;

(g) *If X is finite, then* $P(Y,X) = P(Y,\{\bigwedge X\})$;

(h) *If X is P-normal,* $P(Y \cup \{A \wedge \neg A\},X) = 0$;

(i) $P(Y,\varnothing) = P(Y,\{\neg(A \wedge \neg A)\}) = P(Y,\{\top\})$;

(j) *If* $X \cup Z$ *is P-normal, then X is P-normal;*

(k) *X is P-abnormal if and only if* $P(\{\bot\},X) = 1$;

(l) *If P meets* **RSK**, *then* $P(\{A\},Z) = 1$ *for every statement Z, if* $P(\{A\},\varnothing) = 1$;

(m) $P(S(A),X) = \lim_{n\to\infty} P(\{A_1,\ldots,A_n\},X)$;

(n) $P(I((\forall x)A),X) = \lim_{n\to\infty} P(\{A(t_1/x),\ldots,A(t_n/x)\},X)$.

Proof: (a) By **RS3** and **RS2**, or by **RS3**, **RS2**, and **RSω**.

(b) $P(\varnothing,X) = 1$ by **RS1**; hence $P(Y,X) \leq 1$ by (a).

(c) $P(Z \cup Y,X \cup Z) = 1 \times P(Y,X \cup Z)$ by **RS3** and **RS1**.

(d) By **RS3** and **RS7**.

(e) $P(Y \cup \{A \wedge B\},X)$

$= P(\{A \wedge B\},Y \cup X) \times P(Y,X)$	(**RS3**)
$= P(\{A,B\},Y \cup X) \times P(Y,X)$	(**RS6**)
$= P(Y \cup \{A,B\},X)$	(**RS3**)

(f) $P(Y,X \cup \{A \wedge B\})$

$= P(Y \cup \{A \wedge B\},X \cup \{A \wedge B\})$	((c))
$= P(Y \cup \{A,B\},X \cup \{A \wedge B\})$	((e))
$= P(Y,\{A,B\} \cup X \cup \{A \wedge B\}) \times P(\{A,B\},X \cup \{A \wedge B\})$	(**RS3**)
$= P(Y,\{A,B\} \cup X \cup \{A \wedge B\}) \times 1$	(**RS6**, **RS1**)
$= P(Y,\{A,B\} \cup X \cup \{A \wedge B\}) \times P(\{A \wedge B\},X \cup \{A,B\})$	(**RS1**, **RS6**)
$= P(Y \cup \{A \wedge B\},X \cup \{A,B\})$	(**RS3**)
$= P(Y,X \cup \{A,B\})$	((e), (f))

(g) By (f).

(h) Suppose *X P*-normal. Then

$P(Y \cup \{A \wedge \neg A\},X)$

$= P(Y \cup \{A,\neg A\},X)$	((e))
$= P(Y \cup \{A\},X) - P(Y \cup \{A,A\},X)$	((d))
$= 0$	

(i) $P(Y,\varnothing)$

$= P(Y \cup \{\neg(A \wedge \neg A)\},\varnothing)$	((d), **RS4**, (h))
$= P(Y,\{\neg(A \wedge \neg A)\}) \times P(\{\neg(A \wedge \neg A)\},\varnothing)$	(**RS3**)
$= P(Y,\{\neg(A \wedge \neg A)\}) \times (1 - P(\{A \wedge \neg A\},\varnothing))$	((d), **RS4**)
$= P(Y,\{\neg(A \wedge \neg A)\})$	((h), **RS4**)

(j) Assume $P(Y,X \cup Z) \neq 1$. Then $P(Z \cup Y,X) \neq 1$ by **RS3** and (b), i.e. *X* is *P*-normal.

(k) If *X* is *P*-abnormal, then $P(\{\bot\},X) = 1$ by definition. And if $P(\{\bot\},X) = 1$, then *X* is *P*-abnormal by (h).

(l) Suppose *P* meets **RSK** and assume $P(\{A\},\varnothing) = 1$. Then $P(\{\neg A\},\varnothing) = 0$ by **RS4** and **RS5** and hence $\{\neg A\}$ is *P*-abnormal by **RSK**. If *Z* is *P*-abnormal, $P(\{A\},Z) = 1$ by definition. So, assume that *Z* is *P*-normal. Then $P(\{A \wedge \neg A\},Z) = 0$ by (h), and

$P(\{A\},\{\neg A\} \cup Z) \times P(\{\neg A\},Z) = 0$ by **RS3** and **RS6**. Since $\{\neg A\} \cup Z$ is P-abnormal by (j), $P(\{A\},\{\neg A\} \cup Z) = 1$. Hence $P(\{\neg A\},Z) = 0$, and hence $P(\{A\},Z) = 1$. So, if $P(\{A\},\emptyset) = 1$, $P(\{A\},Z) = 1$, whether or not Z is P-normal.

(m) $P(\mathrm{S}(A),X) = \inf_{Z'} \{P(Z',X) : Z' \subseteq_f \mathrm{S}(A)\}$ by **AS**ω. Let n be an arbitrary positive number. Then $\{A_1,...,A_n\}$ is a finite subset of S(A). Hence

$$\inf_{Z'} \{P(Z',X) : Z' \subseteq_f \mathrm{S}(A)\} \leq \operatorname{limit}_{n\to\infty} P(\{A_1,...,A_n\},X)$$

by (a). Now let Z' be an arbitrary finite subset of S(A). Then there exists a positive n such that $Z' \subseteq \{A_1,...,A_n\}$. Hence

$$\operatorname{limit}_{n\to\infty} P(\{A_1,...,A_n\},X) \leq \inf_{Z'} \{P(Z',X) : Z' \subseteq_f \mathrm{S}(A)\}$$

by (a) again.

(n) Like (m).

Relative Probability Functions for Sets of Statements (Language-Independent Constraints)

RS1 *If* $Y \subseteq X$, *then* $P(Y,X) = 1$

RS2' $0 \leq F_{P_X}(U,V)$

RS3 $P(Z \cup Y,X) = P(Z,Y \cup X) \times P(Y,X)$

RS4 $P(Z,\emptyset) \neq 1$ *for at least one set Z of statements*

RSω $P(Y,X) = \inf_{Y' \subseteq L} \{P(Y',X) : Y' \subseteq_f Y\}$

RS$\wedge$ $P(\{A \wedge B\},X) = P(\{A,B\},X)$

RS$\neg$ $P(\{\neg A\},X) = 1 - P(\{A\},X) + P(\{\bot\},X)$

RS$\bot$ *If* $P(\{\bot\},X) \neq 0$, *then* $P(\{A\},X) = 1$

IND *If X and Y are sets of atomic statements and* $X \cap Y = \emptyset$, *then* $P(\{\bot\},X \cup Y^{\neg}) = 0$

Lemma 16. *Let P be a binary function defined on the finite sets of statements of L which meets* **RS1, RS2', RS3,** *and* **RS4,** *or a function on the sets of statements of L which meets those constraints and* **RS**ω.

(a) $P(\emptyset,X) = 1$;

(b) $P(Z \cup Y,X) \leq P(Y,X)$;

(c) $0 \leq P(Y,X) \leq 1$;

(d) *If* $P(Y,X) = 0$, *then* $P(Z \cup Y,X) = 0$;

(e) $0 \leq 1 - P(Z,X) - P(Y,X) + P(Z \cup Y,X)$;

(f) *If* $P(Y,X) = 1$, *then* $P(Z \cup Y,X) = P(Z,Y \cup X) = P(Z,X)$;

(g) *If* $P(Y,X) = P(Z \cup Y,X)$, *then* $P(Y,X) \leq P(Z,X)$;

(h) *If* $P(Y,X) = 1$ *and* $P(Z,X) \neq 0$, *then* $P(Y,X \cup Z) = 1$;

(i) *If* $Z \subseteq Y$ *and* $P(Y,X) = 1$, *then* $P(Z,X) = 1$;

(j) *If* $P(Y,X) = P(Z,X) = 1$, *then* $P(Z \cup Y,X) = 1$;
(k) $P(Y,X) = 1$ *if and only if* $P(\{A\},X) = 1$ *for every* $A \in Y$;
(l) *If* $Y \subseteq X$, *then* $P(Z \cup Y,X) = P(Z,X)$.

Proof: (a) By **RS1**.

(b) By Lemma 9(d) and **RS2'**.

(c) $0 \leq P(Y,X)$ by Lemma 9(a) and **RS2'**. $P(Y,X) \leq 1$ by (a) and (b).

(d) By (b) and (c).

(e) By induction on card(Y) + card(Z).
Basis: $Y = Z = \varnothing$. Then $0 \leq 1 - P(Z,X) - P(Y,X) + P(Z \cup Y,X)$ by (a).
Inductive Step: card(Y) + card(Z) > 0, say $Z \neq \varnothing$, and let A be any member of Z. Then

$$0 \leq 1 - P(Z - \{A\},X \cup \{A\}) - P(Y,X \cup \{A\}) + P((Z - \{A\}) \cup Y,X \cup \{A\})$$

by the inductive hypothesis and $0 \leq P(\{A\},X)$ by (b). Hence

$$0 \leq P(\{A\},X) - P(Z,X) - P(\{A\} \cup Y,X) + P(Z \cup Y,X)$$

by **RS3**. But

$$0 \leq 1 - P(\{A\},X) - P(Y,X) + P(\{A\} \cup Y,X)$$

by Lemma 9(g), (a), and **RS2'**. So $0 \leq 1 - P(Z,X) - P(Y,X) + P(Z \cup Y,X)$.

(f) Suppose $P(Y,X) = 1$. Then $P(Z \cup Y,X) = P(Z,Y \cup X)$ and $P(Z \cup Y,X) = P(Y,Z \cup X) \times P(Z,X)$ by **RS3**. Hence $P(Z,Y \cup X) \leq P(Z,X)$ by (c). But by (e) $P(Z,X) \leq P(Z \cup Y,X)$ and hence $P(Z,X) \leq P(Z,Y \cup X)$. So $P(Z \cup Y,X) = P(Z,Y \cup X) = P(Z,X)$.

(g) By **RS3** and (c).

(h) Suppose $P(Y,X) = 1$ and $P(Z,X) \neq 0$. Then $P(Z,Y \cup X) = P(Y,Z \cup X) \times P(Z,X)$ by **RS3**. But $P(Z,Y \cup X) = P(Z,X)$ by (f). Hence $P(Y,X \cup Z) = 1$.

(i) Suppose $P(Y,X) =1$ and $Z \subseteq Y$. Then $P(Y,X) = P(Z,Y \cup X) \times P(Y,X)$ by **RS3**. Hence $P(Z,Y \cup X) = 1$. But $P(Z,Y \cup X) = P(Z,X)$ by (f). So $P(Z,X) = 1$.

(j) Suppose $P(Y,X) = P(Z,X) = 1$. Then $P(Z,Y \cup X) = 1$ by (f). Hence $P(Z \cup Y,X) = 1$ by **RS3**.

(k) By (i) and (j) (and **RSω** in case Y is infinite).

(l) By **RS3** and **RS1**.

Lemma 17. *Let P be a function on the sets of statements of* L_P *that meets*
(i) **RS1**, **RS2'**, **RS3**, **RS4**, *and* **RSω**,
(ii) **RS∧**, **RS¬**, *and* **RS⊥**.

Then:
(a) $P(\{A\},X \cup \{\bot\}) = 1$;
(b) $P(Y,X \cup \{\bot\}) = 1$;
(c) $P(\{\bot\} \cup Y,X) = P(\{\bot\},X)$;
(d) $P(\{\neg A\},X) = 1 - P(\{A\},X) + P(\{\bot\},X \cup \{A\}) \times P(\{A\},X)$;

(e) $P(\{\neg A\} \cup Y, X) = P(Y, X) - P(\{A\} \cup Y, X) + P(\{\bot\}, X)$;

(f) $P(\{A \wedge \neg A\}, X) = P(\{\bot\}, X)$;

(g) $P(\{\neg A\}, X) = 1 - P(\{A\}, X) + P(\{B \wedge \neg B\}, X)$;

(h) *If* $P(\{B \wedge \neg B\}, X) \neq 0$, *then* $P(\{A\}, X) = 1$;

(i) $P(\{B \wedge \neg B\}, \varnothing) \neq 1$;

(j) *If* $P(\{\bot\}, X) = 1$, *then* $P(\{\bot\}, X \cup Z) = 1$;

(k) *If* $P(\{\bot\}, X) = 0$ *but* $P(\{\bot\}, X \cup Y) = 1$, *then* $P(Y, X) = 0$;

(l) *If* $P(\{\bot\}, X) = 0$ *and* $P(\{\bot\}, Y) = 1$, *then* $P(Y, X) = 0$;

(m) $P(\{\neg A\} \cup Y, X \cup \{A\}) = P(\{\bot\} \cup Y, X \cup \{A\})$;

(n) $P(\{\neg\neg A\}, X) = P(\{A\}, X)$;

(o) $P(\{A \supset B\}, X) = 1 - P(\{A\}, X) + P(\{B\}, X \cup \{A\}) \times P(\{A\}, X)$;

(p) $P(\{A \vee B\}, X) = P(\{A\}, X) + P(\{B\}, X) - P(\{A, B\}, X)$.

Proof: (a) $P(\{\bot\}, X \cup \{\bot\}) = 1$ by **RS1**. Hence $P(\{A\}, X \cup \{\bot\}) = 1$ by **RS⊥**.

(b) By (a) and Lemma 16(j).

(c) If $P(\{\bot\}, X) = 1$, $P(\{\bot\} \cup Y, X) = 1$ by **RS⊥** and if $P(\{\bot\}, X) \neq 1$, $P(\bot\}, X) = 0$ by **RS⊥** and $P(\{\bot\} \cup Y, X) = 0$ by **RS3**. Hence $P(\{\bot\} \cup Y, X) = P(\{\bot\}, X)$ in either case.

(d) By **RS3** $P(\{\bot\}, X \cup \{A\}) \times P(\{A\}, X) = P(\{A\}, X \cup \{\bot\}) \times P(\{\bot\}, X)$. But $P(\{A\}, X \cup \{\bot\}) = 1$ by (a). Hence $P(\{\neg A\}, X) = 1 - P(\{A\}, X) + P(\{\bot\}, X \cup \{A\}) \times P(\{A\}, X)$ by **RS¬**.

(e) $P(\{\neg A\} \cup Y, X)$

$= P(\{\neg A\}, Y \cup X) \times P(Y, X)$ (**RS3**)

$= [1 - P(\{A\}, Y \cup X) + P(\{\bot\}, Y \cup X)] \times P(Y, X)$ (**RS¬**)

$= P(Y, X) - P(\{A\} \cup Y, X) + P(\{\bot\}, X)$ (**RS3**)

(f) $P(\{A \wedge \neg A\}, X)$

$= P(\{A, \neg A\}, X)$ (**RS∧**)

$= P(\{A\}, X) - P(\{A\}, X) + P(\{\bot\}, X)$ ((e))

$= P((\bot\}, X)$.

(g) By (d) and **RS¬**.

(h) By (d) and **RS⊥**.

(i) By (d) and **RS4**.

(j) If $P(\{\bot\}, X) = 1$, then $P(\{\bot\} \cup Z, X) = 1$ by **RS⊥** and so $P(\{\bot\}, X \cup Z) = 1$ by **RS3**.

(k) Assume $P(\{\bot\}, X) = 0$ and $P(\{\bot\}, X \cup Y) = 1$. Then, by **RS3**, $P(\{\bot\} \cup Y, X) = P(Y, \{\bot\} \cup X) \times P(\{\bot\}, X) = 0$. But $P(\{\bot\} \cup Y, X) = P(Y, X)$, also by **RS3**. Hence $P(Y, X) = 0$.

(l) By (j) and (k).

(m) $P(\{\neg A\} \cup Y, X \cup \{A\})$

$= P(Y, X \cup \{A\}) - P(\{A\} \cup Y, X \cup \{A\}) + P(\{\bot\}, X \cup \{A\})$ ((e))

$= P(\{\bot\}, X \cup \{A\})$ (Lemma 16(f), **RS1**)

(n) By **RS¬**.

(o) For by the definition of '⊃', **RS¬**, **RS∧**, (e), and **RS3**

$P(\{A \supset B\}, X)$

$= P(\{\neg(A \wedge \neg B)\}, X)$

$= 1 - P(\{A \wedge \neg B\}, X) + P(\{\bot\}, X)$

$= 1 - [P(\{A\}, X) - P(\{A, B\}, X) + P(\{\bot\}, X)] + P(\{\bot\}, X)$

$= 1 - P(\{A\}, X) + P(\{B\}, X \cup \{A\}) \times P(\{A\}, X)$.

(p) By the definition of '∨', **RS¬**, and (e)

$P(\{A \vee B\}, X)$

$= P(\{\neg(\neg A \wedge \neg B)\}, X)$

$= 1 - P(\{\neg A \wedge \neg B\}, X) + P(\{\bot\}, X)$

$= 1 - [P(\{\neg B\}, X) - P(\{A, \neg B\}, X) + P(\{\bot\}, X)] + P(\{\bot\}, X)$

$= 1 - [1 - P(\{B\}, X) + P(\{\bot\}, X)] + [P(\{A\}, X) - P(\{A, B\}, X) + P(\{\bot\}, X)$

$= P(\{A], X) + P\{B\}, X) - P(\{A, B\}, X)$.

Intuitionistic Relative Probability Functions for Sets of Statements

RS1 *If* $Y \subseteq X$, $P(Y, X) = 1$

RS2' $0 \leq F_{P_X}(Y, Z)$

RS3 $P(Z \cup Y, X) = P(Z, Y \cup X) \times P(Y, X)$

RS4 $P(Z, \varnothing) \neq 1$ *for at least one set Z of statements*

RSω $P(Y, X) = \inf_{Y'} \{P(Y', X) : Y' \subseteq_f Y\}$

RS∧ $P(\{A \wedge B\}, X) = P(\{A, B\}, X)$

RS∨ $P(\{A \vee B\}, X) = P(\{A\}, X) + P(\{B\}, X) - P(\{A, B\}, X)$

RSI⊃ $P(\{A \supset B\}, X) = \max_{Y \subseteq_f L} \{P(Y, X) : P(\{A\} \cup Y, X) = P(\{A, B\} \cup Y, X)\}$

RSI¬ $P(\{\neg A\}, X) = \max_{Y \subseteq_f L} \{P(Y, X) : P(\{A\} \cup Y, X) = P(\{\bot\}, X)\}^2$

RS⊥ *If* $P(\{\bot\}, X) \neq 0$, *then* $P(\{A\}, X) = 1$

Lemma 18. *Let P be a binary function on the sets of statements of* L_I *that meets*

(i) **RS1**, **RS2'**, **RS3**, **RS4**, *and* **RSω**,

(ii) **RS∧**, **RS∨**, **RSI⊃**, **RSI¬**, *and* **RS⊥**.

Then:

(a) $\max_{Y \subseteq_f L} \{P(Y, X) : P(Y, X) = P(\{A\} \cup Y, X)\} = P(\{A\}, X)$;

(b) $P(\{A \wedge B\} \cup Y, X) = P(\{A, B\} \cup Y, X)$;

(c) $P(Y, X \cup \{A \wedge B\}) = P(Y, X \cup \{A, B\})$;

(d) $P(\{A, \neg A\}, X) = P(\{\bot\}, X)$;

(e) $P(\{\bot\} \cup Y, X) = P(\{\bot\}, X)$;

(f) $\max_{Y \subseteq_f L} \{P(Y,X)\} = 1$;
(g) $P(\{A, A \supset B\}, X) = P(\{A, B, A \supset B\}, X)$;
(h) *If* $P(\{\bot\}, Y \cup X) = 1$, $P(Z \cup Y, X) = P(\{\bot\}, X)$;
(i) *If* $P(\{B\}, \{A\} \cup X) = 1$, *then* $P(\{A \supset B\}, X) = 1$;
(j) $P(\{A \supset A\}, \varnothing) = 1$;
(k) $P(\{\bot\}, \{A \supset A\}) = 0$;
(l) *If X is P-normal,* $P(Y \cup \{A \wedge \neg A\}, X) = 0$;
(m) *If* $X \cup Z$ *is P-normal, then X is P-normal;*
(n) $P(\{A\}, X) = 0$ *if and only if* $P(\{\neg A\}, X) = 1$;
(o) *If* $P(\{A\}, X) = 1$, $P(\{\neg A\}, X) = 0$.

Proof: (a) Since $\{A\}$ meets the condition $P(Y,X) = P(\{A\} \cup Y, X)$ on Y,

$$P(\{A\}, X) \leq \max_{Y \subseteq_f L} \{P(Y,X) : P(Y,X) = P(\{A\} \cup Y, X)\}$$

and if $P(Y,X) = P(\{A\} \cup Y, X)$, then $P(Y,X) \leq P(\{A\}, X)$ by Lemma 16(g).

(b) By **RS3** and **RS**$\wedge$.

(c) $P(Y, X \cup \{A \wedge B\})$

$= P(Y \cup \{A \wedge B\}, X \cup \{A \wedge B\})$ (Lemma 16(f))
$= P(Y \cup \{A, B\}, X \cup \{A \wedge B\})$ ((b))
$= P(Y, \{A, B\} \cup X \cup \{A \wedge B\}) \times P(\{A, B\}, X \cup \{A \wedge B\})$ (**RS3**)
$= P(Y, \{A, B\} \cup X \cup \{A \wedge B\}) \times 1$ (**RS**$\wedge$, **RS1**)
$= P(Y, \{A, B\} \cup X \cup \{A \wedge B\}) \times P(\{A \wedge B\}, X \cup \{A, B\})$ (**RS1, RS**$\wedge$)
$= P(Y \cup \{A \wedge B\}, X \cup \{A, B\})$ (**RS3**)
$= P(Y, X \cup \{A, B\})$ ((b), Lemma 16(f))

(d) $P(\{\neg A\}, X \cup \{A\})$

$= \max_{Y \subseteq_f L} \{Y, X \cup \{A\}) : P(\{A\} \cup Y, X \cup \{A\}) = P(\{\bot, A\} \cup Y, X \cup \{A\})\}$
(**RSI**$\neg$)
$= \max_{Y \subseteq_f L} \{Y, X \cup \{A\}) : P(Y, X \cup \{A\}) = P(\{\bot\} \cup Y, X \cup \{A\})\}$ (Lemma 16(f))
$= P(\{\bot\}, X \cup \{A\})$ ((a))

So, $P(\{A, \neg A\}, X) = P(\{\bot\}, X)$ by **RS3**.

(e) If $P(\{\bot\}, X) = 1$, $P(\{\bot\} \cup Y, X) = 1$ by **RS**$\bot$ and if $P(\{\bot\}, X) \neq 1$, $P(\bot\}, X) = 0$ by **RS**$\bot$ and $P(\{\bot\} \cup Y, X) = 0$ by **RS3**. Hence $P(\{\bot\} \cup Y, X) = P(\{\bot\}, X)$ in either case.

(f) By Lemmas 16(a) and 16(c).

(g) $P(\{A, A \supset B\}, X)$

$= \max_{Y \subseteq_f L} \{P(Y, \{A\} \cup X) : P(\{A\} \cup Y, \{A\} \cup X) = P(\{A, B\} \cup Y, \{A\} \cup X)\} \times P(\{A\}, X)$
(**RS3, RSI**$\supset$)
$= \max_{Y \subseteq_f L} \{P(Y, \{A\} \cup X) : P(Y, \{A\} \cup X) = P(\{B\} \cup Y, \{A\} \cup X)\} \times P(\{A\}, X)$
(Lemma 16(f))
$= P(\{A, B\}, X)$ ((a), **RS3**)

$P(\{A, B, A \supset B\}, X)$

$$= \max_{Y \subseteq_f L} \{P(Y,\{A,B\} \cup X) : P(\{A\} \cup Y,\{A,B\} \cup X) = P(\{A,B\} \cup Y,\{A,B\} \cup X)\} \times P(\{A,B\},X) \quad (\textbf{RS3}, \textbf{RSI}\supset)$$

$$= \max_{Y \subseteq_f L} \{P(Y,\{A,B\} \cup X) : P(Y,\{A,B\} \cup X) = P(Y,\{A,B\} \cup X)\} \times P(\{A,B\},X) \quad \text{(Lemma 16(f))}$$

$$= 1 \times P(\{A,B\},X) \quad ((f))$$

Hence $P(\{A,A \supset B\},X) = P(\{A,B,A \supset B\},X)$.

(h) By **RS3** and (e).

(i) Suppose $P(\{B\},\{A\} \cup X) = 1$. Then $P(\{A\},X) = P(\{A,B\},X)$ by **RS3**. Hence $P(\varnothing,X) \leq P(\{A \supset B\},X)$ by **RSI**$\supset$. Hence $P(\{A \supset B\},X) = 1$ by Lemmas 16(a) and 16(c).

(j) By (i) and **RS1**.

(k) Suppose $P(\{\bot\},\{A \supset A\}) = 1$. Then $P(\{A \supset A\},\varnothing) = P(\{\bot\},\varnothing)$ by (h), which is impossible by (j) and **RS4**.

(l) Suppose X P-normal. Then

$$P(Y \cup \{A \wedge \neg A\},X)$$
$$= P(Y \cup \{A,\neg A\},X) \quad ((b))$$
$$= P(\{\bot\},X) \quad (\textbf{RS3}, (d), (e))$$
$$= 0 \quad (\textbf{RS}\bot)$$

(m) Assume $P(Y,X \cup Z) \neq 1$. Then $P(Z \cup Y,X) \neq 1$ by **RS3** and Lemma 16(c), i.e. X is P-normal.

(n) Suppose $P(\{A\},X) = 0$. Then $P(\{\neg A\},X) = 1$ by **RSI**$\neg$, Lemma 16(d), and (f). Suppose $P(\{\neg A\},X) = 1$. Then $P(\{A\},X) = 0$ by (d), **RS3**, and Lemma 16(f).

(o) By **RSI**$\neg$, Lemma 16(d), and (f).

Valuations in the Open Sets of a Topological Space

T1 $\mathbf{I}(\alpha \cap \beta) = \mathbf{I}\alpha \cap \mathbf{I}(\beta)$

T2 $\mathbf{I}(\alpha) \subseteq \alpha$

T3 $\mathbf{I}(\mathbf{I}(\alpha)) = \mathbf{I}(\alpha)$

T4 $\mathbf{I}(\Xi) = \Xi$

V1 $v(A \wedge B) = v(A) \cap v(B)$

V2 $v(A \vee B) = v(A) \cup v(B)$

V3 $v(\neg A) = \mathbf{I}(\backslash v(A))$

V4 $v(A \supset B) = \mathbf{I}(\backslash v(A) \cup v(B))$,

Lemma 19. *Let $\langle \Xi,\mathbf{I} \rangle$ be a topological space and v a valuation of L_I in the open sets of $\langle \Xi,\mathbf{I} \rangle$. Then:*

(a) $v(\Lambda\varnothing) = \Xi$;

(b) $v(\mathrm{V}\varnothing) = \varnothing$;

(c) *If* $\alpha \subseteq \beta$, *then* $\mathbf{I}(\alpha) \subseteq \mathbf{I}(\beta)$*;*

(d) *If* $\alpha \subseteq \beta$ *and* α *is open, then* $\alpha \subseteq \mathbf{I}(\beta)$*;*

(e) $\text{card}\alpha + \text{card}\beta = \text{card}(\alpha \cup \beta) + \text{card}(\alpha \cap \beta)$.

Proof: (a) $\Lambda\varnothing$ is $A \supset A$. Hence

$\nu(\Lambda\varnothing)$

$= \mathbf{I}(\backslash\nu(A) \cup \nu(A))$ **(V4)**

$= \mathbf{I}(\Xi)$

$= \Xi$ **(T4)**

(b) $\text{V}\varnothing$ is $A \wedge \neg A$. Hence

$\nu(\text{V}\varnothing)$

$= \nu(A) \cap \mathbf{I}(\backslash\nu(A))$ **(V1,V3)**

$= \varnothing$

(c) Suppose $\alpha \subseteq \beta$, i.e. $\alpha \cap \beta = \alpha$. Then $\mathbf{I}(\alpha \cap \beta) = \mathbf{I}(\alpha)$, hence $\mathbf{I}(\alpha) \cap \mathbf{I}(\beta) = \mathbf{I}(\alpha)$ by **T1**, i.e. $\mathbf{I}(\alpha) \subseteq \mathbf{I}(\beta)$.

(d) Suppose $\alpha \subseteq \beta$ and α open, i.e. $\mathbf{I}(\alpha) = \alpha$. Then $\mathbf{I}(\alpha) \subseteq \mathbf{I}(\beta)$ by (c). hence $\alpha \subseteq \mathbf{I}(\beta)$.

Appendix II

Proof that P^2 constitutes a truth-value function

Let P be any measure function, suppose $P(A) > 0$, let $E_1, E_2, ..., E_n, ...$ be the atomic statements of L_P, and define a sequence of statements $\pm E_1, \pm E_2, ..., \pm E_n, ...$ as follows:

If $P(E_1 \wedge A) > 0$ *then* $\pm E_1$ *is* E_1, $\neg E_1$ *otherwise;*

If $P(((\pm E_1 \wedge ... \wedge \pm E_n) \wedge E_{n+1}) \wedge A) > 0$ *then* $\pm E_{n+1}$ *is* E_{n+1}, $\neg E_{n+1}$ *otherwise.*

Further define a sequence of measure functions $P_1, P_2, ..., P_n, ...$ as follows:

$$P_n(B) = P((\pm E_1 \wedge ... \wedge \pm E_n) \wedge B) \,/\, P(\pm E_1 \wedge ... \wedge \pm E_n),$$

where, in case $P(\pm E_1 \wedge ... \wedge \pm E_n) = \infty$, $P((\pm E_1 \wedge ... \wedge \pm E_n) \wedge B) / P(\pm E_1 \wedge ... \wedge \pm E_n)$ equals 0 if $P((\pm E_1 \wedge ... \wedge \pm E_n) \wedge B)$ is finite, and 1 otherwise. Finally, let a unary function P^2 be defined as the limit of this sequence of functions, i.e.

$$P^2(B) = \lim_{n \to \infty} P_n(B).$$

Then P^2 constitutes a truth-value function.

For let B be a statement of L_P and let the atomic statements that occur in B be among $E_1, ..., E_n$. It can be shown by induction on the number c of connectives in B that $P((\pm E_1 \wedge ... \wedge \pm E_n) \wedge B)$ equals either 0 or $P(\pm E_1 \wedge ... \wedge \pm E_n)$ and consequently that $P_n(B)$ equals either 0 or 1.

Basis: c = 0, i.e. B is the atomic statement E_i, where $1 \leq i \leq n$.

Case 1.1: $\pm E_i = E_i = B$. Then $P((\pm E_1 \wedge ... \wedge \pm E_n) \wedge B) = P((\pm E_1 \wedge ... \wedge \pm E_n) \wedge \pm E_i) = P(\pm E_1 \wedge ... \wedge \pm E_n)$ by Lemma 1(q). Hence $P_n(B) = 1$.

Case 1.2: $\pm E_i = \neg E_i = \neg B$. Then $P((\pm E_1 \wedge ... \wedge \pm E_n) \wedge B) = P((\pm E_1 \wedge ... \wedge \pm E_n) \wedge \neg \pm E_i)$ by Lemma 1(i). Hence $P((\pm E_1 \wedge ... \wedge \pm E_n) \wedge B) = 0$ by Lemma 1(r), and so $P_n(B) = 0$.

Inductive Step: c > 0.

Case 2.1: B is $\neg C$. Then $P((\pm E_1 \wedge ... \wedge \pm E_n) \wedge B) = P((\pm E_1 \wedge ... \wedge \pm E_n) \wedge \neg C) = P(\pm E_1 \wedge ... \wedge \pm E_n) - P((\pm E_1 \wedge ... \wedge \pm E_n) \wedge C)$ by **AP3**. But then, by the inductive hypothesis, $P((\pm E_1 \wedge ... \wedge \pm E_n) \wedge B)$ equals either $P(\pm E_1 \wedge ... \wedge \pm E_n)$ or 0. So $P_n(B) = 1$ or $P_n(B) = 0$ and $P_n(B) = 1 - P_n(C)$.

Case 2.2: B is $C \wedge D$. Then $P((\pm E_1 \wedge ... \wedge \pm E_n) \wedge B) = P((\pm E_1 \wedge ... \wedge \pm E_n) \wedge (C \,\&\, D))$. If either $P((\pm E_1 \wedge ... \wedge \pm E_n) \wedge C) = 0$ or $P((\pm E_1 \wedge ... \wedge \pm E_n) \wedge D) = 0$, $P((\pm E_1 \wedge ... \wedge \pm E_n) \wedge (C \wedge D)) = 0$ by Lemmas 1(b), 1(n), and 1(p) in Appendix I. So assume that $P((\pm E_1 \wedge ... \wedge \pm E_n) \wedge C) = P((\pm E_1 \wedge ... \wedge \pm E_n) \wedge D) = P(\pm E_1 \wedge ... \wedge \pm E_n)$. Then $P((\pm E_1 \wedge ... \wedge \pm E_n) \wedge (C \wedge D))$ equals $P((\pm E_1 \wedge ... \wedge \pm E_n) \wedge D)$ by Lemmas 1(e) and 1(j), and hence equals $P(\pm E_1 \wedge ... \wedge \pm E_n)$. So, by the inductive hypothesis, $P((\pm E_1 \wedge ... \wedge \pm E_n) \wedge B)$ equals 0 or $P(\pm E_1 \wedge ... \wedge \pm E_n)$, $P_n(B) = 0$ or $P_n(B) = 1$, and $P_n(B) = P_n(C) \times P_n(D)$.

For $m > n$, $P_m(B) = P_n(B)$ and, in the first instance, $P_{n+1}(B) = P_n(B)$. Indeed if $P((\pm E_1 \wedge ... \wedge \pm E_n) \wedge B) = 0$, $P(((\pm E_1 \wedge ... \wedge \pm E_n) \wedge \pm E_{n+1}) \wedge B) = 0$ by Lemmas 1(e), 1(d), 1(n), and 1(b). And if $P((\pm E_1 \wedge ... \wedge \pm E_n) \wedge B) = P(\pm E_1 \wedge ... \wedge \pm E_n)$, then $P((\pm E_1 \wedge ... \wedge \pm E_n) \wedge \neg B) = 0$ by **AP3**, so that $P((\pm E_1 \wedge ... \wedge \pm E_n) \wedge \pm E_{n+1}) \wedge \neg B) = 0$ by the same reasoning as before, and hence $P((\pm E_1 \wedge ... \wedge \pm E_n) \wedge \pm E_{n+1}) \wedge B) = P(\pm E_1 \wedge ... \wedge \pm E_n) \wedge \pm E_{n+1})$ by **AP3**. It follows that $P_m(B) = P_n(B)$ for every $m \geq n$ and hence that $P^2(B) = P_n(B)$. So P^2 is a unary function with values 0 and 1 such that $P^2(\neg B) = 1 - P^2(B)$ and $P^2(C \mathbin{\&} D) = P^2(C) \times P^2(D)$, i.e. P^2 is a truth-value function.

Since $P((\pm E_1 \wedge ... \wedge \pm E_n) \wedge A) > 0$ for every n, $P_m(A) = 1$ for every m such that the atomic statements that occur in A are among $E_1, ..., E_m$. Therefore $P^2(A) = 1$.

Notes

Chapter 1

1 See Carnap (1950).

2 As long as we are dealing with L_P, the logical notions invoked are the familiar truth-functional ones; e.g. logical equivalence is truth-functional equivalence.

3 See Popper (1938), (1955), and (1959), Appendix *ii, for more on this. Popper himself preferred to think of the arguments of his probability functions (both absolute and relative ones) as *elements*, i.e. statements if you will, but possibly also sets, or properties, or finite classes, etc. We opt for statements, as most students of Popper's functions do, and, later on, sets of statements.

4 The following constraints are equivalent to **AP1–AP6**:

AP1' $0 \leq P(A)$

AP2' $P(A \wedge \neg A) = 0$

AP3' $P(A) + P(\neg A) = 1$

AP4' $P(A \wedge B) + P(A \wedge \neg B) = P(A)$

AP5' $P(A \wedge B) \leq P(B \wedge A)$

AP6' $P(A \wedge (B \wedge C)) \leq P((A \wedge B) \wedge C)$

The two sets of constraints have **AP1**, **AP3**, **AP4**, and **AP5** in common. **AP2'** and **AP3'** are in effect Lemmas 1(f) and 2(a) in Appendix I. So, **AP1–AP6** yield **AP1'–AP6'**. On the other hand, **AP2'** and **AP3'** yield **AP2**, and **AP4'** yields

$$P(A \wedge A) + P(A \wedge \neg A) = P(A).$$

Hence,

$$P(A \wedge A) = P(A)$$

by **AP2'**, and hence **AP6**. So, **AP1'–AP6'** yield **AP1–AP6**.

5 The constraint

If $P(A) = 1$, *then A is logically true,*

equivalent to it by virtue of Lemma 2(a), will also be referred to as **AC**.

6 This result was reported in Leblanc (1982).

7 Rényi remarks in Note 2 of (1955) that a considerably simpler family of functions is gotten by requiring the disjunction of any two members of S_2 to belong to S_2.

8 Inadvertently, as he later admitted to Hugues Leblanc, Carnap in (1952) left out constraint **RC2**, in the absence of which all of his relative probabilities could equal 0. In (1950) he also left out constraint **RC2** but added this one:

$$P(\neg(A \wedge \neg A), B) = 1,$$

which does yield $P(A,A) = 1$.

9 This identity is first mentioned in Field (1977).

10 The infimum of a set of real numbers is the greatest lower bound of those numbers.

11 It is of course due to constraint **RS6** that sets of statements of L_P are interpreted here *conjunctively*.

12 ΛX is again some conjunction of the members of X, while $\mathrm{V}Y$ is a disjunction of the members of Y. $\Lambda\varnothing$ is T and $\mathrm{V}\varnothing$ is $\bot$.

Chapter 2

1 An ω-sequence is an infinite sequence that has a first element, a second one, a third one, and so on ad infinitum.

2 In L_ω the finitary connective '∧' could be dispensed with by treating $A \wedge B$ as short for $\Lambda\langle A,B\rangle$. But, again to stress the fact that L_ω is an extension of L_P, we prefer to retain '∧' as a primitive connective of L_ω. Note that (i) like L_P, L_ω has only $\aleph_0$ atomic statements, but (ii) unlike L_P, it has $2^{\aleph_0}$ compound ones. There are, for example, $2^{\aleph_0}$ different sequences of atomic statements of L_ω and hence already $2^{\aleph_0}$ conjunctions of L_ω of the sort $\Lambda\boldsymbol{A}$, where $\boldsymbol{A}$ is a countable sequence of these atomic statements. The statements of L_P are the finitary statements of L_ω.

3 The need for having the extra conjunct 'B' in a constraint of this kind was first noted by Bendall (1979) in connection with absolute probabilities of universal quantifications. The presence of 'B' ensures that $B \wedge \Lambda\langle A_1, A_2, ..., A_n, ...\rangle$ is P-indiscernible from $\Lambda\langle B, A_1, A_2, ..., A_n, ...\rangle$ in the sense of Section 1 of the previous chapter, reflecting the logical equivalence of $B \wedge \Lambda\langle A_1, A_2, ..., A_n, ...\rangle$ and $\Lambda\langle B, A_1, A_2, ..., A_n, ...\rangle$. See also Note 9.

4 When $\boldsymbol{A}$ is the empty sequence $\langle\,\rangle$, $\underset{n\to\infty}{\text{limit}}\, P(A_1 \wedge ... \wedge A_n)$ should be understood as $P(\neg(A \wedge \neg A))$, A here any atomic statement you please; and when $\boldsymbol{A}$ consists for some k or other larger than 0 of the k statements $A_1, A_2, ...,$ and A_k, it should be understood as $P(A_1 \wedge ... \wedge A_k)$. Like remarks hold concerning $\underset{n\to\infty}{\text{limit}}\, P(A_1 \wedge ... \wedge A_n, B)$ and $\underset{n\to\infty}{\text{limit}}\, P(A, B_1 \wedge ... \wedge B_n)$.

5 Note that, whereas **RPΛ1** is equivalent – given **RPΛ** – to **DL2**, **RPΛ** merely implies **DL1**. The two double-limit formulas besides **DL1** and **DL2**, namely

$$\underset{m\to\infty}{\text{limit}}\ \underset{n\to\infty}{\text{limit}}\, P(B_1 \vee ... \vee B_n, B_1 \vee ... \vee B_m) = 1$$

and

$$\underset{m\to\infty}{\text{limit}}\ \underset{n\to\infty}{\text{limit}}\, P(B_1 \wedge ... \wedge B_m, B_1 \wedge ... \wedge B_n) = 1$$

are trivial consequences of **RP1–RP7**. **DL1** and **DL2**, by the way, are independent of each other; and, given **RP1–RP7** *alone*, each of

$$\underset{m\to\infty}{\text{limit}}\ \underset{n\to\infty}{\text{limit}}\, P(B_1 \vee ... \vee B_m, B_1 \vee ... \vee B_n)$$

and

$$\underset{m\to\infty}{\text{limit}}\ \underset{n\to\infty}{\text{limit}}\, P(B_1 \wedge ... \wedge B_n, B_1 \wedge ... \wedge B_m)$$

equals in view of Lemma 5 in Appendix I either 0 or 1 for any given sequence $\langle B_1, B_2, ..., B_n, ...\rangle$.

6 See Theorem 18 of Chapter 5, Section 4.

7 For a study of the two consequence relations see Leblanc, Roeper, Thau, and Weaver (1991). But note that what we call here the *substitutional sense* is called there the *instantial sense*.

8 See Leblanc, Roeper, Thau, and Weaver (1991).

9 Frequently used, e.g. by Carnap and Popper, has been the constraint

$$P((\forall x)A) = \underset{n\to\infty}{\text{limit}}\, P(A_1 \wedge ... \wedge A_n).$$

But Bendall reported in (1979) that this constraint is not sufficient and that wanted instead is **AP∀**, i.e.

$$P((\forall x)A \wedge B) = \underset{n\to\infty}{\text{limit}}\, P((A_1 \wedge ... \wedge A_n) \wedge B).$$

For an early account of probability functions for quantificational languages see also Gaifman in (1964).

10 This is in essence the approach taken by Gaifman (1964).

11 This result is an analogue of the Löwenheim-Skolem Theorem in first-order logic. Indeed, the present proof can be turned into a proof of that theorem by assuming that P_α is a 2-valued Kolmogorov function.

12 $t_1, t_2, ..., t_n, ...$ are again the terms in $\mathcal{D}_0 \cup \mathcal{D}_\omega$.

Chapter 3

1 As explained in Chapter 1, T is an arbitrary but fixed statement of the kind $C \vee \neg C$ and $\perp$ is $C \wedge \neg C$.

2 Note that when B is not logically false, $P_2(B) \neq 0$ by **AC**.

3 Note that if L has infinitely many atomic statements, then, for every Popper function P_1 that is not a Carnap one, there are sure to exist statements D_1 and D_2 such that

(i) D_1 and D_2 are not logically false,

(ii) D_2 does not logically imply D_1, and

(iii) $P_1(D_1) = P_1(D_2) = 0$.

4 Note that if $P_1(B \wedge \neg D_1) = 0$, then $P_1(B) = 0$ by the hypothesis on D_1, **AP3**, and Lemma 1(a).

5 In light of Note 3 this reduces to the result that if L has infinitely many atomic statements, then those among Popper's relative probability functions that are neither Carnap 1952 functions nor Kolmogorov ones relativise those among Popper's absolute probability functions on L that are not Carnap ones.

Chapter 4

1 As in earlier chapters, T is an arbitrary but fixed statement of the kind $C \vee \neg C$ and $\perp$ is a statement of the kind $C \wedge \neg C$.

2 This proof is due to Spohn (1986).

Chapter 5

1 Two atomic statements A and B are said to be stochastically independent under an absolute probability function P for L_P if $P(A \wedge B) = P(A) \times P(B)$. The formula $P(X \cup Y) \times P(X \cap Y) = P(X) \times P(Y)$ generalises the concept of stochastic independence.

2 Strictly speaking, the functions P_1 and P_2 are not absolute probability functions for L_P, if L_P is understood to have denumerably many atomic statements, but for a sublanguage of L_P which has but two atomic statements. However, it is easy to extend these functions to the whole of L_P by stipulating that $P_1(E) = P_2(E) = 1$ for every atomic statement E other than E_1 and E_2 or, what comes to the same, by treating every such atomic statement E as if it were a logical truth.

3 See Note 2.

4 The restriction of P to L_P is sure to meet constraints **RP1–RP7** and hence does constitute a relative probability function for L_P.

Chapter 6

1 Strictly speaking, the functions P_1 and P_2 are not absolute probability functions for L_P, if L_P is understood to have denumerably many atomic statements, but for a sublanguage of L_P which has but three atomic statements. However, it is easy to extend these functions to the whole of L_P by stipulating that $P_1(E) = P_2(E) = 1$ for every atomic statement E other than E_1, E_2, and E_3 or, what comes to the same, by treating every such atomic statement E as if it were a logical truth.

Chapter 7

1 Many of the considerations in this chapter are drawn from Roeper and Leblanc (1995).

2 Cf. Shoesmith and Smiley (1978).

3 The term *designated* is traditionally used in connection with many-valued logics in order to single out one 'truth-value' from the rest as corresponding to the truth-value *true* in the 2-valued case.

4 Gentzen (1934–35).

5 See Shoesmith and Smiley (1978).

6 For proof take the following simple example of a language with just two statements A and B. The two functions F_1 and F_2, meeting constraints **F1–F3**, are distinct but agree on their values for $\langle\{A\},\varnothing\rangle$ and $\langle\{B\},\varnothing\rangle$.

	$F_1(X,Y)$	Y: $\varnothing$	$\{A\}$	$\{B\}$	$\{A,B\}$
	$\varnothing$	1	1/2	1/2	1/4
X	$\{A\}$	1/2	1	1/4	1
	$\{B\}$	1/2	1/4	1	1
	$\{A,B\}$	1/4	1	1	1

	$F_2(X,Y)$	Y: $\varnothing$	$\{A\}$	$\{B\}$	$\{A,B\}$
	$\varnothing$	1	1/2	1/2	1/3
X	$\{A\}$	1/2	1	1/6	1
	$\{B\}$	1/2	1/6	1	1
	$\{A,B\}$	1/3	1	1	1

7 See Gentzen (1934–35) or Gentzen (1969). Our constraints correspond to rules of the type used by Gentzen, but not in all cases to his actual rules.

8 See the *Hauptsatz* in Gentzen (1934–35).

9 While the constraint **ASC** is not autonomous, **FIND** allows us to characterise Carnap absolute probability functions autonomously.

10 See Carnap (1950).

11 The Gentzen constraints for '$\forall$' are

MVR *If* $X \vdash Y \cup S(A)$, *then* $X \vdash Y \cup \{\forall A\}$

MVL *If* $\{A_i\} \cup X \vdash Y$ *for* $i = 1, 2, \ldots,$ *then* $\{\forall A\} \cup X \vdash Y$

12 We do not consider the language $L_{Q=}$ in this section. But results paralleling those for L_Q to be found in this section can be obtained for $L_{Q=}$ as well.

13 We draw here on Leblanc, Roeper, Thau, and Weaver (1991).

Chapter 8

1 Cf. Shoesmith and Smiley (1978)

2 Cf. Shoesmith and Smiley (1978), p.82 ff.

3 As before, $Y^{\neg}$ is the set of statements whose members are the negations of the members of Y.

4 Similarly, Carnap 1950 relative probability functions can be characterised autonomously by **IND** and **RSK**.

5 $\wedge\varnothing = \top$; hence $\varnothing \sim_P \{\top\}$.

6 A set S of statements is maximally consistent if S is logically consistent and $S \cup \{A\}$ is inconsistent for any A not a member of S.

7 Two-valued Kolmogorov functions are also discussed in Chapter 6.

Chapter 9

1 Actually, only intuitionistic relative probability functions seem to have attracted attention up till now. The aim is then to meet a corresponding soundness and completeness condition. See Morgan and Leblanc (1983), van Fraassen (1981 I), van Fraassen (1981 II), Morgan and Mares (1995).

2 See Chapter 1, Section 4.

3 Cf. Rasiowa and Sikorski (1970), Chapter IX.

4 In particular, $\backslash\varnothing = \Xi$ and $\backslash\Xi = \varnothing$.

Chapter 10

1 By virtue of Lemma 18(e) **RSI¬** can be obtained from **RSI⊃**, if one reads '$\neg A$' as '$A \supset \perp$', something that is often done.

Appendix I

1 A stronger result can be had to the effect that there exists a statement B of L_P such that for every absolute probability function P of L_P and statement A of L_P

$$P(A \wedge B) = P(A) \times P(B).$$

2 By virtue of Lemma 18(e) **RSI¬** can be obtained from **RSI⊃**, if one reads '$\neg A$' as '$A \supset \perp$'.

Bibliography

Bendall, K. (1979), "Belief-Theoretic Formal Semantics for First-Order Logic and Probability", *Journal of Philosophical Logic* **8**, 375–397.

Byrne, L., (1946), "Two Brief Formulations of Boolean Algebra", *Bulletin of the American Mathematical Society* **52**, 269–272.

Carnap, R. (1942), *Introduction to Semantics*, Cambridge, Mass.: Harvard University Press.

Carnap, R. (1950), *Logical Foundations of Probability*, Chicago: The University of Chicago Press.

Carnap, R. (1952), *The Continuum of Inductive Methods*, Chicago: The University of Chicago Press.

Császár, A. (1955), "Sur la Structure des Espaces de Probabilité Conditionelle", *Acta Mathematica Academiae Scientiarum Hungaricae* **6**, 337-361.

De Finetti, B. (1937), "La prévision: Ses lois logiques, ses sources subjectives", *Annales de l'Institut Henri Poincaré* **7**, 1-68.

Dorn, G. H. W. (1992–93), "Popper's Laws of Excess of the Probability of the Conditional over the Conditional Probability", *Conceptus* **68**, 3–61.

Field, H. H. (1977), "Logic, Meaning, and Conceptual Role", *The Journal of Philosophy* **74**, 379-409.

Gaifman, H. (1964), "Concerning Measures on First Order Calculi", *Israel Journal of Mathematics* **2**, 1–18.

Gentzen, G. (1934–35), "Untersuchungen über das logische Schliessen", *Mathematische Zeitschrift* **39**, 176-210, 405–431.

Gentzen, G. (1969), *The Collected Papers of Gerhard Gentzen*, edited and translated by M. Szabo, Studies in Logic and the Foundations of Mathematics, North-Holland, Amsterdam.

Harper, W. L. (1983), "A Conditional Belief Semantics for Free Quantification Logic with Identity", *Essays in Epistemology and Semantics*, Haven Publications, New York.

Harper, W. L., Leblanc, H., and van Fraassen, B. C. (1983), "On Characterizing Popper and Carnap Probability Functions", *Essays in Epistemology and Semantics*, Haven Publications, New York.

Karp, C. (1964), *Languages with Expressions of Infinite Length*, Amsterdam: North-Holland Publishing Company.

Keisler, H. J. (1971), *Modal Theory for Infinitary Logic*, North-Holland Publishing Company, Amsterdam-London.

Keynes, J. (1952), *A Treatise on Probability*, MacMillan, London.

Kolmogorov, A. N. ((1933) 1950), *Foundations of the Theory of Probability*. Translated by N. Morrison. Originally published as "Grundbegriffe der Wahrscheinlichkeitsrechnung", *Ergebnisse der Mathematik*, **2**: 1-61.

Leblanc, H. (1961), "The Algebra of Logic and the Theory of Deduction", *The Journal of Philosophy* **58**, 553–558.

Leblanc, H. (1976), *Truth-Value Semantics*, Amsterdam: North-Holland Publishing Company.

Leblanc, H. (1982), "Popper's 1955 Axiomatization of Absolute Probability", *Pacific Philosophical Quarterly* **63**, 133–145.

Leblanc, H. (1983a), "Alternatives to Standard First-Order Semantics", in D. Gabbay and F. Guenthner (eds.), *Handbook of Philosophical Logic, Volume I: Elements of Classical Logic*, Dordrecht: D. Reidel Publishing Company, 189–274.

Leblanc, H. (1983b), "Probability Functions and their Assumption Sets - the Singulary Case", *Journal of Philosophical Logic* **12**, 379-402.

Leblanc, H. and Morgan, C. G. (1984), "Probability Functions and their Assumption Sets - the Binary Case", *Synthese* **60**, 91–106.

Leblanc, H. and Roeper, P. (1986), "Absolute Probability Functions: A Recursive and Autonomous Account", in W. Leinfellner and F. N. Wuketits (eds.), *The Tasks of Contemporary Philosophy*. Vienna: Hölder-Pichler-Tempsky, pp. 39-49.

Leblanc, H. and Roeper, P. (1989), "On Relativizing Kolmogorov's Absolute Probability Functions", *Notre Dame Journal of Formal Logic* **30**, 485-512.

Leblanc, H. and Roeper, P. (1990), "What are Absolute Probabilities a Function of?", in J. M. Dunn and A. Gupta (eds.), *Truth or Consequences: Essays in Honor of N. D. Belnap*. Dordrecht: Kluwer Academic Publishers, pp. 307-325.

Leblanc, H. and Roeper, P. (1992), "Probability Functions: The Matter of their Recursive Definability", *Philosophy of Science 59*, 372–388.

Leblanc, H. and Roeper, P. (1993), "Getting the Constraints on Popper's Probability Functions Right", *Philosophy of Science* **60**, 151–167.

Leblanc, H., Roeper, P., Thau, M., and Weaver, G. (1991), "Henkin's Completeness Proof: Forty Years Later", *Notre Dame Journal of Formal Logic* **32**, 212-232.

Lewis, D. K. (1976), "Probabilities of Conditionals and Conditional Probability", *Philosophical Review* **85**, 279–315. Reprinted in Lewis, *Philosophical Papers*, Volume 2, Oxford: Oxford University Press, 1986, 133–152.

Miller, D. and Popper, K. (1986), "Deductive Dependence", *Actes IV Congrés Català de Lògica*, Universitat Politècnica de Catalunya & Universitat de Barcelona, 21–29.

Morgan, C. G. and Leblanc, H. (1983), "Probabilistic Semantics for Intuitionistic Logic", *Notre Dame Journal of Formal Logic* **24**, 161–180.

Morgan, C. G. and Mares, E. D. (1995), "Conditionals, Probability, and Non-Triviality", *Journal of Philosophical Logic* **24**, 455–467.

Popper, K. R. (1938), "A Note on Probability", *Mind, N. S.* **47**, 275–279. Also *The Logic of Scientific Discovery*, Appendix *ii, Basic Books, New York, 318–322.

Popper, K. R. (1955), "Two Autonomous Axiom Systems for the Calculus of Probabilities", *The British Journal for the Philosophy of Science* 6, 51–57.

Popper, K. R. (1959), *The Logic of Scientific Discovery*, Basic Books, New York.

Popper, K. R. (1963), "Creative and Non-creative Definitions in the Calculus of Probability", *Synthese* **15**, 167–186.

Popper, K. R. and Miller, D. (1994), "Contributions to the Formal Theory of Probability", *Patrick Suppes: Scientific Philosopher*, vol. 1, Kluwer Academic Publishers, pp. 3–23.

Rasiowa, H. and Sikorski, R. (1970), *The Mathematics of Metamathematics*, PWN – Polish Scientific Publishers, Warszawa.

Rényi, A. (1955), "On a New Axiomatic Theory of Probability", *Acta Mathematica Academiae Scientiarum Hungaricae* **6**, 286-335.

Rényi, A. (1964), "Sur les Espaces Simples des Probabilités Conditionnelles", *Annales de l'Institut Poincaré, N Série, Section B1*: 3-21.

Roeper, P. and Leblanc, H. (1991), "Indiscernibility and Identity in Probability Theory", *Notre Dame Journal of Formal Logic* **32**, 1-46.

Roeper, P. and Leblanc, H. (1995), "Consequence and Confirmation", *Notre Dame Journal of Formal Logic* **36**, 341-363.

Rosenbloom, P. C. (1950), *The Elements of Mathematical Logic*, New York: Dover.

Shoesmith, D.J. and Smiley, T.J. (1978), *Multiple-conclusion Logic*, Cambridge: Cambridge University Press.

Spohn, W. (1986), "The Representation of Popper Measures", *Topoi* **5**, 69-74.

Stalnaker, R. (1970), "Probability and Conditionals", *Philosophy of Science* **37**, 64-80.

van Fraassen, B.C. (1976), "Representation of Conditional Probabilities", *Journal of Philosophical Logic* **5**, 417-430.

van Fraassen, B.C. (1981 I), "Probabilistic Semantics Objectified: II. Implication in Probabilistic Model Sets", *Journal of Philosophical Logic* **10**, 371–394.

van Fraassen, B.C. (1981 II), "Probabilistic Semantics Objectified: I. Postulates and Logics", *Journal of Philosophical Logic* **10**, 495–510.

Waismann, F. (1930–31), "Logische Analyse des Wahrscheinlichkeitsbegriffs", *Erkenntniss* **1**, 228–248.

Wittgenstein, L. (1922), *Tractatus Logico-Philosophicus*. London: Routledge & Kegan Paul Ltd.

Index

Index of Constraints